D1070712

Topics in Fracture and Fatigue

Frank A. McClintock (ca. 1980)

This book is dedicated to Frank A. McClintock, Professor of Mechanical Engineering at the Massachusetts Institute of Technology, in recognition of his pioneering contributions to the understanding of the mechanics and mechanisms of ductile fracture and fatigue crack propagation.

A.S. Argon
Editor

Topics in Fracture and Fatigue

With 152 Illustrations

Springer-Verlag
New York Berlin Heidelberg London Paris
Tokyo Hong Kong Barcelona Budapest

A.S. Argon
Department of Mechanical Engineering
Massachusetts Institute of Technology
Cambridge, MA 02139-4307
USA

Library of Congress Cataloging-in-Publication Data
Topics in fracture and fatigue / edited by A.S. Argon.
p. cm.
Includes bibliographical references and index.
ISBN 0-387-97833-X. — ISBN 3-540-97833-X
1. Fracture mechanics. 2. Materials—Fatigue. I. Argon, Ali S.
TA409.T65 1992
620.1′126—dc20 92-10375

Printed on acid-free paper.

Production managed by Hal Henglein; manufacturing supervised by Robert Paella.
Camera-ready copy prepared by the editor.
Printed and bound by Edwards Brothers, Inc., Ann Arbor, MI.
Printed in the United States of America.

9 8 7 6 5 4 3 2 1

ISBN 0-387-97833-X Springer-Verlag New York Berlin Heidelberg
ISBN 3-540-97833-X Springer-Verlag Berlin Heidelberg New York

Preface

The phenomenon of fracture, now widely recognized to be of central importance in governing the useful service life of engineering structures of all types, was not always considered as such in physics and materials science. Even today the study of fracture is considered by some as a field in which no self-respecting scientist should be seen working. The principal reason for this attitude has been the widespread empiricism that has been associated with the field. While the practicing engineer has found this empiricism quite useful in coping with often bewildering material response characteristics, this has discouraged the development of more fundamental and lasting approaches.

A new departure occurred in the understanding of fracture with the pioneering work of Griffith in 1920 and 1924, who not only associated fracture and strength with cracks, but also stated the conditions for fracture in brittle solids in terms of crack growth criteria. Considerable awareness existed in the field in the late 40's that Griffith's conditions did not apply well to many ductile solids, and that these materials appeared to be coming apart more gradually by internal cavitation. While the effect of plastic deformation on the crack tip environment was studied by many workers in the late 50's and early 60's, the definitive mechanistic developments of dealing with these new problems on the microstructural scale began with McClintock's pioneering work of 1968 on ductile fracture by plastic cavitation.

The stimulation of this work and related modeling studies of others resulted in a series of rapid fundamental developments in both crack tip mechanics and material response characteristics on the microstructural scale that has continued very actively up to the present and is currently referred to often as "micromechanics." McClintock has contributed importantly to this field at every turn, starting with his elastic-plastic mode III crack tip field solution (1957), to local field models of growing mode III cracks (1970), material model for ductile cavitation (1968), growth of cavities in the environment of a plastically blunted mode I crack (1969), fully plastic slip line field studies of crack tip environments in pure mode and mixed mode deformation (1971), mechanics and mechanisms of fatigue crack growth in mode III (1956), fatigue crack growth by quasi-homogeneous damage accumulation (1963), and fatigue crack growth by irreversible crack tip distortions (1967). The contributions extended also to studies of fundamental material response in the form of strength of high angle grain boundaries (1965), brittle intergranular fractures among 2-D hexagonal grains (1974), and included atomistic modeling studies on emission of dislocations from mode III cracks (1969, unpublished), and generalized ductile cavitation loci (1970, unpublished). A short annotated list of these contributions and some others can be found at the end of this book in Chapter 10.

This volume was assembled on the occasion of the retirement of McClintock in June of 1991, and is the outcome of a mini-international symposium held at Endicott House, the MIT Conference Estate at that time. Its purpose is to assess some of the important developments in the field of ductile fracture and fatigue to which McClintock contributed significantly. The select group of contributors to this volume include a number of active practitioners in this important field, who individually have made contributions as pioneering and lasting as those of McClintock himself. They discuss developments varying from atomistic level responses of crack tips, to elastic-plastic crack tip fields, to cavitation models for ductile fracture and intergranular creep fracture, and include models of fracture in composites, and mechanisms of fatigue fracture. While the coverage of these topics is not at all complete or uniform, they all incorporate unique mechanistic perspectives that collectively furnish examples of the best thinking in this field and in both style and approach point out the future directions in further development.

I gratefully acknowledge the financial support of the National Science Foundation, Mechanics and Materials Program in the Engineering Directorate, under Grant 9108983-MSS, that has made this book possible. I am also grateful to Ms. Mary Toscano for her expert help, not only in the preparation of the basic manuscript, but also for her perseverance in extracting manuscripts from some of the illustrious contributors and for the preparation of the index.

A.S. Argon
Cambridge, MA
December, 1991

Contents

List of Contributors

A. S. Argon
Dept. of Mechanical Eng.
Mass. Institute of Technology
Cambridge, MA 02139–4307
USA

G. E. Beltz
Div. of Applied Sciences
Harvard University
Cambridge, MA 02138
USA

A. G. Evans
College of Engineering
University of California
Santa Barbara, CA 93106
USA

J. W. Hancock
Dept. of Mechanical Eng.
University of Glasgow
Glasgow, Scotland

K. J. Hsia
Dept. of Theo. & Applied Mech.
University of Illinois
Urbana, IL 61801
USA

J. W. Hutchinson
Div. of Applied Sciences
Harvard University
Cambridge, MA 02138
USA

F. A. McClintock
Dept. of Mechanical Eng.
Mass. Institute of Technology
Cambridge, MA 02139–4307
USA

R. M. McMeeking
College of Engineering
University of California
Santa Barbara, CA 93106
USA

K. J. Miller
Dept. of Mechanical Eng.
University of Sheffield
Sheffield, England

A. Needleman
Div. of Engineering
Brown University
Providence, RI 02912
USA

D. M. Parks
Dept. of Mechanical Eng.
Mass. Institute of Technology
Cambridge, MA 02139–4307
USA

A. Pineau
Centre des Materiaux
Ecole des Mines
Paris, France

J. R. Rice
Div. of Applied Sciences
Harvard University
Cambridge, MA 02138
USA

Y. Sun
Div. of Applied Sciences
Harvard University
Cambridge, MA 02138
USA

V. Tvergaard
Department of Solid Mechanics
The Tech. University of Denmark
Lyngby, Denmark

F. Zok
College of Engineering
University of California
Santa Barbara, CA 93106
USA

1

Peierls Framework for Dislocation Nucleation from a Crack Tip

J. R. Rice, G. E. Beltz, and Y. Sun

ABSTRACT Dislocation nucleation from a stressed crack tip is analyzed based on the Peierls concept, in which a periodic relation between shear stress and atomic shear displacement is assumed to hold along a slip plane emanating from a crack tip. This approach allows some small slip displacement to occur near the tip in response to small applied loading and, with increase in loading, the incipient dislocation configuration becomes unstable and leads to a fully formed dislocation which is driven away from the crack. An exact solution for the loading at that nucleation instability was developed using the J-integral for the case when the crack and slip planes coincide (Rice, 1992). Solutions are discussed here for cases when they do not. The results were initially derived for isotropic materials and some generalizations to take into account anisotropic elasticity are noted here. Solutions are also given for emission of dissociated dislocations, especially partial dislocation pairs in fcc crystals. The level of applied stress intensity factors required for dislocation nucleation is shown to be proportional to $\sqrt{\gamma_{us}}$ where γ_{us}, the *unstable stacking* energy, is a new solid state parameter identified by the analysis. It is the maximum energy encountered in the block-like sliding along a slip plane, in the Burgers vector direction, of one half of a crystal relative to the other. Approximate estimates of γ_{us} are summarized, and the results are used to evaluate brittle versus ductile response in fcc and bcc metals in terms of the competition between dislocation nucleation and Griffith cleavage at a crack tip. The analysis also reveals features of the near-tip slip distribution corresponding to the saddle point energy configuration for cracks that are loaded below the nucleation threshold, and some implications for thermal activation are summarized. Additionally, the analysis of dislocation nucleation is discussed in connection with the emission from cracks along bimaterial interfaces, in order to understand recent experiments on copper bicrystals and copper/sapphire interfaces, and we discuss the coupled effects of tension and shear stresses along slip planes at a crack tip, leading to shear softening and eased nucleation.

1.1 Introduction

Armstrong (1966) and Kelly et al. (1967) advanced the viewpoint of brittle versus ductile response as the competition between Griffith cleavage and plastic shear at a crack tip. The latter proposed that the response of a crystal or grain boundary should be treated by comparing the ratio of the largest tensile stress to the largest shear stress close to a crack tip with the ratio of the ideal cleavage stress to the ideal shear stress. Armstrong compared the applied stress necessary to meet the Griffith condition with the stress to shear apart a dislocation dipole near a crack tip, and thereby noted the importance of the dimensionless combination $\gamma_s/\mu b$ (γ_s = surface energy, μ = shear modulus, b = magnitude of the Burgers vector) as an index of how relatively easy it was for the shear process to occur before cleavage. Subsequently Rice and Thomson (1974) specifically modeled the shear process as the nucleation of a dislocation from a stressed crack tip. The Rice and Thomson approach made use of elasticity solutions for a fully formed dislocation (i.e., a dislocation with slip equal to the Burgers vector $\boldsymbol{b}$ of some complete or partial lattice dislocation) and a core cut-off parameter had to be introduced to derive a nucleation criterion. Their analysis showed, likewise, the importance of large $\gamma_s/\mu b$ and also of low core energy (large r_c/b, where r_c is the core cut-off radius in their analysis) for ductile response.

Recent treatments of the Rice-Thomson model have evolved to characterizing the crack-tip competition in terms of the parameters G_{cleave}, the energy release rate for cleavage and G_{disl}, the energy release rate associated with the emission of a single dislocation on a slip plane emanating from the crack tip. In its original form, the Rice-Thomson model treated dislocation emission by considering the stability of a straight dislocation line or a semicircular dislocation loop; both proceeded by assuming the existence of a freshly generated dislocation at a relatively small distance (turning out to be less than a few atomic spacings) away from the crack tip, on a slip plane which intersects the crack front. A drawback to this procedure, as well as the Peierls-type model to be discussed shortly, is that the analysis may be straightforwardly applied only to cases in which the slip plane(s) intersect the crack front. Following Mason (1979), however, we may envision a scenario in which dislocations are emitted when a moving crack front undergoes local deviations which bring it into line with a potentially active slip plane. Another drawback to the Rice-Thomson treatment is that it involves the core cutoff radius, an uncertain parameter. Here, following a suggestion by Argon (1987), the Peierls (1940) concept is used in an analysis of dislocation formation at a crack tip. That is, a periodic relation is assumed to hold between shear stress and sliding displacement along a crystal slip plane emanating from a crack tip, and a solution is then derived for the critical external loading which corresponds to dislocation nucleation.

A first report on this approach has been given by Rice (1992). We follow the text and format of that work closely here, adopting entire sections where appropriate. We also enlarge on the development to review exact numerical solutions, anisotropic effects, coupled shear and tension, thermal activation, and bimaterial interface cracks.

With the results derived in the Peierls framework, we shall have no further need for introduction of the poorly defined core cut-off at a crack tip in analyzing nucleation phenomena. Indeed, the results show that no feature resembling a fully formed dislocation is present at the crack tip prior to the instability. The instability begins a slip event leading to a fully formed dislocation which moves away from the crack tip. Prior to the instability there exists only an incipient dislocation in the form of a nonlinear shear distribution along a slip plane, with maximum deformation equivalent to a slip at the crack tip of generally less than a half of that of the fully formed dislocation. This agrees with the suggestion by Argon (1987) that nucleation instability should occur at a slip less than that of the dislocation which ultimately emerges. The Peierls concept has also been used in a recent analysis of dislocation nucleation by Schoeck (1991). His analysis was somewhat more approximate and did not uncover the exact solution for nucleation within the Peierls framework that is derived here. The results by Rice (1992) identified a new solid state parameter, denoted γ_{us} and called the *unstable stacking* energy, which characterizes the resistance to dislocation nucleation.

1.2 Description of Model

Suppose that a crack tip intersects one of the possible slip planes in a ductile crystal (Fig. 1.1(a)). The question addressed is that of what loading of the cracked solid suffices to nucleate a dislocation from the tip, assuming that cleavage decohesion does not occur first. By adopting the Peierls (1940) concept, the shear stress τ along the slip plane is regarded as a (periodic) function of the slip displacement δ along it. Thus, the problem addressed consists of an externally loaded solid containing a crack with traction-free surfaces, and with the additional boundary condition that the shear traction τ must be a function of the slip displacement δ along a plane of displacement discontinuity emanating from the crack tip. For the present we assume that there is a discontinuity of slip displacement only along that plane. More precise models in which there are discontinuities in both shear and opening displacement (the latter relating to dilatancy of an atomic array during large shear and, also, to the presence of tensile stress σ across the slip plane) are discussed by Beltz and Rice (1991, 1992a) and Sun et al. (1992). Hence, if $\boldsymbol{s}$ and $\boldsymbol{n}$ are unit vectors in the slip direction and normal to the slip plane, then $\delta = u_s^+ - u_s^-$ where $u_s = \boldsymbol{s} \cdot \boldsymbol{u}$; $\boldsymbol{u}$ is the displacement vector and $+, -$ refer to the two sides of the slip plane with $\boldsymbol{n}$ pointing

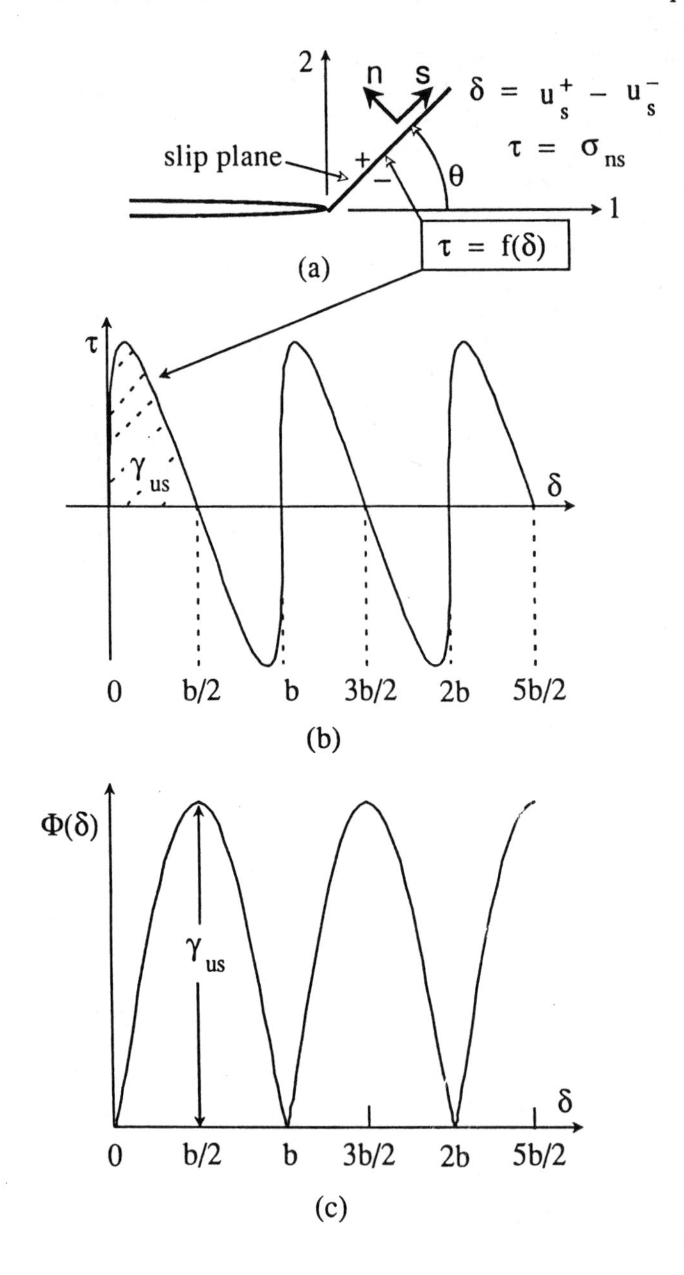

FIGURE 1.1. (a) Crystal slip plane emanating from crack tip. (b) Periodic relation between stress and shear displacement discontinuity (see discussion in connection with next figure to understand basis for vertical tangent at zero slip). (c) Energy associated with slip discontinuity.

from − to +. Within the present simplification, other components of $\boldsymbol{u}$ are continuous. Also, $\tau = n_\alpha \sigma_{\alpha\beta} s_\beta = \sigma_{ns}$ where $\sigma_{\alpha\beta}$ is the stress tensor.

The τ *vs.* δ relation is assumed to have a form like in Fig. 1.1(b), i.e., a periodic function with period b equal to the Burgers vector of a full dislocation, and with an axis crossing in-between, at $b/2$ in lattices with simple symmetry. Ways of estimating the form of the relation, and why it has been drawn with a vertical tangent at $\delta = 0$ and b, are discussed below; adaptations of concepts so as to deal with complex dislocations having stacking faults or anti-phase boundaries are discussed in a later section. Weertman (1981) analyzed a similar model but with the τ *vs.* δ relation in the form of a rectangular wave.

The result derived by Rice (1992), exactly for a special geometry and approximately in all cases, is that dislocation nucleation occurs under critical crack tip stress intensity factors which scale with $\sqrt{\gamma_{\mathrm{us}}}$. Here γ_{us}, the unstable stacking energy, is identified in Fig. 1.1(b) as the area under the τ *vs.* δ curve between $\delta = 0$ and $\delta = b/2$ (more generally, γ_{us} is the area between $\delta = 0$ and the first δ at which $\tau = 0$ again). Figure 1.1(c) shows the energy per unit area of the slip plane, $\Phi = \int \tau d\delta$. Thus γ_{us} is the maximum value of Φ. We may take the viewpoint that the same τ *vs.* δ relation could be used to describe the block-like shear, along a slip plane, of one half of a perfect lattice relative to the other. Hence Φ (or, more accurately, a related energy Ψ introduced below) corresponds to the γ-energy plot of Vitek (1968) and Vitek et al. (1972) and γ_{us}, the maximum value of Φ (and of Ψ) along the slip path, is the energy barrier to be overcome in block-like shear.

To understand the τ *vs.* δ relation, consider the states of shear of an initially rectangular lattice illustrated in Fig. 1.2. The relative shear displacement of the central pair of planes is denoted Δ; these are separated by distance h and are the pair of planes which will ultimately be displaced a lattice distance b. Lattice configurations (a) to (d), starting at the lower left and going clockwise, correspond to point (a) to (d) on the τ versus Δ curve. All the configurations shown are homogeneous in the direction of the shear displacement, but not perpendicular to it. When sufficiently sheared, like in (c) and (d), there exist configurations in which the lattice is not homogeneously strained, like it is in (b), but rather for which the central pair of planes corresponds to a Δ along the descending part of the τ versus Δ relation, while the crystal outside is stressed at the same level at an amount of shear corresponding to the rising part of the curve. Position (d) corresponds to the unstressed but unstable equilibrium state for which the central pair of lattice planes are displaced by $b/2$ while the crystal outside is unstrained. This is the unstable stacking configuration and the work to create it (area under τ versus Δ between $\Delta = 0$ and $b/2$) is γ_{us}, the same γ_{us} of Figs. 1.1(b) and (c), as explained next.

Although the configurations considered in Fig. 1.2 are homogeneous in the direction of shear, we follow Peierls (1940) in applying the τ versus Δ

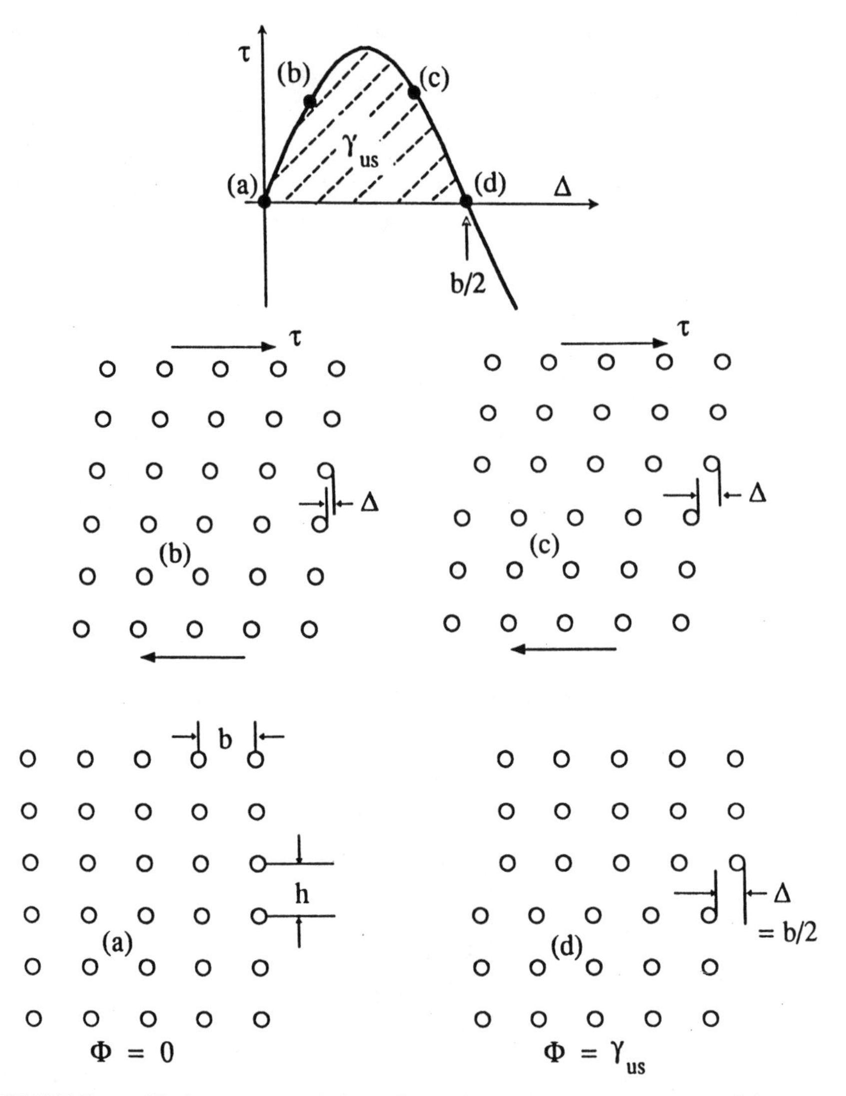

FIGURE 1.2. Various states of shear for a simple cubic lattice; state (d) shows the unstable stacking configuration, with energy γ_{us} per unit area of slip plane.

relation locally to states of inhomogeneous shear like along the slip plane in Fig. 1.1(a). Since that inhomogeneous shear is modeled here as a displacement discontinuity, of amount δ, along a cut of zero thickness in an elastic continuum, it is sensible to identify δ not with Δ, which denotes relative displacement of points a distance h apart, but rather to write $\Delta = \delta + h\tau/\mu$ so that relative displacement Δ of atomic planes at spacing h is composed of the discontinuity δ on the mathematical cut plus an additional amount due to elastic shearing by amount τ/μ over a distance h perpendicular to the cut; μ is the shear modulus. Thus, if $\tau = F(\Delta)$, of period b, describes the τ versus Δ relation of Fig. 1.2, where $F(0) = 0$ and $\mu = hF'(0)$, then the τ versus δ relation, $\tau = f(\delta)$, is given parametrically by $\tau = F(\Delta)$ and $\delta = \Delta - h\tau/\mu = \Delta - F(\Delta)/F'(0)$. This means that the resulting $\tau = f(\delta)$ is of period b and that $f'(\delta)$ is unbounded at $\delta = 0$, as illustrated in Fig. 1.1(b). The transformation from Δ to δ as displacement variable preserves the area, namely $\gamma_{\rm us}$, under the τ *vs.* displacement curve between the origin and the next zero of τ. An energy $\Psi(\Delta)$ may be defined from $\tau d\Delta = d\Psi$; it is the form in which an energy of sheared configurations has been calculated from atomic models (e.g., Vitek, 1968; Vitek et al., 1972; Yamaguchi et al., 1981; Sun et al., 1991). Given that the energy $\Phi(\delta)$ of Fig. 1.1(c) satisfies $\tau d\delta = d\Phi$, the relation $\delta = \Delta - h\tau/\mu$ shows that $d\Phi = d\Psi - h\tau d\tau/\mu$ and thus that $\Phi(\delta) = \Psi(\Delta) - h\tau^2(\Delta)/2\mu$.

The simplest case of a $\tau = F(\Delta)$ relation is the Frenkel sinusoidal function

$$\tau = (\mu b/2\pi h)\sin(2\pi\Delta/b), \tag{1.1}$$

in which case,

$$\delta = \Delta - (b/2\pi)\sin(2\pi\Delta/b) \tag{1.2}$$

and the energies Ψ and Φ are

$$\Psi = (\mu b^2/2\pi^2 h)\sin^2(\pi\Delta/b), \text{ and } \Phi = (\mu b^2/2\pi^2 h)\sin^4(\pi\Delta/b). \tag{1.3}$$

In this case $\gamma_{\rm us}$, which is the common maximum of Φ and Ψ, is given by $\gamma_{\rm us} = \mu b^2/2\pi^2 h$, an estimate that will be considered subsequently along with others. The plots in Figs. 1.1(b), 1.1(c) and 1.2 have been drawn based on the Frenkel sinusoid.

1.3 Analysis of Simplified Geometry with Coincident Crack and Slip Planes

This section repeats the major result of Rice (1992). While geometries like in Fig. 1.1(a), typically loaded by tensile, or predominantly tensile, forces relative to the crack plane are of primary interest, the problem posed there

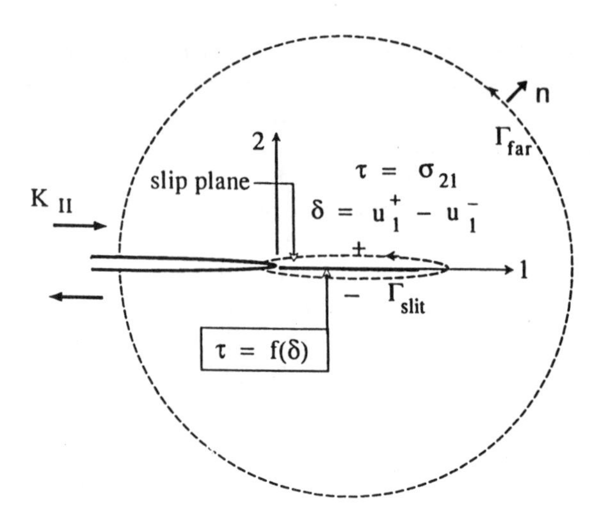

FIGURE 1.3. Coincident crack and slip plane, mode II loading.

is solvable only by numerical methods (Beltz and Rice, 1991a,b). Tensile loading of that configuration of Fig. 1.1(a) causes high shear stress τ (at least, when slip δ is precluded) along any slip plane in the general range of, say, $\theta = 30°$ to $120°$; the mode I crack tip field has highest shear stress along $\theta \approx 70°$. Some of the same features of the configuration of Fig. 1.1(a), namely, shear along a highly stressed plane emanating from the crack tip, are preserved in the simplified configuration of Fig. 1.3, for which an exact solution will be derived. In that simplified case, the most stressed slip plane is assumed to be coplanar with the crack ($\theta = 0$), with $\boldsymbol{s}$ in the x_1 direction, so emerging dislocations are of edge character relative to the tip, and the external loading is by in-plane (mode II) shear. A nearly identical analysis may be followed when $\boldsymbol{s}$ is in the x_3 direction, so that emerging dislocations are of screw type relative to the tip, and loading is by anti-plane shear (mode III).

Along the prolongation of the crack into the slip plane in Fig. 1.3, we have $\delta = u_1^+ - u_1^-$ and $\tau = \sigma_{21}$, where $\tau = f(\delta)$ like in Fig. 1.1(b); u_2 and u_3 are continuous there. Recognizing that this configuration is being analyzed as a simplified analog of more realistic tensile-loaded cases like in Fig. 1.1(a), we do not extend applicability of the τ versus δ relation back onto the crack faces in Fig. 1.3 but, rather, assume that the crack faces are traction free ($\sigma_{2j} = 0, j = 1, 2, 3$).

The crack is assumed to be sufficiently long that any region near its tip where significant slip develops, prior to unstable dislocation nucleation, is assumed to be of negligible length compared to crack length and other overall dimensions of the cracked solid, such as distance to boundaries and to points of external force application. In that case it suffices to consider the crack as a semi-infinite slit in an unbounded solid, with all loadings applied at infinitely remote distance so that all we need consider is the singular crack tip stress field, characterized by stress intensity factors, $K_{\text{I}}, K_{\text{II}}, K_{\text{III}}$, that the loadings would induce in the linear elastic model of the actual solid. At present, only K_{II} is assumed to be non zero, such that the stress field ahead of the crack tip $(x_2 = 0, x_1 > 0)$ in the linear elastic model of our solid, when restrained against slip δ, is $\sigma_{21} \equiv \tau_0 = K_{\text{II}}/\sqrt{2\pi r}, \sigma_{22} = \sigma_{23} = 0$, with $r = x_1$, and the Irwin energy release rate G is, in the isotropic case,

$$G = (1-\nu)K_{\text{II}}^2/2\mu \tag{1.4}$$

where μ is the shear modulus and ν the Poisson ratio.

We now follow a similar argument to that used by Rice (1968 a,b), based on the path-independent J integral, in proof of the equivalence of the Griffith criterion $G = 2\gamma_s (\gamma_s =$ surface energy) for tensile crack growth under mode I loading to the criterion derived from the tensile-decohesion analog of the model described so far here (i.e., from a model in which σ_{22} is a function of opening displacement, $u_2^+ - u_2^-$, along the prolongation of the crack plane, with that function increasing to a maximum and then diminishing to zero at large opening displacements, such that its integral from 0 to ∞ is $2\gamma_s$). That same equivalence was demonstrated earlier by Willis (1967), using integral representations of the linear elastic solution for the field outside the decohesion zone, and the Willis method was also adapted by Rice (1992) to the present analysis of shear dislocation emission at a crack tip. See Eshelby (1970), Rice (1987), and Rice and Wang (1989) for related discussions.

The J integral is

$$J = \int_\Gamma [n_1 W(\nabla \boldsymbol{u}) - n_\alpha \sigma_{\alpha\beta} \partial u_\beta / \partial x_1]\, ds \tag{1.5}$$

where W is the strain energy density, $\sigma_{\alpha\beta} = \partial(W(\Delta \boldsymbol{u}))/\partial(\partial u_\beta/\partial x_\alpha)$ is the stress tensor, s is arc length, and, here, $\boldsymbol{n}$ is the unit outward normal to the path Γ, where Γ starts on the lower crack surface, surrounds the crack tip and any slip zone in its vicinity, and ends on the upper crack surface. The integral is independent of path when evaluated for any 2D solution $\boldsymbol{u}(\boldsymbol{x})$ of the elastostatic equilibrium equations $\partial\sigma_{\alpha\beta}/\partial x_\alpha = 0$, at least when the elastic properties are invariant to translation in the x_1 direction. The path independence applies not only for conventional stable elastic solutions corresponding to a minimum of the energy functional $U[\boldsymbol{u}(\boldsymbol{x})]$, but also to 2D fields $\boldsymbol{u}(\boldsymbol{x})$ corresponding to other extremals of $U[\boldsymbol{u}(\boldsymbol{x})]$ such as saddle-point configurations, of interest for activation over energy barriers; the field equations $\partial\sigma_{\alpha\beta}/\partial x_\alpha = 0$ are satisfied at all extrema. Here U is the energy of the stressed solid per unit distance along the crack front.

Since J has the same value for all paths which do not traverse the crack or slip zone ahead of it, we can advantageously evaluate J on two contours,

Γ_{far} and Γ_{slit}; Γ_{far} lies far from the crack tip and the nonlinear perturbation of the linear elastic field due to the incipient slip process near the tip, whereas Γ_{slit} coincides with the upper and lower surfaces of the slit lying ahead of the crack tip on which the displacement u_1 is discontinuous by (variable) amount δ. The value of J on Γ_{far} will depend only on the remote linear elastic field characterized by K_{II} and, as is well known in that case, the result is $J = G$. The value along Γ_{slit} can be written as

$$\begin{aligned} J &= -\int_0^{\infty} \sigma_{21}\partial(u_1^+ - u_1^-)/\partial x_1 dx_1 = -\int_0^{\infty} \tau\partial\delta/\partial x_1 dx_1 \\ &= \int_0^{\delta_{\text{tip}}} \tau d\delta \equiv \Phi(\delta_{\text{tip}}) \end{aligned} \tag{1.6}$$

where δ_{tip} is the slip displacement discontinuity at the crack tip. Since J is independent of path, the two evaluations must agree and hence the amount of slip at the crack tip associated with any static solution must satisfy

$$G \equiv (1-\nu)K_{\text{II}}^2/2\mu = \Phi(\delta_{\text{tip}}). \tag{1.7}$$

For anisotropic solids the same result applies but with $(1-\nu)/2\mu$ replaced with the appropriate compliance factor from the Stroh (1958) and Barnett and Asaro (1972) results relating G to K_{II}.

Thus as the applied K_{II} and hence G increases from zero, one first follows the rising branch of the $\Phi(\delta)$ function of Fig. 1.4, having solutions for δ_{tip} like that illustrated at point A. Such values of δ_{tip} [$= \delta(r)$ at $r = 0$] are reasonably assumed to correspond to functions $\delta(r)$ that give minima of $U[\delta(r)]$ and that represent an incipient, but not yet fully formed, dislocation at the crack tip. It is evident that no static solution can exist when G exceeds γ_{us}, the maximum value of Φ, and hence the incipient dislocation configuration discussed loses stability at

$$G \equiv (1-\nu)K_{\text{II}}^2/2\mu = \gamma_{\text{us}}, \tag{1.8}$$

which therefore corresponds to nucleation of a full dislocation. The slip δ_{tip} at the crack tip when instability is reached is well short of that (namely, b) for a full dislocation, and corresponds to $b/2$ in lattices with simple symmetry. Thus no feature resembling a fully formed dislocation is actually present at the crack tip prior to the instability at which the full dislocation is nucleated.

As further shown in Fig. 1.4, the equation $G = \Phi(\delta_{\text{tip}})$ for $G < \gamma_{\text{us}}$ has multiple roots, illustrated by solution points A, C, A', A'', ... Points A', A'', etc. have a clear interpretation as corresponding to incipient dislocation configurations after one, two, etc. full dislocations have already been formed from the crack tip. Since A and A' may be presumed to correspond to stable solutions, minimizing $U[\mathbf{u}(\mathbf{x})]$, we should expect there to be a

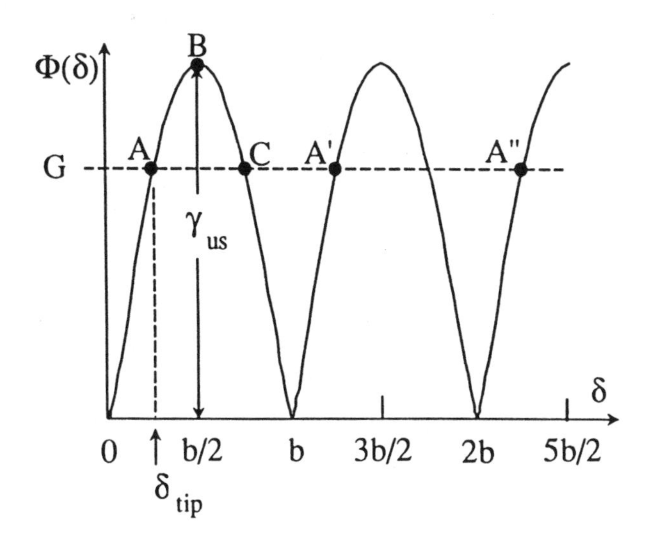

FIGURE 1.4. Solution for the slip displacement at the crack tip, for stable solution A (and A', A'', ... corresponding to one, two, or more previously emitted dislocations), and for 2D saddle-point configuration C.

saddle-point configuration between theses two states, also an extrema of $U[\boldsymbol{u}(\boldsymbol{x})]$. That saddle-point configuration evidently has a slip δ_{tip} at the crack tip given by point C in Fig. 1.4, and hence we are able to calculate an important feature of the activated configuration, at least in a 2D treatment. This is of limited use because the actual saddle-point configuration, defining the activation energy for an analysis of thermally assisted dislocation nucleation when $G < \gamma_{\text{us}}$, will involve a 3D elastic field associated with a localized outward protrusion of slip from the stable 2D incipient dislocation distribution corresponding to point A. Further discussion of activated states in dislocation nucleation is deferred to Section 1.13.

The same analysis as above may be followed for a crack tip loaded under mode III conditions and for which the slip direction $\boldsymbol{s}$ is in the x_3 direction, so that the emerging dislocation is of screw type. We now identify τ as σ_{23} and δ as $u_3^+ - u_3^-$. The above equations hold with K_{II} replaced with K_{III}, and with $(1-\nu)$ replaced by 1, so that the nucleation condition is then

$$G \equiv K_{\text{III}}^2/2\mu = \gamma_{\text{us}}. \tag{1.9}$$

At this point we want to extend the results to nucleation of general dislocations, combining both edge and screw components, at crack tips under general mixed-mode loading. Also, we want to model the nucleation of dislocations in dissociated form, with first one partial dislocation nucleating,

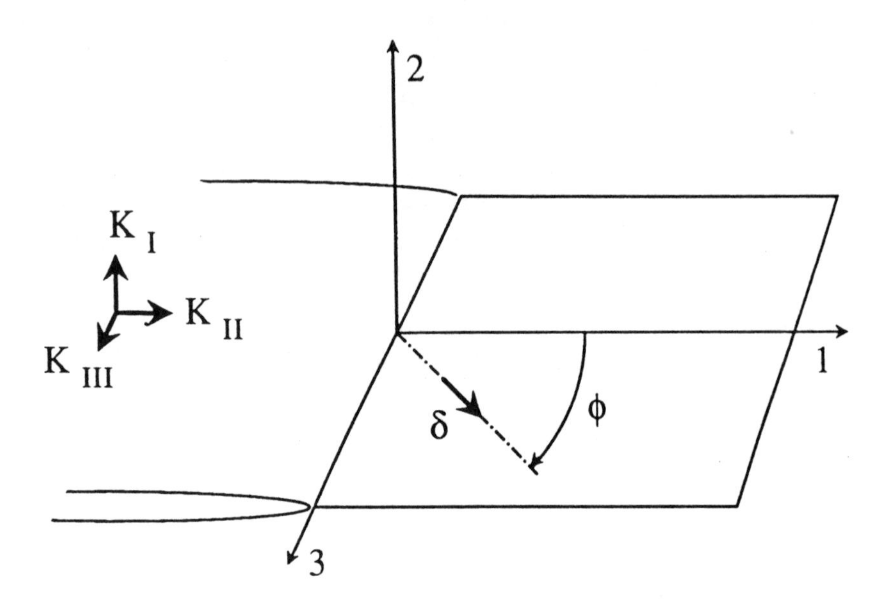

FIGURE 1.5. General mixed mode loading. Relative displacement along the slip plane assumed to follow a constrained path of pure sliding, without opening, along slip direction at angle ϕ.

leaving a faulted plane behind it, and then the remainder of the dislocation nucleating under increased external loading (e.g. , fcc metals in which partial dislocations on {111} planes are separated by stacking faults, and ordered alloys in which superlattice dislocations dissociate into partials separated by an antiphase boundary). Reasonably exact results (Rice, 1992) are given next within a "constrained-path" approximation, that is already tacit in the results presented so far.

1.4 Results for General Shear Loading, Coincident Crack and Slip Planes

Suppose now that the solid of Fig. 1.3 is loaded in combined mode I, II and III, Fig. 1.5, so that stresses on the slip plane in the absence of any relaxation would be $\sigma_{21} = K_{II}/\sqrt{2\pi r}, \sigma_{22} = K_{I}/\sqrt{2\pi r}, \sigma_{23} = K_{III}/\sqrt{2\pi r}$. In general the displacement discontinuity on the slip plane could have components in both the 1, 2 and 3 directions, $\delta_1 = u_1^+ - u_1^-, \delta_2 = u_2^+ - u_2^-$ and $\delta_3 = u_3^+ - u_3^-$. The energy $\Phi^*(\delta_1, \delta_2, \delta_3)$ of the slip plane is now related to the stresses by $\sigma_{2\alpha} = \partial\Phi^*(\delta_1, \delta_2, \delta_3)/\partial\delta_\alpha$ and an application of the J integral paralleling that in the previous section shows that solutions of the static elastic equations for this case must have relative displacements

$(\delta_{1\text{tip}}, \delta_{2\text{tip}}, \delta_{3\text{tip}})$ at the crack tip satisfying

$$G \equiv \left[(1-\nu)\left(K_{\mathrm{I}}^2 + K_{\mathrm{II}}^2\right) + K_{\mathrm{III}}^2\right]/2\mu = \Phi^*\left(\delta_{1\text{tip}}, \delta_{2\text{tip}}, \delta_{3\text{tip}}\right). \qquad (1.10)$$

This condition, however, does not let us determine a nucleation condition since now there are too many degrees of freedom at the tip.

A solution can be found if we make the assumptions that the relative motion along the slip plane is pure shear so that opening $\delta_2 = 0$ (an improvement is to consider relaxed paths along which $\partial\Phi^*/\partial\delta_2 = 0$), and that a certain direction or more generally that a certain set of crystallographically equivalent directions within a slip plane are far less resistant to shear than are any other directions. Such directions would, of course, coincide with the observed slip directions $\boldsymbol{s}$, i.e., the directions of Burgers vectors $\boldsymbol{b}$. Calculations from atomic models (Yamaguchi et al., 1981; Sun et al., 1991) of slip plane energies for different directions of shear do indeed show very large differences in energy; see Section 1.9. Thus let the angle ϕ denote the angle of the easy slip direction on the slip plane, where ϕ is measured from the x_1 axis (Fig. 1.5) so that $\phi = 0$ corresponds to an edge dislocation, whereas $\phi = \pi/2$ corresponds to a screw dislocation, relative to the crack front. When there are several such directions, we shall regard ϕ as denoting the first such direction to meet the nucleation condition, derived below, under the given ratio of K_{III} to K_{II} loading.

We now make the approximation that the resistance to slip along directions other than ϕ, and the resistance to tensile opening, is so great that we can regard the relative displacement as being constrained to a pure slip path at angle ϕ, so that

$$\delta_1 = \delta\cos\phi, \delta_2 = 0, \text{ and } \delta_3 = \delta\sin\phi \qquad (1.11)$$

where δ is the slip along direction ϕ. Thus if

$$\tau = \sigma_{21}\cos\phi + \sigma_{23}\sin\phi \qquad (1.12)$$

denotes the resolved shear stress in the slip direction, we may assume as boundary condition along the slip plane that τ is related to δ like in Fig. 1.1(b), and that Φ of Fig. 1.1(c) is given as before as $\Phi = \int \tau d\delta$. Because of the constraint on the relative displacements, it will no longer be the case that the slip process relaxes the stress singularity at the crack tip. Thus, in addition to the K_{I}, K_{II} and K_{III} characterizing the remotely applied loading, we will also have a non-zero stress intensity factors $K_{\mathrm{I(tip)}}$, $K_{\mathrm{II(tip)}}$ and $K_{\mathrm{III(tip)}}$ remaining at the crack tip at $x_1 = 0$.

Evaluation of the J integral along the path Γ_{far} gives

$$J = G \equiv \left[(1-\nu)\left(K_{\mathrm{I}}^2 + K_{\mathrm{II}}^2\right) + K_{\mathrm{III}}^2\right]/2\mu \qquad (1.13)$$

whereas in evaluating the contribution along Γ_{slit} we now have to include the contribution from the crack tip singularity, thus getting

$$\begin{aligned} J &= \left[(1-\nu)\left(K_{\text{I(tip)}}^2 + K_{\text{II(tip)}}^2\right) + K_{\text{III(tip)}}^2\right]/2\mu - \int_0^\infty \sigma_{2\alpha}\partial\delta_\alpha/\partial x_1 dx_1 \\ &= \left[(1-\nu)\left(K_{\text{I(tip)}}^2 + K_{\text{II(tip)}}^2\right) + K_{\text{III(tip)}}^2\right]/2\mu + \Phi(\delta_{\text{tip}}) \end{aligned} \quad (1.14)$$

where it has been noted that $\sigma_{2\alpha}\partial\delta_\alpha/\partial x_1 = \tau\partial\delta/\partial x_1 = \partial\Phi(\delta)/\partial x_1$ in view of the constraint on the slip path. The following conditions may be brought to bear: Since τ is bounded at the tip,

$$K_{\text{II(tip)}}\cos\phi + K_{\text{III(tip)}}\sin\phi = 0. \quad (1.15)$$

Also, by using the separate mode I, II and III solutions for the effect of slip on alteration of the stress intensity factors, we have

$$\begin{aligned} &[K_{\text{I}} - K_{\text{I(tip)}}, K_{\text{II}} - K_{\text{II(tip)}}, K_{\text{III}} - K_{\text{III(tip)}}] \\ &\quad = -\frac{\mu}{\sqrt{2\pi}(1-\nu)}\int_0^\infty \frac{d}{d\rho}[\delta_2(\rho), \delta_2(\rho), (1-\nu)\delta_3(\rho)]\frac{d\rho}{\sqrt{\rho}}. \end{aligned} \quad (1.16)$$

When the above constraints on the δ_α are used, this gives

$$K_{\text{I(tip)}} = K_{\text{I}}, \text{ and } (1-\nu)\sin\phi(K_{\text{II}} - K_{\text{II(tip)}}) - \cos\phi(K_{\text{III}} - K_{\text{III(tip)}}) = 0. \quad (1.17)$$

We may therefore solve for $K_{\text{II(tip)}}$ and $K_{\text{III(tip)}}$ as

$$(K_{\text{II(tip)}}, K_{\text{III(tip)}}) = \frac{(\sin\phi, -\cos\phi)}{\cos^2\phi + (1-\nu)\sin^2\phi}[(1-\nu)\sin\phi K_{\text{II}} - \cos\phi K_{\text{III}}], \quad (1.18)$$

and when we substitute these results into the two expressions for J above, and equate the expressions, one finds after a little manipulation that the slip δ_{tip} at the crack tip is given by

$$\frac{1-\nu}{2\mu}\frac{(\cos\phi K_{\text{II}} + \sin\phi K_{\text{III}})^2}{\cos^2\phi + (1-\nu)\sin^2\phi} = \Phi(\delta_{\text{tip}}). \quad (1.19)$$

This coincides with the results of the last section for mode II loading in emission of an edge ($\phi = 0$) dislocation and for mode III loading in emission of a screw ($\phi = \pi/2$). Since the maximum of Φ is γ_{us}, the nucleation criterion is therefore

$$\cos\phi K_{\text{II}} + \sin\phi K_{\text{III}} = \sqrt{\frac{2\mu}{1-\nu}[\cos^2\phi + (1-\nu)\sin^2\phi]\gamma_{\text{us}}} \quad (1.20)$$

(assuming that the left side is positive; a minus sign should precede the right side if not). The combination

$$K \equiv \cos\phi K_{\mathrm{II}} + \sin\phi K_{\mathrm{III}} \tag{1.21}$$

which enters the criterion has an evident interpretation as the intensity factor for the resolved shear stress along the slip direction.

Anisotropy: We may readily generalize the above discussions to take into account anisotropic elasticity (Sun and Rice, 1992). Evaluation of the J integral along the path Γ_{far} yields

$$J = G = \Lambda_{\alpha\beta} K_\alpha K_\beta \tag{1.22}$$

where $\boldsymbol{K} = (K_1, K_2, K_3) = (K_{\mathrm{II}}, K_{\mathrm{I}}, K_{\mathrm{III}})$ and $\Lambda_{\alpha\beta}$ is the appropriate matrix (derived by methods of Stroh, 1958, and Barnett and Asaro, 1972) for the anisotropic material. $\Lambda_{\alpha\beta}$ is real, symmetric, and positive definite, and is a function of the elastic constants of the material; it has the dimension of compliance and generally contains off-diagonal elements. Assume that a cohesive zone with slip $\delta(r)$ along a constrained path as above is developed to relax the singular stress field near the tip in response to the loading. There still exist residual singular stress components, which are described by the stress intensity factors $K_{\alpha(\mathrm{tip})}, \alpha = 1,2,3; \boldsymbol{K}_{(\mathrm{tip})} = (K_{1(\mathrm{tip})}, K_{2(\mathrm{tip})}, K_{3(\mathrm{tip})})$. The J integral evaluated along Γ_{slit} contains two contributions; one is from the crack tip singularity and the other is due to the energy $\Phi(\delta_{\mathrm{tip}})$,

$$J = \Lambda_{\alpha\beta} K_{\alpha(\mathrm{tip})} K_{\beta(\mathrm{tip})} + \Phi(\delta_{\mathrm{tip}}) \tag{1.23}$$

We define $\boldsymbol{s}(\phi) = (\cos\phi, 0, \sin\phi)$ so that $\delta_\alpha(r) = \delta(r) s_\alpha(\phi)$. The anisotropic derivation for the emission criterion proceeds analogously to that of the isotropic case. Further details may be found in Sun and Rice (1992) who use $s_\alpha(\phi) K_{\alpha(\mathrm{tip})} = 0$, corresponding to Eq. 1.15, and also the anisotropic analog of Eq. 1.16, expressing each $K_\alpha - K_{\alpha(\mathrm{tip})}$ as a similar integral operator on $[\Lambda^{-1}]_{\alpha\beta}\delta_\beta$, to derive the nucleation condition

$$s_\alpha(\phi) K_\alpha = \sqrt{\gamma_{\mathrm{us}} p(\phi)} \tag{1.24}$$

where we have defined $p(\phi) = [\Lambda^{-1}]_{\alpha\beta} s_\alpha(\phi) s_\beta(\phi)$.

Some tendency for dilatant opening across a lattice plane ($\delta_2 \neq 0$) must, in general, accompany shear. A particular embedded-atom model for iron, used in molecular dynamics simulations by Cheung (1990), provides an example for which the constrained-path approximation with $\delta_2 = 0$ is not so good, in that high tensile stress across slip planes at a crack tip noticeably reduce the resistance to dislocation emission (Cheung et al., 1991). These features require a more detailed formulation including numerical solution of coupled integral equations for the distribution of the δ's. The coupling of dilatant opening and shear has been analyzed based on such numerical solutions in work by Beltz and Rice (1991, 1992a) and Sun et al. (1992) to be discussed, and confirm the conclusions of Cheung et. al (1991) for their model of α-iron.

An approximate account of such tension-shear coupling can be made by simply interpreting γ_{us} as the unstable stacking energy for *relaxed* shear, along a path with δ_2 chosen to make $\partial\Phi^*(\delta_1, \delta_2, \delta_3)/\partial\delta_2 = 0$. Such may be formally justified within an alternative "constrained path" approximation by observing that the J integral conservation applies not only to the entire elastic field but also to each crack tip mode individually. Thus, we can equate the expressions for J given by the right sides of Eqs. 1.13 and 1.14, but with the K_{I} and $K_{\mathrm{I(tip)}}$ terms deleted and understanding that now α in Eq. 1.24 ranges over just 1 and 3. Hence, choosing as the constrained path that with $\partial\Phi^*/\partial\delta_2 = 0$, but restricting $\delta_1 = \delta\cos\phi$ and $\delta_3 = \delta\sin\phi$ as above, we re-derive Eq. 1.20 with γ_{us} now interpreted as the relaxed value. The procedure does a good job of describing coupled tension-shear results as will be discussed in Section 1.7.

1.5 Nucleation of Dissociated Dislocations, Coincident Crack and Slip Planes

The parts of this section based on isotropic elasticity also follow Rice (1992). Suppose that a complete lattice dislocation in a certain crystal is composed of two partial dislocations with respective Burgers vectors $\boldsymbol{b}_A$ and $\boldsymbol{b}_B$, where these share the same slip plane and are separated by a faulted portion of slip plane with energy γ_{sf} (*stacking fault* energy) per unit area. We continue with the simplification that the crack plane and slip plane are coincident as in Fig. 1.5, and make the constrained-path approximation for each partial dislocation individually. Thus partial dislocation A is created by slip δ_A from 0 to b_A along a definite direction at angle ϕ_A (the first of the different possible partial dislocation directions on the slip plane to meet the nucleation condition, under the prevailing $K_{\mathrm{III}}/K_{\mathrm{II}}$ ratio), and then partial dislocation B can come into existence by slip δ_B from 0 to b_B at angle ϕ_B (taken to be the most favorable of the allowed crystal directions for continuation of slip as a second partial). For $\{111\}$ planes in fcc lattices, with partials of Burgers vectors in $< 211 >$ directions summing to complete $< 110 >$ dislocations, ϕ_B and ϕ_A differ by 60°. (Anderson (1986) previously analyzed partial dislocation nucleation within the Rice-Thomson framework.)

Energy functions Φ for the two partials are shown in Fig. 1.6. The first slip over b_A carries the energy Φ_A from zero, through the peak at γ_{us}, and to a residual state of energy γ_{sf}; the next slip starts with energy Φ_B at γ_{sf}, goes through the same peak γ_{us}, and returns to zero after slip b_B, a complete dislocation having then been formed. Let

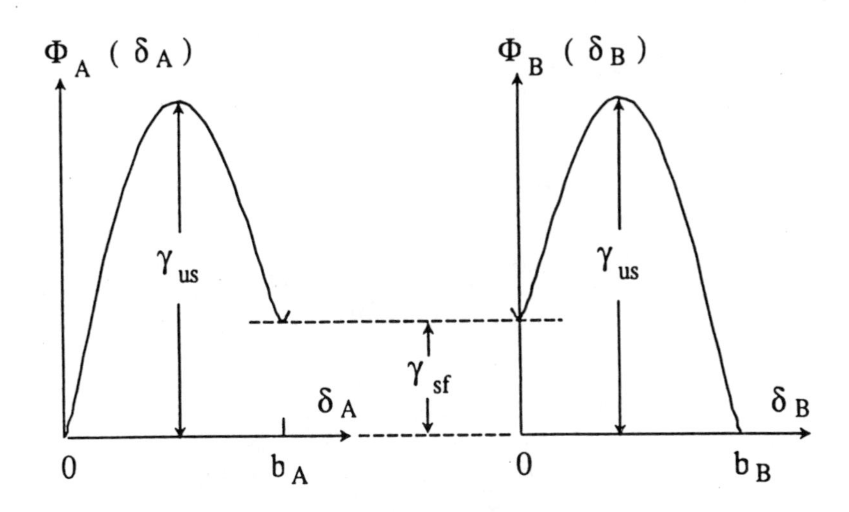

FIGURE 1.6. Energy versus slip for two partial dislocation which combine to form a complete lattice dislocation; γ_{sf} is the energy per unit area of the stacking fault.

$$K_A = K_{\mathrm{II}} \cos \phi_A + K_{\mathrm{III}} \sin \phi_A, \; K_B = K_{\mathrm{II}} \cos \phi_B + K_{\mathrm{III}} \sin \phi_B. \quad (1.25)$$

The analysis of the previous section shows that first partial nucleates when

$$K_A = K_{A\mathrm{crit}} \equiv \sqrt{\frac{2\mu}{1-\nu}[\cos^2 \phi_A + (1-\nu)\sin^2 \phi_A]\gamma_{\mathrm{us}}}. \quad (1.26)$$

However, the fully formed partial dislocation which emerges, of Burgers vector b_A, leaves a faulted plane of energy γ_{sf} behind it and thus is not swept indefinitely far away by the stress field but instead remains in the vicinity of the crack tip. Let r_A be the position of the core of that partial dislocation. It is determined by equilibrium between Peach-Koehler configurational forces; that due to the applied stress field $K_A/\sqrt{2\pi r}$ must balance the sum of the dislocation image force due to the presence of the stress-free crack surface (Rice and Thomson, 1974) and the force γ_{sf} tending to annihilate the fault. (We can treat this defect as a classical, singular line dislocation, without considering its spread-out core, since it will be seen that r_A is typically very large compared to b_A). Thus r_A is the (largest) root of

$$K_A b_A/\sqrt{2\pi r_A} = \gamma_{sf} + \mu b_A^2[\cos^2 \phi_A + (1-\nu)\sin^2 \phi_A]/4\pi(1-\nu)r_A, \quad (1.27)$$

from which one finds that

$$\frac{\mu b_A}{(1-\nu)\sqrt{2\pi r_A}} = \frac{K_A[1-\sqrt{1-(\gamma_{sf}/\gamma_{us})(K_{A\text{crit}}/K_A)^2}}{\cos^2\phi_A + (1-\nu)\sin^2\phi_A} \tag{1.28}$$

(the combination on the left will be needed shortly) and that

$$\frac{r_A}{b_A} = \frac{(K_{A\text{crit}}/K_A)^2[\cos^2\phi_A + (1-\nu)\sin^2\phi_A](\mu b_A/\gamma_{us})}{4\pi(1-\nu)\left[1-\sqrt{1-(\gamma_{sf}/\gamma_{us})(K_{A\text{crit}}/K_A)^2}\right]^2}. \tag{1.29}$$

The last expression, to be used after nucleation ($K_A \geq K_{A\text{crit}}$), defines a rapidly increasing function of K_A. It is least when $K_A = K_{A\text{crit}}$, and then gives the position r_A to which the partial jumps just after nucleation. Later we will see estimates of $\mu b_A/\gamma_{us}$ ranging from 25 to 40 for fcc metals, and $\gamma_{sf} \approx \gamma_{us}/4$ to $\gamma_{us}/2$ seems to be representative (smaller values give larger r_A). These lead, for $\phi_A = 0$ and $\nu = 0.3$, to $r_A/b \approx 30$ to 250. This is an overestimate, in that it neglects lattice friction against motion of the partial dislocation.

The simplest way to address emission of the second partial is to note that the first partial dislocation has the effects of (i) modifying the K_{II} and K_{III} at the tip (say, to values K_{II}^* and K_{III}^*), and (ii) resetting the energy of the unslipped state from zero to γ_{sf}, Fig. 1.6, so that the peak energy to be surmounted for the instability leading to dislocation B to occur is reduced from γ_{us} to $\gamma_{us} - \gamma_{sf}$. With those factors taken into account, we can just use the result of the last section so that, at instability,

$$\begin{aligned} K_B^* &\equiv K_{II}^*\cos\phi_B + K_{III}^*\sin\phi_B \\ &= \sqrt{\frac{2\mu}{1-\nu}\left[\cos^2\phi_B + (1-\nu)\sin^2\phi_B\right](\gamma_{us}-\gamma_{sf})} \end{aligned} \tag{1.30}$$

(When we take into account the expressions for K_{II}^* and K_{III}^*, given next, the same result could be derived, alternatively, by applying the J integral, in the style of the last section, to the entire dislocated array, partial dislocation A, the associated stacking fault zone, and incipient partial dislocation B.)

The expressions for K_{II}^* and K_{III}^* are derivable from Eq. 1.26 as

$$K_{II}^* = K_{II} - \mu b_A\cos\phi_A/(1-\nu)\sqrt{2\pi r_A},\ K_{III}^* = K_{III} - \mu b_A\sin\phi_A/\sqrt{2\pi r_A}. \tag{1.31}$$

Using Eq. 1.28 for the latter terms, the quantity K_B^* entering the criterion for nucleation of the second partial is given by

$$K_B^* = K_B - \eta K_A + \eta\sqrt{K_A^2 - \frac{2\mu}{1-\nu}\gamma_{sf}[\cos^2\phi_A + (1-\nu)\sin^2\phi_A]} \quad (1.32)$$

where

$$\eta = \frac{\cos\phi_A\cos\phi_B + (1-\nu)\sin\phi_A\sin\phi_B}{\cos^2\phi_A + (1-\nu)\sin^2\phi_A}. \quad (1.33)$$

Anisotropy. The treatment of dissociated partial dislocations in an anisotropic elastic medium proceeds analogously with the isotropic case (Sun and Rice, 1992). The emission of the first partial occurs when the nucleation condition is reached in anisotropic medium,

$$K_A \equiv s_\alpha(\phi_A)K_\alpha = K_{A\text{crit}} = \sqrt{\gamma_{us}p(\phi_A)}. \quad (1.34)$$

The emitted partial dislocation is treated as a line defect; its stable equilibrium position is found from the condition that the force exerted on it vanishes:

$$f_A = K_A b_A/\sqrt{2\pi r_A} - \gamma_{sf} + f_r = 0. \quad (1.35)$$

The image force f_r on the partial dislocation itself is given by Rice (1985) as

$$f_r = -\frac{b_A^2[\Lambda^{-1}]_{\alpha\beta}s_\alpha(\phi_A)s_\beta(\phi_A)}{8\pi r_A}. \quad (1.36)$$

Hence,

$$\frac{b_A}{2\sqrt{2\pi r_A}} = \frac{K_A\left[1-\sqrt{1-(\gamma_{sf}/\gamma_{us})(K_{A\text{crit}}/K_A)^2}\right]}{p(\phi_A)} \quad (1.37)$$

The stress intensity factor is now shielded by the emitted first partial dislocation

$$K_\alpha^* = K_\alpha - b_A[\Lambda^{-1}]_{\alpha\beta}s_\beta(\phi_A)/2\sqrt{2\pi r_A}. \quad (1.38)$$

Combining Eq. 1.25 with Eq. 1.38 gives

$$K_B^* = s_\alpha(\phi_B)K_\alpha^* = K_B - \eta(\phi_B,\phi_A)K_A + \eta(\phi_B,\phi_A)\sqrt{K_A^2 - \gamma_{sf}p(\phi_A)} \quad (1.39a)$$

where

$$\eta(\phi_B,\phi_A) = \frac{[\Lambda^{-1}]_{\alpha\beta}s_\alpha(\phi_B)s_\beta(\phi_A)}{[\Lambda^{-1}]_{\alpha\beta}s_\alpha(\phi_A)s_\beta(\phi_A)} = \frac{[\Lambda^{-1}]_{\alpha\beta}s_\alpha(\phi_B)s_\beta(\phi_A)}{p(\phi_A)}. \quad (1.39b)$$

The emission of the second partial dislocation occurs when

$$K_B^* = \sqrt{(\gamma_{us} - \gamma_{sf})p(\phi_B)}. \quad (1.40)$$

The resulting nucleation criterion (now considering isotropic materials only) is a little complex to study in general, but it takes a simpler form in a special case of considerable interest for fcc metals, in which $\phi_A = 0°$ and $|\phi_B| = 60°$ (+ or - chosen according to the sign of $K_{\rm III}$). The 0° partial will be the first nucleated only if $K_B < (\sqrt{4-3\nu}/2)K_A$, which is equivalent to $|K_{\rm III}| < (\sqrt{4-3\nu}-1)K_{\rm II}/\sqrt{3}$ or, for $\nu = 0.3$, to $|K_{\rm III}| < 0.44 K_{\rm II}$, a condition which is now assumed to hold. The first partial nucleates when

$$K_{\rm II} = \sqrt{2\mu\gamma_{\rm us}/(1-\nu)} \tag{1.41}$$

and the condition for nucleation of the second, given above, now simplifies to

$$\sqrt{3}|K_{\rm III}| + \sqrt{K_{\rm II}^2 - 2\mu\gamma_{\rm sf}/(1-\nu)} = \sqrt{2\mu(4-3\nu)(\gamma_{\rm us}-\gamma_{\rm sf})/(1-\nu)}. \tag{1.42}$$

Since this equation takes effect only after $K_{\rm II}$ reaches the value to nucleate the first partial, the $K_{\rm II}$ which enters it will always be at least as large as that of Eq. 1.26, and hence the quantity of which the square root is taken is always positive since (since $\gamma_{\rm us} > \gamma_{\rm sf}$).

Three possibilities exist, depending on $K_{\rm III}$: (i) When $K_{\rm III}$ is zero or sufficiently small, $K_{\rm II}$ must be increased to nucleate the second partial. (ii) For $|K_{\rm III}|$ greater than a certain limit $K_{\rm III(sp)}$ given below, the second partial nucleates spontaneously once the first has formed; no increase in $K_{\rm II}$ is then required. (iii) And for $|K_{\rm III}|$ yet larger, the analysis ultimately becomes untenable because, instead, the $|\phi| = 60°$ partial nucleates first, and we have to start from the beginning, interchanging A and B.

The greatest $K_{\rm II}$ to nucleate the second partial results when $K_{\rm III} = 0$, in which case

$$K_{\rm II} = \sqrt{2\mu[(4-3\nu)\gamma_{\rm us} - 3(1-\nu)\gamma_{\rm sf}]/(1-\nu)} \tag{1.43}$$

When $\nu = 0.3$ and $\gamma_{\rm sf} = \gamma_{\rm us}/3$, this is 55% higher than the $K_{\rm II}$ to nucleate the first dislocation. The required increase in $K_{\rm II}$ diminishes to zero when $|K_{\rm III}| = K_{\rm III(sp)}$, where

$$K_{\rm III(sp)} = \sqrt{2\mu(\gamma_{\rm us}-\gamma_{\rm sf})/(1-\nu)}\left(\sqrt{4-3\nu}-1\right)\sqrt{3} \tag{1.44}$$

is calculated by setting $K_{\rm II}$ equal to that to nucleate the first partial. For the numerical values above, $K_{\rm III(sp)}$ is 0.36 times the $K_{\rm II}$ to nucleate the first partial. The range of $K_{\rm III}$ for which there is spontaneous nucleation of the 60° partial persists up to a limit given by the same expression as for $K_{\rm III(sp)}$ but with $(\gamma_{\rm us} - \gamma_{\rm sf})$ increased to $\gamma_{\rm us}$; beyond that limit, it is the 60° partial which nucleates first.

Nucleation by the partial mechanism discussed here is considered again in the next section, where slip planes at angle $\theta \neq 0$ are considered.

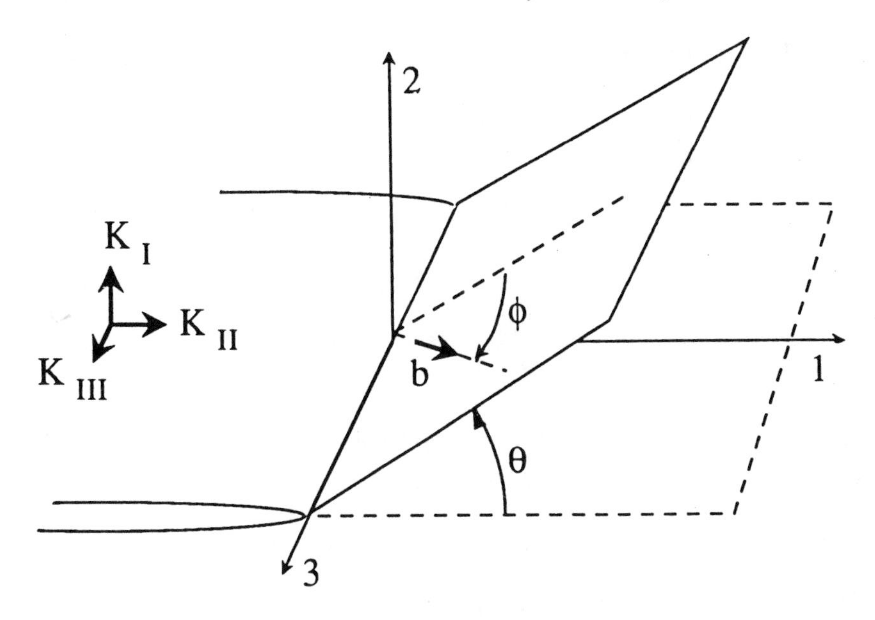

FIGURE 1.7. Slip plane inclined at angle θ with the prolongation of the crack plane; slip direction inclined at angle ϕ with the normal to the crack tip.

1.6 Approximate Nucleation Condition, Slip Plane Not Coincident with Crack Plane

In general the most highly stressed slip plane will make a non-zero angle θ relative to the crack plane, like in Fig. 1.7, and the Burgers vector direction along that plane will make an angle ϕ with a line drawn perpendicular to the crack tip, similar to Fig. 1.5.

Effective stress intensity factor concept. Rice (1992) proposed the following approximate treatment of nucleation when $\theta \neq 0$. Suppose that the solid is loaded so as to induce a general set of intensity factors K_{I}, K_{II} and K_{III} at the crack tip. The in-plane and anti-plane shear stress components acting along the slip plane, according to the linear elastic solution, are

$$\sigma_{\theta r} = \left[K_{\text{I}} f_{\text{I}}(\theta) + K_{\text{II}} f_{\text{II}}(\theta)\right] / \sqrt{2\pi r}, \ \sigma_{\theta 3} = K_{\text{III}} f_{\text{III}}(\theta) / \sqrt{2\pi r}. \tag{1.45}$$

where, for the isotropic case,

$$\begin{aligned} f_{\text{I}}(\theta) &= \cos^2(\theta/2)\sin(\theta/2), \ f_{\text{II}}(\theta) = \cos(\theta/2)[1 - 3\sin^2(\theta/2)], \\ f_{\text{III}}(\theta) &= \cos(\theta/2). \end{aligned} \tag{1.46}$$

The form of these results motivate the notion of *effective* mode II and mode III intensity factors along the slip plane at angle θ. These are defined as

$$K_{\mathrm{II}}^{\mathrm{eff}} = K_{\mathrm{I}} f_{\mathrm{I}}(\theta) + K_{\mathrm{II}} f_{\mathrm{II}}(\theta), \quad K_{\mathrm{III}}^{\mathrm{eff}} = K_{\mathrm{III}} f_{\mathrm{III}}(\theta). \tag{1.47}$$

The same idea can be applied to anisotropic elasticity; the set of effective stress intensity factors K_α^{eff} are defined by writing

$$\sigma_{\theta\alpha}(r) = K_\alpha^{\mathrm{eff}} / \sqrt{2\pi r} \tag{1.48}$$

with $\alpha = r, \theta, z$ from the general loading $\boldsymbol{K} = (K_1, K_2, K_3) = (K_{\mathrm{II}}, K_{\mathrm{I}}, K_{\mathrm{III}})$ on the main crack,

$$K_\alpha^{\mathrm{eff}} = F_{\alpha\beta}(\theta) K_\beta \tag{1.49}$$

where $F_{\alpha\beta}(\theta)$ appear in the near tip expression for $\sigma_{\theta\alpha}$ under mode β loading. Equations in the form of Eq. 1.45 and 1.47 are sufficient for orthotropic elasticity with principal axes along the x_1, x_2, x_3 system, in the following approximations, with Eqs. 1.46 then replaced by results appropriate to orthotropic elasticity, since in-plane deformation is then decoupled from anti-plane deformation.

As a simple approximation, we may now assume that the nucleation conditions derived for $\theta = 0$ in all the earlier sections of the paper apply as well to an inclined slip plane, $\theta \neq 0$, when we replace K_{II} and K_{III} in expressions earlier in the paper with the effective intensity factors $K_{\mathrm{II}}^{\mathrm{eff}}$ and $K_{\mathrm{III}}^{\mathrm{eff}}$ above. Thus the basic nucleation condition of Eq. 1.20 for a complete dislocation in an isotropic material becomes, approximately when $\theta \neq 0$,

$$\begin{aligned} &[f_{\mathrm{I}}(\theta) K_{\mathrm{I}} + f_{\mathrm{II}}(\theta) K_{\mathrm{II}}] \cos\phi + f_{\mathrm{III}}(\theta) K_{\mathrm{III}} \sin\phi \\ &\quad = \sqrt{\frac{2\mu}{1-\nu} \left[\cos^2\phi + (1-\nu)\sin^2\phi\right] \gamma_{\mathrm{us}}} \end{aligned} \tag{1.50}$$

and corresponding results are given shortly for nucleation of a dissociated dislocation.

We treat anisotropic elasticity in a parallel fashion to the isotropic case. The proper $\Lambda_{\alpha\beta}$ matrix, denoted by $\Lambda_{\alpha\beta}^{(\theta)}$ for the crack extension force G_r for a crack extending along the radial direction ($G_r = \Lambda_{\alpha\beta}^{(\theta)} K_\alpha^{\mathrm{eff}} K_\beta^{\mathrm{eff}}$, where now α, β range over r, θ, z.) is

$$\Lambda_{\alpha\beta}^{(\theta)} = R_{\alpha\gamma} \Lambda_{\gamma\delta} R_{\beta\delta} \tag{1.51}$$

where $\boldsymbol{R}$ is the rotation matrix

$$\boldsymbol{R} = \begin{bmatrix} \cos\theta & \sin\theta & 0 \\ -\sin\theta & \cos\theta & 0 \\ 0 & 0 & 1 \end{bmatrix} \tag{1.52}$$

The dislocation nucleation criterion is written the same way for K_α^{eff} as for the coplanar case, so that

$$s_\alpha(\phi)F_{\alpha\beta}(\theta)K_\beta = \sqrt{\gamma_{\text{us}}p(\phi,\theta)} \tag{1.53}$$

with

$$p(\phi,\theta) = \Lambda_{\alpha\beta}^{(\theta)^{-1}} s_\alpha(\phi)s_\beta(\phi) \tag{1.54}$$

and where now $\mathbf{s}(\phi)$ has components ($\cos\phi$, 0, $\sin\phi$) in the (r, θ, z) directions.

Exact analysis for $\theta \neq 0$. By contrast, the exact (but neglecting tension-shear coupling) route to determining a nucleation condition would involve enforcing equilibrium along the inclined slip plane, as achieved for nucleation in the edge mode in the isotropic case by satisfying the following integral equation:

$$\tau(r) = \frac{K_{\text{I}}f_{\text{I}}(\theta) + K_{\text{II}}f_{\text{II}}(\theta)}{\sqrt{2\pi r}} - \frac{\mu}{2\pi(1-\nu)}\int_0^\infty \frac{d\delta(\rho)}{d\rho}g_{11}(r,\rho;\theta)d\rho \tag{1.55}$$

where $g_{11}(r,\rho;\theta)$ reduces to $\sqrt{\rho/r}/(r-\rho)$ when $\theta = 0$, and where $\tau = f(\delta(r))$, e.g., $\tau = (\mu/2\pi)\sin(2\pi\Delta/b)$ where $\delta = \Delta - (h/2b)\sin(2\pi\Delta/b)$ for the Frenkel model. Solutions of Eq. 1.55 have been found in connection with the dislocation emission problem (Beltz, 1991); it has been found that there is a critical value of $K_{\text{II}}^{\text{eff}}$ beyond which no solutions exist, and that value is taken as the nucleation value. The critical $K_{\text{II}}^{\text{eff}}$ thus found has some θ dependence, in contrast to the prediction of Eq. 1.50; see below.

For pure mode I loading of an isotropic solid, in which case $G = (1-\nu)K_{\text{I}}^2/2\mu$, the above approximate criterion given by Eq. 1.50 reduces to

$$G = 8\frac{1+(1-\nu)\tan^2\phi}{(1+\cos\theta)\sin^2\theta}\gamma_{\text{us}} \tag{1.56}$$

for dislocation nucleation. Figure 1.8 gives a comparison of the approximate result in Eq. 1.56 with the numerical results from the solution of Eq. 1.55 based on the Frenkel form. There is reasonable agreement and the results become identical in the limit of small θ. However, G as estimated by Eq. 1.56, based on the K^{eff} concept, must be reduced by 11% when $\theta = 45°$, to agree with the exact result based on Eq. 1.55, by 16% when $\theta = 55°$, and by 26% when $\theta = 90°$.

Dislocation nucleation versus cleavage. The G for dislocation emission (as given by its approximation in Eq. 1.56, in order to expedite the algebraic manipulations to follow) may be compared to

$$G = 2\gamma_s \tag{1.57}$$

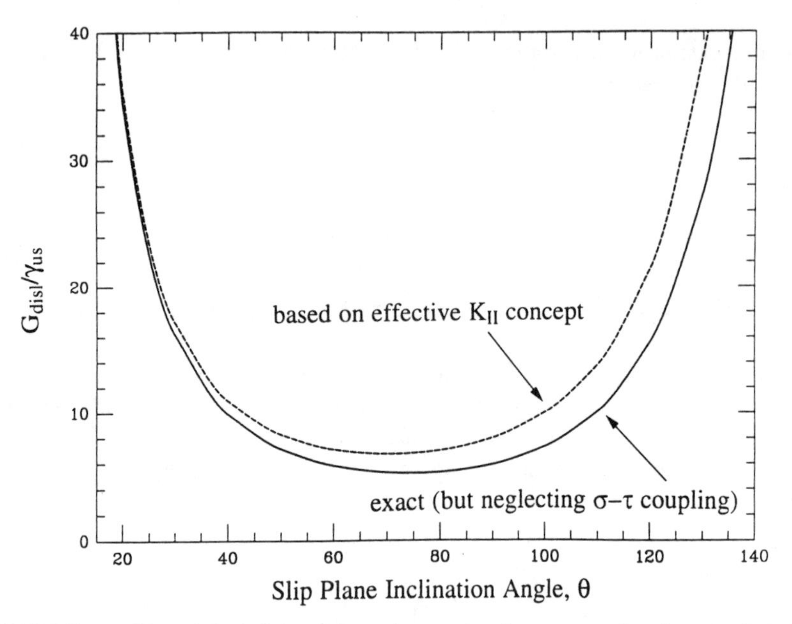

FIGURE 1.8. The critical G for dislocation nucleation versus slip plane inclination angle. The solid line is based on an exact solution (but neglects shear-tension coupling); the dashed line is calculated based on the K^{eff} concept.

(γ_s = surface energy) for Griffith cleavage. Hence crack tip blunting by dislocation nucleation should occur before conditions for Griffith cleavage decohesion are met if the latter G exceeds the former, which happens for the isotropic solid when

$$\frac{\gamma_s}{\gamma_{\text{us}}} > 4\frac{1 + (1-\nu)\tan^2\phi}{(1+\cos\theta)\sin^2\theta}. \tag{1.58}$$

Cleavage occurs before the tip can blunt when the inequality is reversed. (Given the discussion in connection with Fig. 1.8, the number on the right side of Eq. 1.58 should be reduced by an amount ranging from approximately 10% when $\theta = 45°$ to 25% when $\theta = 90°$, if we are to have a more accurate estimate. We use the simpler formulae based on Eqs. 1.50, 1.53, and 1.56 here and in Section 1.10, except where noted otherwise).

The critical $\gamma_s/\gamma_{\text{us}}$ ratio is, however, usually quite sensitive to deviations from pure mode I loading. For example, if x and z denote fractional shear loadings, defined by writing $K_{\text{II}} = xK_{\text{I}}$ and $K_{\text{III}} = zK_{\text{I}}$, then the inequality to be met for emission before cleavage is (Rice, 1992)

$$\frac{\gamma_s}{\gamma_{\text{us}}} > \frac{4[1 + x^2 + z^2/(1-\nu)][1 + (1-\nu)\tan^2\phi]}{(1+\cos\theta)[\sin\theta + (3\cos\theta - 1)x + 2z\tan\phi]^2} \tag{1.59}$$

For a pure mode I loaded crack tip in an *anisotropic* elastic medium, the condition of dislocation emission before crack extension is,

$$\frac{\gamma_s}{\gamma_{us}} > \frac{\Lambda_{22} s_\alpha(\phi) \Lambda_{\alpha\beta}^{(\theta)^{-1}} s_\beta(\phi)}{2[s_\alpha(\phi) F_{\alpha 2}(\theta)]^2} \tag{1.60}$$

Equation 1.60 specializes for orthotropic elasticity, with principal axes aligned with the x_1, x_2 and x_3 coordinate system, as

$$\frac{\gamma_s}{\gamma_{us}} > \beta, \quad \text{where } \beta = \frac{[\Lambda_2 \cos^2\theta/\Lambda_1 + \sin^2\theta + \Lambda_2 \tan^2\phi/\Lambda_3]}{2 f_{\rm I}(\theta)^2} \tag{1.61}$$

and this generalizes for tensile loading with some fractional shear loadings as

$$\beta = \frac{(x^2\Lambda_1/\Lambda_2 + 1 + z^2\Lambda_3/\Lambda_2)[\Lambda_2 \cos^2\theta/\Lambda_1 + \sin^2\theta + \Lambda_2 \tan^2\phi/\Lambda_3]}{2[x f_{\rm II}(\theta) + f_{\rm I}(\theta) + z f_{\rm III}(\theta)]^2} \tag{1.62}$$

Here Λ_i (i = 1, 2, 3) are the diagonal components of the $\Lambda_{\alpha\beta}$ matrix discussed in connection with Eq. 1.22 and $f_{\rm I}(\theta)$, $f_{\rm II}(\theta)$ and $f_{\rm III}(\theta)$ are the appropriate functions defined in Eq. 1.45 for a singular crack tip field but here based upon orthotropic elasticity instead of isotropic elasticity (Sun and Rice, 1992).

Consider a case of interest for bcc solids: a crack on a {100} plane with tip along a < 100 > type direction, so as to intersect a {110} slip plane on which < 111 > slip can occur. In that case, $\theta = 45°$ and $\phi = \arctan(1/\sqrt{2}) = 35.3°$. Thus, for pure mode I loading and using isotropic expressions with $\nu = 0.3$, Eq. 1.58 predicts $\gamma_s/\gamma_{us} > 6.3$ for dislocation nucleation to occur before Griffith cleavage, but according to Eq. 1.59 the required ratio reduces nearly by a half, to $\gamma_s/\gamma_{us} = 3.5$, when $K_{\rm II}$ and $K_{\rm III}$ are just 10% of $K_{\rm I}$ (i.e., $x = z = 0.1$). Implications for specific solids are discussed in Section 1.10, after reviewing some estimates of γ_{us} in Section 1.9.

Partial dislocation pair. For the nucleation of dissociated dislocations with $\theta \neq 0$, we consider a geometry of interest for fcc solids, with a crack on a {100} plane and tip along a < 110 > direction, and assume that the most stressed {111} slip plane is that at $\theta = 54.73°$, and that the loading is such that the first partial to nucleate involves slip along the < 211 > direction at $\phi_A = 0°$ with the second at $\phi_B = 60°$. Then K_A and K_B of the earlier discussion of dissociated dislocations can be replaced by $K_A^{\rm eff}$ and $K_B^{\rm eff}$, defined like K_A and K_B in Eqs. 1.39a, 1.39b but in terms of $K_{\rm II}^{\rm eff}$ and $K_{\rm III}^{\rm eff}$. For the special θ and ϕ's considered, these quantities are

$$K_A^{\rm eff} = 0.363(K_{\rm I} + 0.897 K_{\rm II}), \; K_B^{\rm eff} = 0.769 K_{\rm III} + 0.5 K_A^{\rm eff}. \tag{1.63}$$

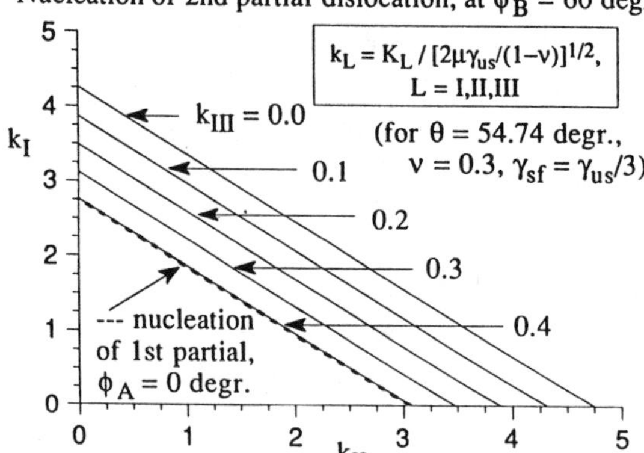

FIGURE 1.9. Combinations of $K_{\rm I}$, $K_{\rm II}$ and $K_{\rm III}$ for nucleation of the second of two partial dislocations in a fcc crystal with crack on $\{100\}$ plane, with tip along $< 110 >$ direction so that the relevant $\{111\}$ slip plane is at $\theta = 54.74^\circ$; the first partial is assumed to nucleate with $\phi = 0^\circ$ and the second with $\phi = 60^\circ$.

It is assumed that $K_{\rm II} \geq 0$ and $K_{\rm III} \geq 0$. If not, the same phenomena will occur relative to $\theta = -54.73^\circ$ if $K_{\rm II} < 0$ and to $\phi_B = -60^\circ$ if $K_{\rm III} < 0$, so $K_{\rm II}$ and $K_{\rm III}$ here can be interpreted as $|K_{\rm II}|$ and $|K_{\rm III}|$.

Reading from the earlier results, interpreted approximately in terms of the effective shear stress intensity factors, the 0° partial will indeed be the first to nucleate when $K_{\rm III} < 0.179(K_{\rm I} + 0.897K_{\rm II})$, and the nucleation condition (from Eq. 1.26) for that first partial is

$$K_{\rm I} + 0.897K_{\rm II} = 2.75\sqrt{2\mu\gamma_{\rm us}/(1-\nu)}. \tag{1.64}$$

This is shown as the dashed line in Fig. 1.9, which is analogous to the mixed mode nucleation diagrams of Lin and Thomson (1986). The nucleation condition for the second partial, at $\phi_B = 60^\circ$, is then, from Eqs. 1.30, 1.32, 1.33, 1.63,

$$\begin{aligned} 0.769K_{\rm III} \quad &+ \quad 0.5\sqrt{[0.363(K_{\rm I} + 0.897K_{\rm II})]^2 - \frac{2\mu}{1-\nu}\gamma_{\rm sf}} \\ &= \quad \sqrt{\frac{(4-3\nu)\mu}{2(1-\nu)}(\gamma_{\rm us} - \gamma_{\rm sf})}. \end{aligned} \tag{1.65}$$

For pure mode I loading, this is

$$K_{\rm I} = 2.75\sqrt{2\mu[(4-3\nu)\gamma_{\rm us} - 3(1-\nu)\gamma_{\rm sf}]/(1-\nu)}, \tag{1.66}$$

or $K_I = 4.26\sqrt{2\mu\gamma_{us}/(1-\nu)}$ when $\nu = 0.3$ and $\gamma_{sf} = \gamma_{us}/3$. The combined loading result is plotted in Fig. 1.9, based on $\nu = 0.3$ and $\gamma_{sf} = \gamma_{us}/3$, for various values of K_{III}. The nucleation condition is extremely sensitive to K_{III}: While the numerical factor 2.75 in Eq. 1.64 above increases to 4.26 for nucleation of the second partial when $K_{III} = 0$, that factor is reduced back to 2.75 (so that there is spontaneous nucleation of the second partial) when K_{III} is increased so that an analogously defined numerical factor for K_{III} reaches only 0.404.

Emission of partial dislocations before cleavage. From Eqs. 1.64 and 1.57, the first partial will nucleate before the Griffith cleavage condition is met, under pure mode I loading of the fcc configuration considered, if $\gamma_s/\gamma_{us} > 3.8$. Since $\phi_A = 0$, this result is insensitive to mode III loading, at least as long as $|K_{III}| < 0.179K_I$ so that the $\phi = 0°$ partial is actually the first to nucleate. If there is also a 10% mode II loading ($x = 0.1$), the inequality changes somewhat, to $\gamma_s/\gamma_{us} > 3.2$.

Under pure mode I loading, the second partial, and hence the complete fcc dislocation, nucleates before Griffith cleavage if, from Eqs. 1.66 and 1.57,

$$\gamma_s/\gamma_{us} > 3.8[4 - 3\nu - 3(1-\nu)\gamma_{sf}/\gamma_{us}], \tag{1.67}$$

which is $\gamma_s/\gamma_{us} > 9.1$ when $\nu = 0.3$ and $\gamma_{sf} = \gamma_{us}/3$. However, as anticipated from the discussion above, this result is extremely sensitive to small shear mode contributions, especially in mode III. Thus for loading with K_{II} and K_{III} both 10% of K_I, and with $\nu = 0.3$ and $\gamma_{sf} = \gamma_{us}/3$, the inequality becomes $\gamma_s/\gamma_{us} > 4.2$, so that there is a reduction to less than a half of the γ_s/γ_{us} value required for nucleation prior to cleavage under pure mode I loading.

Because of the strong sensitivity to shear loadings illustrated here, and in the earlier bcc discussion, it should rather commonly be the case that dislocations emerge from (nominally) tensile loaded cracks in solids which violate the γ_s/γ_{us} requirement for ductility under pure tensile loading by as much as, say, a factor of 2.

Anisotropy: We consider emission of paired partial dislocations in fcc metals within the anisotropic elasticity framework, adopting the same crack geometry and slip plane as in the isotropic case. That is, the crack is on the (001) plane growing along the $[\bar{1}10]$ direction; the slip plane which contains the [110] crack front and which is the easiest plane on which slip can occur under tensile loading, is the $(1\bar{1}1)$ plane. The slip plane makes an angle of $\theta = 54.7°$ with the crack plane.

We consider the paired partials, (1/6) $[\bar{1}12]$ as the first Shockley partial and (1/6) $[\bar{2}\bar{1}1]$ as the second partial. The condition for the first partial emission under pure mode K_I loading is expressed by Eq. 1.61, and the resulting parameter β is listed in Table 1.1 for fcc metals. The first partial slip is along the r direction, $\phi_A = 0°$.

The anisotropic case can be discussed similarly to the isotropic for emission of the second partial, based on the K_α^{eff} concept and the appropriate $\Lambda_{\alpha\beta}^{(\theta)}$ matrix for that inclined slip plane and Eq. 1.34 to 1.40 of the last section for partial dislocation when $\theta = 0°$. For the same geometry as in the isotropic case, $\phi_A = 0°$, and $\phi_B = 60°$, $\eta(\phi_A, \phi_B) = 1/2$ from Eq. 1.39b. The nucleation condition for the second partial, after Eq. 1.40, is thus

$$(\sqrt{3}/2)K_z^{\text{eff}} + (1/2)\sqrt{K_r^{\text{eff}\,2} - \gamma_{\text{sf}}p(\phi_A, \theta)} = \sqrt{(\gamma_{\text{us}} - \gamma_{\text{sf}})p(\phi_B, \theta)} \quad (1.68)$$

For pure tensile loading, the loading required for nucleation of the second partial is thus

$$K_{\text{I}} = \sqrt{4(\gamma_{\text{us}} - \gamma_{\text{sf}})p(\phi_B, \theta) + \gamma_{\text{sf}}p(\phi_A, \theta)}/f_{\text{I}}(\theta) \quad (1.69)$$

The condition of second partial emission before Griffith cleavage condition is met is again expressed in the inequality form

$$\frac{\gamma_s}{\gamma_{\text{us}}} > \beta \quad (1.70)$$

where now

$$\beta = \frac{\Lambda_{22}[4(1 - \gamma_{\text{sf}}/\gamma_{\text{us}})p(\phi_B, \theta) + \gamma_{\text{sf}}p(\phi_A, \theta)/\gamma_{\text{us}}]}{2[f_{\text{I}}(\theta)]^2} \quad (1.71)$$

The equation above further simplifies, when taking $\gamma_{\text{sf}}/\gamma_{\text{us}} = 1/3$ as in estimates for the isotropic case, to

$$\beta = \frac{\Lambda_{22}[8p(\phi_B, \theta) + p(\phi_A, \theta)]}{6[f_{\text{I}}(\theta)]^2} \quad (1.72)$$

The values for various fcc metals are listed in Table 1.1 with γ_{sf} taken as $\gamma_{\text{us}}/3.7$.

1.7 Effects of Tension Across the Slip Plane; Shear-Tension Coupling

The procedures for determining G discussed thus far have not explicitly considered effects of tensile stress across the slip plane. Cheung et al. (1990), and earlier Argon (1987), have argued that shear softening by high tensile stress is a critical element in dislocation nucleation. We can model such effects by broadening the framework to include coupled shear and tension (Beltz and Rice, 1991, 1992a). In the calculations to be discussed, the tensile stress across the slip plane, in the absence of slip, is assumed to follow the well-known fit, with energy proportional to $-(L + \Delta_\theta)\exp(-\Delta_\theta/L)$, to the universal bonding correlation of Rose et. al. (1983). Here Δ_θ is the opening and Δ_r the shear displacement between neighboring atomic planes. The parameter L has been suggested as scaling with the Thomas-Fermi screening length; here it can be loosely interpreted as the characteristic length associated with the decohesion process (σ reaches its maximum, at $\Delta_r = 0$, when $\Delta_\theta = L$). An analytical form for the shear and normal stress on a slip plane as functions of the relative atomic shear Δ_r and opening Δ_θ, which combines the Frenkel relation and the universal bonding correlation, has been proposed by Beltz and Rice (1991, 1992a) and is given by the following equations:

$$\tau = A(\Delta_\theta) \sin\left(\frac{2\pi\Delta_r}{b}\right) \tag{1.73a}$$

$$\sigma = \left[B(\Delta_r)\Delta_\theta - C(\Delta_r)\right] e^{-\Delta_\theta/L} \tag{1.73b}$$

where

$$A(\Delta_\theta) = \frac{\pi\gamma_{\text{us}}^{(u)}}{b} - \frac{2\pi\gamma_s}{b}\left\{q\left(1 - e^{-\Delta_\theta/L}\right) - \left(\frac{q-p}{1-p}\right)\frac{\Delta_\theta}{L}e^{-\Delta_\theta/L}\right\} \tag{1.74a}$$

$$B(\Delta_r) = \frac{2\gamma_s}{L^2}\left\{1 - \left(\frac{q-p}{1-p}\right)\sin^2\left(\frac{\pi\Delta_r}{b}\right)\right\} \tag{1.74b}$$

$$C(\Delta_r) = \frac{2\gamma_s}{L}\frac{p(1-q)}{1-p}\sin^2\left(\frac{\pi\Delta_r}{b}\right) \tag{1.74c}$$

and

$$q = \frac{\gamma_{\text{us}}^{(u)}}{2\gamma_s}, \; p = \frac{\Delta_\theta^*}{L} \tag{1.75}$$

and where Δ_θ^* is the value of Δ_θ after shearing to the state $\Delta_r = b/2$ under conditions of zero tension, $\sigma = 0$ (i.e., relaxed shearing). The form of these equations is consistent with the existence of a potential $\Psi = \Psi(\Delta_r, \Delta_\theta)$, with $\sigma = \partial\Psi/\partial\Delta_\theta$ and $\tau = \partial\Psi/\partial\Delta_r$, and Ψ is the same as the potential introduced by Needleman (1990) when $p = q$. Here $\gamma_{\text{us}}^{(u)}$ denotes the unstable stacking energy for *unrelaxed* shear ($\Delta_\theta = 0$). It has been determined that coupling effects may be approximated by using a modified form of γ_{us}, denoted $\gamma_{us}^{(u*)}$ and defined as $\int \tau d\Delta_r$ along the path from $\Delta_r = 0$ to $\Delta_r = b/2$ with Δ_θ *fixed* at the value Δ_θ^*. In terms of the above quantities $\gamma_{us}^{(u*)}$ may be written as

$$\frac{\gamma_{\text{us}}^{(u*)} - \gamma_{\text{us}}^{(u)}}{\gamma_{\text{us}}^{(u)}} = -\frac{(p^2 - q)e^{-p} + q(1-p)}{q(1-p)} \tag{1.76}$$

The parameter p, referred to here as the "dilation parameter," as well as q and L/b have been estimated so as to be consistent with results of various

TABLE 1.1. Partial Dislocation Nucleation; Anisotropic Formulation fcc metals [(001) cracks growing along [$\bar{1}$10] with slip on (1$\bar{1}$1) plane, (1/6) [$\bar{1}$12] as the first partial and (1/6) [$\bar{2}$11] as the second partial]. The β value, for the inequality $\gamma_s/\gamma_{us} > \beta$ in order that the first partial and second partial be emitted before the Griffith cleavage condition is reached under pure tensile loading, is listed. These β values are based on the K_{eff} concept.

Solid	β for first partial emission	β for second partial emission*
isotropic, $\nu = 0.3$	3.8	9.1
Ag	5.33	11.1
Al	4.05	9.23
Au	5.04	11.0
Cu	5.55	11.8
Ir	4.44	10.85
Ni	5.29	12.0
Pb	5.52	10.8
Pt	4.35	9.33

TABLE 1.2. Comparison of the Critical G for Emission on Inclined Slip Planes

System	$G/\gamma_{us}^{(u*)}$ (based on K_{II}^{eff})	$G/\gamma_{us}^{(u*)}$ (slip only; no coupling)	$G/\gamma_{us}^{(u*)}$ (full $\sigma-\tau$ coupling)	% Red., (K_{II}^{eff} to full $\sigma-\tau$ coupling)
EAM-Fe (bcc)				
($p = 0.214, q = 0.158, L/b = 0.204$)				
$\theta = 45°$	9.37	8.34	7.80	17%
$\theta = 90°$	8.00	5.92	6.33	21%
EAM-Al (fcc)				
($p = 0.140, q = 0.0855, L/b = 0.279$)				
$\theta = 54.7°$	7.61	6.43	6.33	17%
$\theta = 90°$	8.00	5.92	6.21	22%
EAM-Ni (fcc)				
($p = 0.132, q = 0.0879, L/b = 0.271$)				
$\theta = 54.7°$	7.61	6.43	6.27	18%
$\theta = 90°$	8.00	5.92	6.15	23%

embedded atom models by Sun et al. (1992). Representative values are shown in Table 1.2. For general in-plane loadings, the equilibrium discussed in connection with Eq. 1.55 for an incipient dislocation of edge type is now described by a pair of coupled integral equations:

$$\begin{aligned}\tau(\Delta_r, \Delta_\theta) = & \frac{K_{\mathrm{I}} f_{\mathrm{I}}(\theta) + K_{\mathrm{II}} f_{\mathrm{II}}(\theta)}{\sqrt{2\pi r}} \\ & - \frac{\mu}{2\pi(1-\nu)} \int_0^\infty g_{11}(r,\rho;\theta) \frac{d\delta_r(\rho)}{d\rho} d\rho \\ & - \frac{\mu}{2\pi(1-\nu)} \int_0^\infty g_{12}(r,\rho;\theta) \frac{d\delta_\theta(\rho)}{d\rho} d\rho \end{aligned} \tag{1.77a}$$

$$\begin{aligned}\sigma(\Delta_r, \Delta_\theta) = & \frac{K_{\mathrm{I}} f_{\mathrm{I}}^{\sigma}(\theta) + K_{\mathrm{II}} f_{\mathrm{II}}^{\sigma}(\theta)}{\sqrt{2\pi r}} \\ & - \frac{\mu}{2\pi(1-\nu)} \int_0^\infty g_{21}(r,\rho;\theta) \frac{d\delta_r(\rho)}{d\rho} d\rho \\ & - \frac{\mu}{2\pi(1-\nu)} \int_0^\infty g_{22}(r,\rho;\theta) \frac{d\delta_\theta(\rho)}{d\rho} d\rho \end{aligned} \tag{1.77b}$$

where the functions $f^{\sigma}(\theta)$ relate the applied stress intensity factors to the normal stress $\sigma = \sigma_{\theta\theta}$ across a slip plane in the linear elastic solution, and where $\delta_r = \Delta_r - h\tau/\mu$, $\delta_\theta = \Delta_\theta - L^2\sigma/2\gamma_s$. We show results for three choices of the parameters p, q, and L/b. Each set has been chosen to approximately fit the potential $\Psi(\Delta_r, \Delta_\theta)$ associated with Eqs. 1.68, 1.69 to the corresponding potential found numerically by atomic calculations (Sun, 1991), using the Embedded Atom Method, for the block-like relative motion of one half of a crystal relative to the other by shear and opening across a lattice slip plane. This has been done for $\{110\} < 111 >$ slip in an EAM model of α-Fe and also for $\{111\} < 211 >$ slip, corresponding to emission of the first Shockley partial, in EAM models of Al and Ni; some further details relating to the atomic calculations are given in the next section.

Figures 1.10(a), 1.10(b), and 1.10(c) show results for pure edge dislocation nucleation for the three different materials when the slip plane is taken to be coplanar with the crack plane (i.e., $\theta = 0$). The critical G for emission or cleavage, whichever occurs first along the slip plane, is plotted as a solid line as a function of the loading phase angle Ψ, defined such that $\tan\Psi = K_{\mathrm{II}}/K_{\mathrm{I}}$ (e.g., pure shear corresponds to $\Psi = 90°$ and pure tension corresponds to $\Psi = 0°$). The Griffith condition $G = 2\gamma_s$ is reproduced at small Ψ. Figure 1.10(a) is for the slip system $\{110\} < 111 >$ in iron, and Figs. 1.10(b) and 1.10(c) are for the slip system $\{111\} < 211 >$ in aluminum and nickel, respectively (i.e., the emission of the first Shockley partial). The dashed line in each figure gives results when no tension

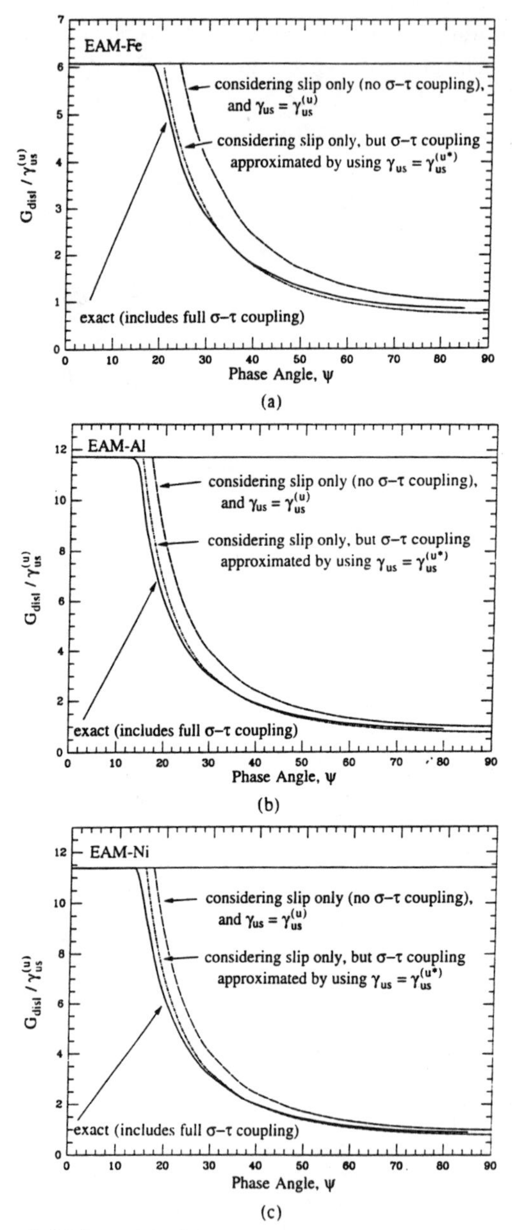

FIGURE 1.10. (a) The critical G for emission of a full dislocation in (EAM) α-Fe versus the loading phase angle ($\tan \Psi = K_{II}/K_{I}$) when the slip plane is coplanar with the crack plane ($\theta = 0$). The solid line is based on an exact numerical solution of Eqs. 1.77 which takes into account coupling between tension and shear. The dashed line is based on a calculation which only considers slip, for which the exact result is then given by Eq. 1.8, and uses $\gamma_{us} = \gamma_{us}^{(u)}$. The dash/dotted line is based on the same calculation but with $\gamma_{us} = \gamma_{us}^{(u*)}$. (b) and (c) are the same except they are for a partial dislocation in EAM-Al and EAM-Ni, respectively.

effects are included and we consider slip only. It corresponds to setting $\Delta_\theta = \delta_\theta = 0$ and ignoring the second integral equation, in which case the solution for the nucleation condition is exactly (since $\theta = 0$) that given by Eq. 1.8 with γ_{us} identified as $\gamma_{us}^{(u)}$, the unrelaxed value. The alternating dash/dot line in each figure corresponds to Eq. 1.8 with γ_{us} identified as the modified value, $\gamma_{us}^{(u*)}$, which will be seen to provide an approximate way of dealing with coupled shear and tension effects.

For all three cases, the approximation based on simply using $\gamma_{us}^{(u*)}$ is seen to be quite good when $\theta = 0$. It is less accurate when used, in conjunction with the K^{eff} concept of Section 1.6, to deal with typical cases of interest when $\theta \neq 0$, although the largest source of error is with the K^{eff} concept, as already discussed in connection with Fig. 1.8. Table 1.2 gives comparisons of the G values for a few special cases in the same materials involving inclined slip planes. The first column gives G based on the K^{eff} concept, and the second column gives G as predicted by the numerical solution of Eq. 1.55, i.e., only slip is taken into account. For both of these methods, γ_{us} is identified with $\gamma_{us}^{(u*)}$. The third column gives G as calculated from the numerical solution of Eqs. 1.77a and 1.77b, in an analysis which thus fully considers tension-shear coupling; the results are normalized to $\gamma_{us}^{(u*)}$ as given by Eq. 1.76. The reduction of G, from its value given by the K^{eff} concept, that occurs when coupling effects are taken into account is expressed as a percentage in this table; these effects appear to be appreciable: reductions of the critical G for emission are in the range of 17–18% for $\theta = 45°$ or 54.7° and 21–23% for $\theta = 90°$.

Inspection of the final two columns of Table 1.2 shows, however, that the approximation based on $\gamma_{us}^{(u*)}$ is also quite good for inclined slip planes, *assuming that the approximation uses the G based on the calculation which considers slip only.* The error in this approximation shows no clear trend; and ranges from ±1.6% to ±6.9%. This justifies the earlier statement that the major source of error in this approximation for inclined slip planes is due to the K^{eff} concept. These considerations are important when addressing the ductile versus brittle behavior of crystals, as will be taken up in Section 1.10.

1.8 Width of the Incipient Dislocation Zone at Instability

The width of the incipient dislocation zone at the moment of instability is also of interest. It will be seen that the width at a crack tip is, at the moment of instability, a moderately broad feature compared to a lattice spacing, thus making more appropriate the use of the Peierls concept. Indeed, Peierls (1940) laments towards the end of his paper that the dislocation core size which he calculated, for an isolated dislocation in an otherwise

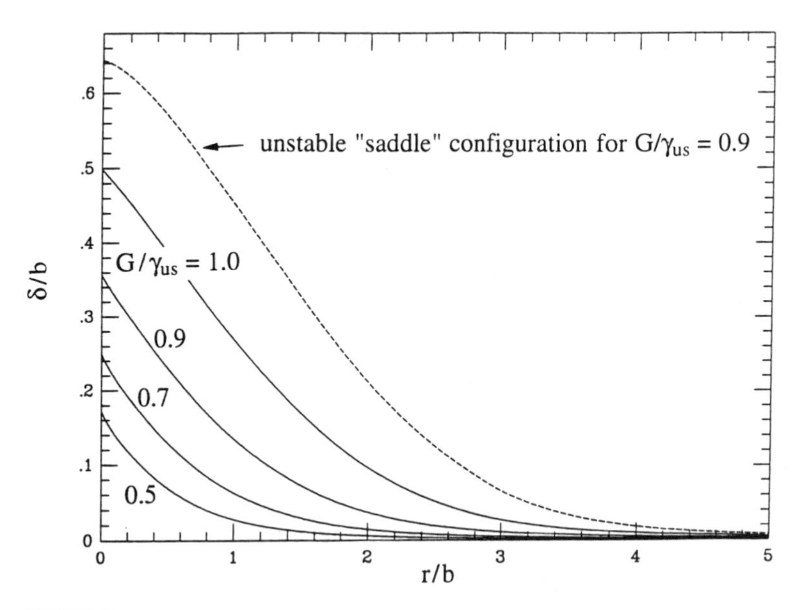

FIGURE 1.11. Displacement profiles at various levels of applied energy release rate up to instability for a pure mode II shear crack in an isotropic material, assuming $\nu = 0.3$ and $h = b$. The dashed line is an unstable "saddle" configuration corresponding to a load of $0.9G_{\mathrm{crit}}$.

perfect lattice, was sufficiently narrow compared to b that the concept of a continuously distributed core displacement, amenable to analysis by continuum elasticity, becomes problematical. The results for nucleation at a crack tip appear to be more favorable.

The core width at instability can be estimated from a full numerical solution of Eq. 1.55. Such solutions have been carried out by Beltz and Rice (1991) based on a $\tau = f(\delta)$ relation obtained from the Frenkel sinusoid for the case when $\theta = 0$ and various ratios of applied K_{I} and K_{II}; here we consider the case when $K_{\mathrm{I}} = 0$, i.e., pure shear. Solutions are shown for $h = b$ and $\nu = 0.3$ in Fig. 1.11 at various load levels up to instability. The characteristic width over which $\delta(r)$ is appreciable is roughly $(2-3)b$.

To quantitatively compare core widths, we may make a comparison with Peierls' width, also based on the Frenkel form, of $h/2(1-\nu)$ (Hirth and Lothe, 1982) for an isolated dislocation. This width is the distance over which τ diminishes from its peak value to its unstable zero value at $\delta = \Delta = b/2$, i.e., $b/4 < \Delta < b/2$, corresponding to $(\pi - 2)b/4\pi < \delta < b/2$. The half width for an isolated dislocation (assuming $b = h$, to be consistent with the conditions under which the integral equation is solved) is about $0.71b$; applying the same definition to the incipient dislocation (loaded at instability) gives a half-width of approximately $2.05b$, an increase by a factor of 2.9.

Nabarro (1947) solved the problem corresponding to that of Peierls for the case of two coplanar dislocations of opposite sign, attracting one another and subjected to a stress just sufficing to hold them in unstable equilibrium, in an otherwise perfect lattice. This is a nice analog of the problem of dislocation nucleation from a crack tip, particularly when we recall the Rice and Thomson (1974) result that the self force on a line dislocation at distance r from a crack tip is the attractive force caused by an oppositely signed dislocation lying at distance $2r$ away in an uncracked, otherwise perfect solid. Like what we infer here, Nabarro's (1947) results show that the core widens considerably from the Peierls size as the two dislocations are brought close to one another.

1.9 Estimates of the Unstable Stacking Energy, γ_{us}

Frenkel estimates: The simplest estimate of γ_{us} is based on the Frenkel sinusoid. This is rewritten here, for shear relative to atomic planes spaced by h, as

$$\tau = (\mu_{slip} b_{eff}/2\pi h)\sin(2\pi\Delta/b_{eff}) \tag{1.78}$$

to emphasize that the modulus, μ_{slip}, should be that for shear relative to the slip system, and given as $\mu_{slip} = (c_{11} - c_{12} + c_{44})/3$ for the fcc and bcc crystal slip systems considered here. Also the Burgers vector is replaced by an effective value, b_{eff}, to emphasize that in some cases the $\Delta(= b_{eff}/2)$ at maximum energy γ_{us}, i.e., at the unstable zero of τ, may not coincide with $b/2$. Thus

$$\gamma_{us(Frenkel)} = \mu_{slip} b_{eff}^2/2\pi^2 h \tag{1.79}$$

and there is no distinction to be made in this simple model between relaxed ($\sigma = 0$) and unrelaxed $\Delta_\theta = 0$) values. The result is shown in the dimensionless form $\gamma_{us(Frenkel)}/\mu_{slip} b$ as the first numerical column of Table 1.3 for partial dislocation on $\{111\}$ planes in fcc solids and for complete dislocation on two common slip planes, $\{110\}$ and $\{211\}$, in bcc solids. For the fcc and first bcc case $b_{eff} = b$ (where, consistently with earlier use, in the fcc case b corresponds to that of a Shockley partial). However, the Frenkel model is expected to give a poor representation of the $\tau = F(\Delta)$ relation for shear on the $\{211\}$ plane in bcc (Vitek et al., 1972), especially for shear in the twinning direction on that plane, in which direction it is possible that slip energy Φ (or Ψ) has a local maximum corresponding to the twinned structure, as it climbs towards γ_{us}. The geometry of shear in the anti-twinning direction (Paxton et al., 1991) seems somewhat simpler and the Frenkel model might apply approximately with the Δ at γ_{us} reduced from $b/2$ to a value perhaps as low as $b/3$. Thus, for that case, b_{eff} is given a range $2b/3$ to b in Table 1.3, resulting in the $\gamma_{\text{us(Frenkel)}}$ range shown.

To go beyond these simple estimates we require models of atomic potentials in solids. In principle, the energy γ_{us} could be determined by a quantum mechanical computations, based on (electron) density functional theory in the local density approximation, of the ground state energy of the configuration for which one half of a lattice is rigidly shifted relative to the other along a slip plane, so as to coincide with the unstable stacking (like in configuration (d) in Fig. 1.2). The analysis of such atomic geometries seems consistent with the present level of development of density functional computations.

For the present it is necessary to be content with empirical atomic models. A recently developed class of these, going beyond pair potentials and thus avoiding Cauchy symmetry of crystal moduli, have been formulated within the *Embedded Atom Method* (Daw and Baskes, 1984) and have found extensive applications to solid state phenomena, including interfacial structure and deformation and fracture. A few results for γ_{us} based on such models are now summarized.

Embedded Atom Models: Such embedded atom models as have been introduced seem to lead to lower estimates of γ_{us} than does the Frenkel model. The results will be different for direct shear with no relaxation in the direction normal to the slip plane (the most commonly available case), and for relaxed shear for which the lattice spacing h is allowed to dilate during shear so as to keep zero normal stress. As we have seen, the latter case is the most relevant one for use in the simplified nucleation criterion (e.g., Fig. 1.10 and Table 1.2, comparing 2nd and 3rd columns).

Cheung (1990) (see also Cheung et al., 1991) employed an embedded atom model for bcc Fe and, from plots of his potential for $\{110\} < 111 >$ shear, we may infer that $\gamma_{\text{us(EAM)}} = 0.44$ (relaxed) to 0.52 (unrelaxed) J/m^2. The dimensionless $\gamma_{\text{us(EAM)}}/\mu_{\text{slip}}b$ is entered for Fe in the second numerical column of Table 1.3 where, here and next, μ_{slip} is the slip system shear modulus that is consistent with the embedded atom potentials used.

Sun et al. (1991, 1992) have done similar calculations based on embedded atom models for $\{111\} < 211 >$ shears forming partial dislocation in fcc metals. These are for the respective cases of Al modeled by the potentials of Hoagland et al. (1990) and Foiles and Daw (1987), and Ni by the potentials of Foiles et al. (1986). These unrelaxed results are $\gamma_{\text{us(EAM)}} = 0.092 J/m^2$ for Al and $0.260 J/m^2$ for Ni; both numbers correspond to nearly the same $\gamma_{\text{us(EAM)}}/\mu_{\text{slip}}b$, of 0.026 as entered in Table 1.3. Relaxed $\gamma_{\text{us(EAM)}}$ values are also shown and are 87% and 86% that of the unrelaxed $\gamma_{\text{us(EAM)}}$ for Ni and Al, respectively. This is close to the 85% of unrelaxed $\gamma_{\text{us(EAM)}}$ found for the EAM α-Fe model.

The modified values of $\gamma_{\text{us(EAM)}}/\mu_{\text{slip}}b$ cited for Fe, Ni and Al are all of the order of 53% to 55% of the corresponding $\gamma_{\text{us(Frenkel)}}/\mu_{\text{slip}}b$. Thus, for

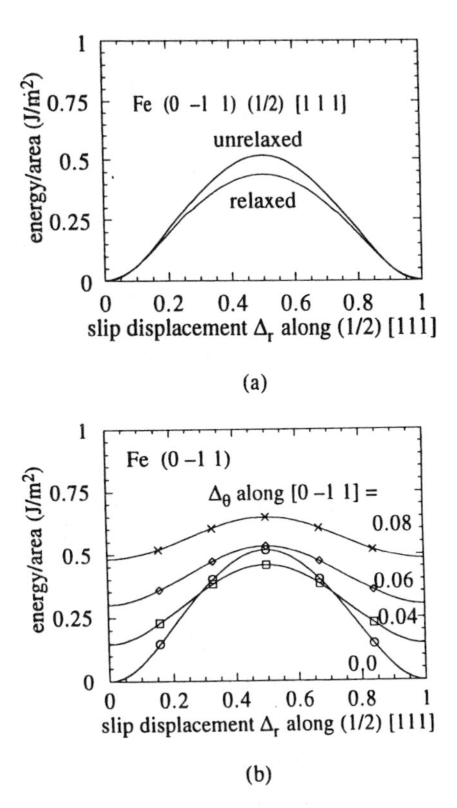

FIGURE 1.12. (a) The slip plane potential energy Ψ as function of slip displacement Δ_r along $(1/2)[111]$ on the $(0\bar{1}1)$ plane in EAM-Fe. (b) The slip plane potential energy Ψ as function of slip displacement Δ_r along $(1/2)[111]$ and opening displacement Δ_θ along $[0\bar{1}1]$ on the $(0\bar{1}1)$ plane in EAM-Fe.

later purposes (Table 1.4) in dealing with $\{111\}$ plane partial dislocations in a large class of fcc solids and with $\{110\}$ plane dislocations in a large class of bcc solids, for most of which embedded atom model results for $\gamma_{\rm us}$ are not available, the rough estimate $\gamma_{\rm us(EAM)} = 0.54\gamma_{\rm us(Frenkel)}$ is used in all cases. This improves upon the estimate $\gamma_{\rm us(EAM)} = 0.7\gamma_{\rm us(Frenkel)}$ made in a table similar to Table 1.4 by Rice (1992).

We now present results for the energy Ψ as function of block-like translational displacements $\{\Delta_r, \Delta_z, \Delta_\theta\}$, calculated based on the embedded atom method potential for α-Fe (Harrison, et al. 1990; Cheung, 1990) and Ni (Foiles and Daw, 1987). The relative positions of atoms in the two blocks are held fixed for each slip configuration. As before, Δ_r is edge-like slip, Δ_θ is opening, and Δ_z is screw-like slip. The energy surface for Δ_r displacement along $(1/2)$ $[111]$ in the $(0\bar{1}1)$ plane in α-Fe $[\Psi$ *vs.* $\Delta_r]$ is shown in Fig. 1.12(a) (also see Cheung, 1990) for $\Delta_\theta = 0$. The Ψ *vs.* Δ_r curve for relaxed conditions, along a path satisfying $\partial\Psi/\partial\Delta_\theta = 0$, is also shown, and it

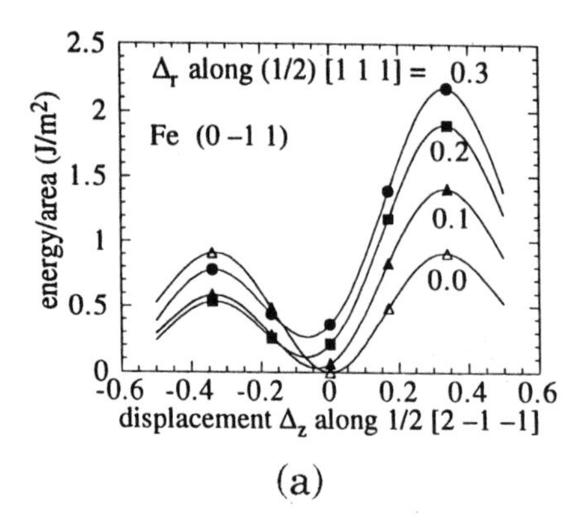

(a)

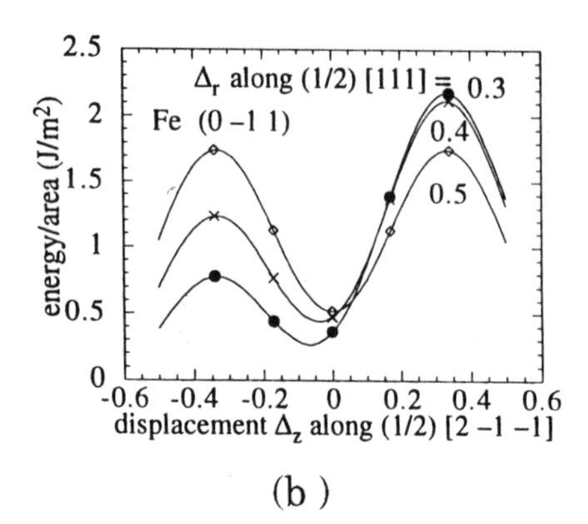

(b)

FIGURE 1.13. The potential energy Ψ as function of slip displacement Δ_r along $(1/2)[111]$ and Δ_z along $(1/2)[2\bar{1}\bar{1}]$ on the $(0\bar{1}1)$ plane in EAM-Fe. (a) shows curves for Δ_r from 0.0 to 0.3; (b) shows curves for Δ_r from 0.3 to 0.5.

has a maximum at $\gamma_{us} = 0.44 J/m^2$ as noted above. The maximum slope is $\tau_{max} = 6.41$ GPa. The energy Ψ *vs.* slip displacement Δ_r along $(1/2)$ $[111]$ at various opening displacements Δ_θ along $[0\bar{1}1]$, is shown in Fig. 1.12(b). The maximum stress along the pure opening direction is $\sigma_{max} = 25.3$ GPa. The ratio $\sigma_{max}/\tau_{max} = 3.95$ for Fe.

The energy Ψ *vs.* slip displacement Δ_z along $(1/2)$ $[2\bar{1}\bar{1}]$, which is perpendicular to Δ_r, for the Fe model is shown in Fig. 1.13(a) and 1.13(b), with opening Δ_θ kept at zero. The saddle-like path first deviates from the Δ_r direction (i.e., along $\boldsymbol{b} = (1/2)$ $[111]$) toward $(1/2)$ $[\bar{2}11]$ and then gradually returns to be parallel to the Δ_r direction. Along the direction perpendicular to the saddle like path direction, the energy Ψ increases much more rapidly than along the saddle-like path direction, as seen in Figs. 1.13(a)

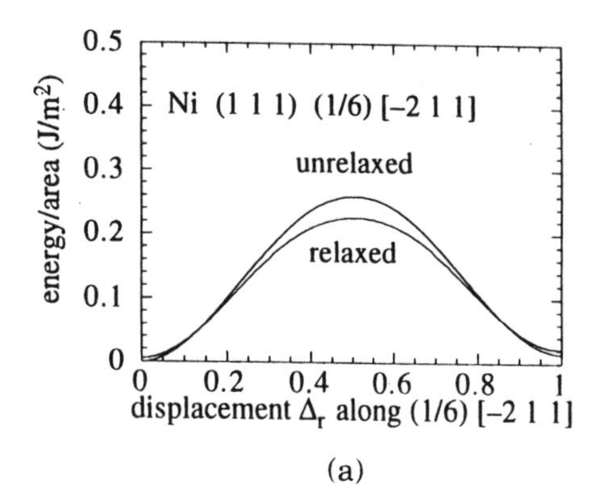

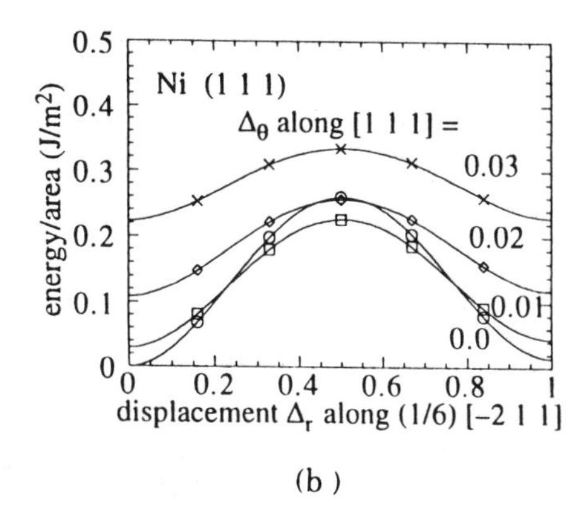

FIGURE 1.14. (a) The potential energy Ψ as function of slip displacement Δ_r along $(1/6)[\bar{2}11]$ on the (111) plane in EAM-Ni. (b) The potential energy Ψ as function of slip displacement Δ_r along $(1/6)[\bar{2}11]$ and opening displacement Δ_θ along $[111]$ on the (111) plane in EAM-Ni.

and 1.13(b). This sort of geometry of the energy surface is, of course, the basis of the constrained path approximation discussed earlier.

The energy surface for block-like Δ_r displacement along $(1/6)$ $[\bar{2}11]$ (i.e. the partial route) in the (111) plane for the EAM model of Ni [Ψ vs. Δ_r] is shown in Fig. 1.14(a), for the unrelaxed condition when $\Delta_\theta = 0$ and the relaxed condition. The relaxed γ_{us}, is $0.226 J/m^2$. The maximum slope is $\tau_{\max} = 5.54$ GPa. The energy Ψ *vs.* slip displacement Δ_r along $(1/6)$ $[\bar{2}11]$, at various opening displacements Δ_θ along $[111]$, is shown in Fig. 1.14(b). The maximum slope along the pure opening direction is $\sigma_{\max} = 28.2$ GPa. The ratio $\sigma_{\max}/\tau_{\max} = 5.09$ for Ni.

The energy Ψ *vs.* slip displacement Δ_z along $(1/2)$ $[0\bar{1}1]$, at various Δ_r,

TABLE 1.3. Estimates of $\gamma_{us}/\mu_{slip}b$

Solid	Frenkel Sinusoid $(b_{eff}^2/2\pi^2 bh)$	Embedded-Atom Models, Block-Like Shear	Density Functional, Homogeneous Simple Shear Strain $(W_{max}h/\mu_{slip}b)$
(1) fcc, partial dislocations, $<211>\{111\}$, $b = a_0/\sqrt{6}, h = a_0/\sqrt{3}, b_{eff} = b$:			
Al	0.036	0.026(u), 0.022(r),0.019($u*$)	0.042(r), 0.043(u)
Cu	0.036	—	0.042(u)
Ir	0.036	—	0.034(r), 0.043(u)
Ni	0.036	0.026(u), 0.023(r),0.020($u*$)	—
(2) bcc, $<111>\{110\}$, $b = \sqrt{3}a_0/2, h = a_0/\sqrt{2}, b_{eff} = b$:			
Fe	0.062	0.045(u),0.038(r),0.032($u*$)	—
(3) bcc, $<111>\{211\}$, $b = \sqrt{3}a_0/2, h = a_0/\sqrt{6}, b_{eff} = 2b/3$ to b :			
Cr	0.048-0.108	—	0.069(u)
Mo	0.048-0.108	—	0.056(u)
Nb	0.048-0.108	—	0.093(u)
V	0.048-0.108	—	0.100(u)
W	0.048-0.108	—	0.060(u)

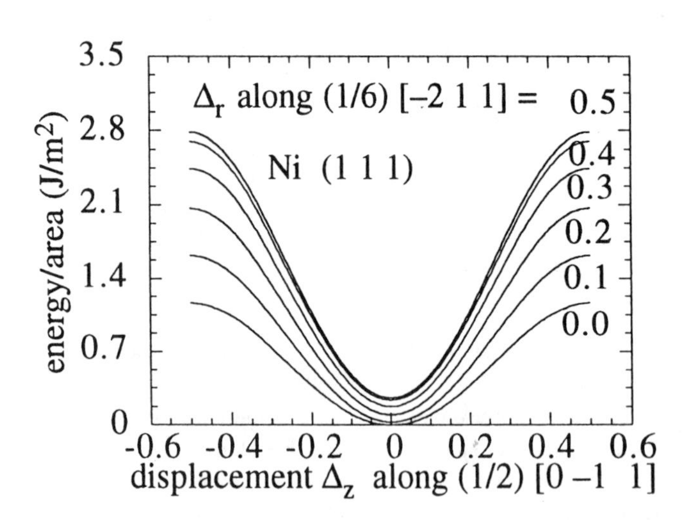

FIGURE 1.15. The potential energy Ψ as function of slip displacement Δ_r along $(1/6)[\bar{2}11]$ and Δ_z along $(1/2)[0\bar{1}1]$ on the (111) plane in EAM-Ni.

TABLE 1.4. Material Properties and γ_s/γ_{us} Ratios

Solid	$\gamma_s(T=0)$ (J/m^2)	μ_{slip} (GPa)	b (nm)	$\gamma_{us(Frenkel)}$ (J/m^2)	$\frac{\gamma_s}{\gamma_{us(Frenkel)}}$	$\frac{\gamma_s}{\gamma_{us(EAM)}}$
fcc metals:						
Ag	1.34	25.6	0.166	0.15	8.8	16.1
Al	1.20	25.1	0.165	0.15	8.1	14.6
Au	1.56	23.7	0.166	0.14	11.0	20.1
Cu	1.79	40.8	0.147	0.22	8.3	15.1
Ir	2.95†	198.	0.156	1.1	2.7	4.8
Ni	2.27	74.6	0.144	0.39	5.9	10.7
Pb	0.61	7.27	0.201	0.053	11.6	21.1
Pt	2.59	57.5	0.160	0.33	7.8	14.3
bcc metals:						
Cr	2.32	131.	0.250	2.0	1.1	2.1
Fe	2.37	69.3	0.248	1.1	2.2	4.1
K	0.13†	1.15	0.453	0.032	4.0	7.4
Li	0.53†	3.90	0.302	0.073	7.3	13.2
Mo	2.28	131.	0.273	2.2	1.0	1.9
Na	0.24†	2.43	0.366	0.055	4.4	7.9
Nb	2.57	46.9	0.286	0.83	3.1	5.6
Ta	2.90	62.8	0.286	1.1	2.6	4.7
V	2.28†	50.5	0.262	0.82	2.8	5.1
W	3.07	160.	0.274	2.7	1.1	2.1
diamond cubic:						
C	5.79†	509.	0.145	2.7	2.2	4.0
Ge	1.20†	49.2	0.231	0.41	2.9	5.4
Si	1.56†	60.5	0.195	0.42	3.7	6.7

Notes: † means γ_s is based on correlation with formation energy; Tyson (1975).
$\mu_{slip} = (c_{11} - c_{12} + c_{44})/3$.
$b = b_{partial} = a_0 < 211 > /6$ for fcc and diamond cubic; $b = a_0 < 111 > /2$ for bcc.
$\gamma_{us(Frenkel)} = 0.036\mu_{slip}b$ for fcc and diamond cubic; $\gamma_{us(Frenkel)} = 0.062\mu_{slip}b$ for bcc.
$\gamma_{us(EAM)}$ equated to $0.54\gamma_{us(Frenkel)}$ based on recent calculations of modified values, $\gamma_{us}^{(u*)}$, summarized here for EAM models of Al, Ni, and Fe. This is a change from Rice (1992) who used the estimate $\gamma_{us(EAM)} = 0.7\gamma_{us(Frenkel)}$.

for Ni is shown in Fig. 1.15, with opening Δ_θ kept at zero. The saddle-like path is strictly along the Δ_r direction. Along the direction perpendicular to the Δ_r direction, the energy Ψ increases much more rapidly than along the Δ_r direction. The constrained path approximation is thus very well justified in this case, more so than for Fe above.

Density functional theory: No directly relevant calculation for the block-like shear of one part of a metal crystal relative to another seems yet to have been reported based on quantum mechanics via density functional theory. However, such calculations appear to be feasible, as Duesbery et al. (1991) have reported energy surfaces for shear of Si along $\{111\}$ planes in a manner corresponding to the introduction of an intrinsic stacking fault. The Duesbery et al. (1991) work also shows that empirical potentials, as available for Si, may agree reasonably with the quantum mechanical calculations for one direction of shear but poorly for another direction on the same crystal plane.

Paxton et al. (1991) used density functional theory in the local approximation to analyze stress-strain relations of homogeneously strained crystals, in fcc cases corresponding to simple shear parallel to $\{111\}$ planes in $< 211 >$ type directions, and for bcc cases to simple shear parallel to $\{211\}$ planes in $< 111 >$ type directions. These are shears leading to twinning transformations (in the softer direction of shear in each case). Paxton et al. report the maximum stress and also the maximum strain energy (say, $W_{\max}$, on a unit volume basis) encountered for simple shear in the twinning direction and in the opposite, or anti-twinning direction. The strain energy maximum, $W_{\max}$, is a rough analog of γ_{us}. Both correspond to maximum energies along a shear path, but for block-like shear of one half the lattice relative to the other in the case of γ_{us} (like in Fig. 1.2, illustration d), and for homogeneous simple shear strain of the entire lattice in the case of $W_{\max}$.

Rice (1992) formed a quantity somewhat like γ_{us} from $W_{\max}$ in the following way: Since $W_{\max}$ is the maximum energy per unit volume in simple shear strain, $W_{\max}h$ is the maximum energy per unit area of slip plane associated with an interplanar separation h. This could be considered comparable to γ_{us} and thus the final column in Table 1.3 shows $W_{\max}h/\mu_{\rm slip}b$ based on $W_{\max}$ from Paxton et al. (1991) and using experimental $\mu_{\rm slip}$ values (expected to correspond within about 10% of those estimated from the density functional calculations; Paxton, private communication, 1991); $W_{\max}$ for the twinning sense is used for the fcc partial dislocation comparisons, and in the anti-twinning sense, suggested in Paxton et al. (1991), for complete $\{211\}$ bcc dislocation comparisons. It is interesting that these values seem approximately compatible with the Frenkel estimates.

The experimental values for $\mu_{\rm slip}$ used in the last column of Table 1.3 (and in Table 1.4) are from Hirth and Lothe (1982) and, if not there, from Brandes (1983) or Anderson (1986). Lattice parameters a_o, used to evaluate b, are from Ashcroft and Mermin (1976).

1.10 Ductile Versus Brittle Crack Tip Response

In using the results of this paper to discuss ductile versus brittle response, in the sense of asking whether conditions for dislocation nucleation will or will not be met prior to Griffith cleavage, it is well to keep the following factors in mind.

(a) Dislocation nucleation is a process susceptible to thermal activation. The analysis given thus far here is, essentially, of temperature $T = 0$ response. The critical K's for nucleation will be reduced somewhat at finite T. The Peierls concept gives a route to treat thermally activated nucleation and some related concepts have already been uncovered in the J integral analysis of the crack tip shear (Fig. 1.4, point C) in the 2D saddle point configuration of $\delta(r)$. The fuller evaluation of the activation energy for dislocation nucleation is not yet complete, but we give some preliminary results on it later. While the K level for dislocation nucleation in some finite waiting time can, in principle, be reduced arbitrarily by increase in T (some solids may melt before there is any substantial reduction), it is interesting that the K for cleavage cannot be reduced arbitrarily and always has the Griffith level (at that T) as a lower bound. Thus increase of T should generally ease dislocation nucleation more than cleavage, and favor ductility. Our considerations in the rest of this section are for low T, when thermal activation is not an important factor.

(b) The present analysis of dislocation nucleation is approximate in many respects, and thus it will be difficult to draw definitive conclusions on ductile versus brittle response in the several borderline cases that arise. We have attained a good understanding of limits to the K^{eff} approximation, and of coupled tension-shear effects, thus far only in the isotropic case. Most importantly, perhaps, we have no very reliable estimates of γ_{us}; the $\gamma_{\mathrm{us(Frenkel)}}$ and $\gamma_{\mathrm{us(EAM)}}$ values of Table 1.3 may contain large errors. Also, reliable values of γ_{sf}, needed in the fcc cases, are not available for most solids.

(c) Dislocation processes not directly associated with nucleation from a crack tip may actually control brittle versus ductile response in many cases. For example, in soft solids with a high density of mobile dislocation, it may never be possible to build up enough stress at a crack tip to meet either a Griffith cleavage or a dislocation nucleation criterion, so the issue of which requires the greater local K value becomes irrelevant. Also, in solids for which dislocation mobility is low, easy nucleation of dislocations from a crack tip does not necessarily imply relaxation of stresses; cleavage may occur because such dislocations cannot move readily enough away from the crack tip so as to relax stress in its vicinity.

Accepting these limitations, consider Table 1.4. Estimates of the surface energy γ_s at $T = 0$, based on measurements that have been extrapolated to low temperature or, where noted by the cross, on correlations thus established with formation energies, are shown in the first column based on Tyson (1975). Shear moduli μ_{slip} and b are also shown (b is for a Shockley partial dislocation in the fcc and diamond cubic cases, and for a complete dislocation in the bcc cases), and the Frenkel estimate $\gamma_{us(Frenkel)} = \mu_{slip}b^2/2\pi^2 h$ is calculated from them, as $0.036\mu_{slip}b$ for partial dislocations in fcc metals and (very uncertainly) in diamond cubic solids, and $0.062\mu_{slip}b$ for complete dislocations on the $\{110\}$ plane in bcc metals.

We can therefore calculate the ratios of γ_s/γ_{us} shown in the last two columns of Table 1.4, based respectively on $\gamma_{us(Frenkel)}$ and $\gamma_{us(EAM)}$, with the latter approximated as $0.54\gamma_{us(Frenkel)}$ based on that being close to the modified values calculated for the three EAM models we have examined (Table 1.3) for Ni, Al, and α-Fe.

To recall now the conclusions drawn in Section 1.7, it was shown that the dislocation nucleation condition is met before that for Griffith cleavage, for the $\{100\}$ cracks considered, if, under pure mode I loading, and using an isotropic elastic model, $\gamma_s/\gamma_{us} > 9.1$ (fcc) or 6.3 (bcc). Those numbers were based on the K^{eff} approximation and were used by Rice (1992) in discussing ductile versus brittle response. We are now in a somewhat better position to estimate these limits, using corrections based on Fig. 1.8 and Table 1.2. Those corrections have been worked out only for edge dislocations, which is the appropriate case for the bcc geometry, but the second partial involved for the fcc geometry has a considerable screw component and the correction in that case is less certain. Here we provisionally use a 17% reduction in both cases, as suggested by results in Table 1.2, so that the condition for nucleation before cleavage, in the isotropic elastic case, is approximately

$$\gamma_s/\gamma_{us} > 7.6(\text{fcc}) \text{ or } 5.2 \text{ (bcc)}, \tag{1.80}$$

Both required ratios were strongly reduced by small deviations from pure mode I, the fcc case most. For example, with both shear mode stress intensity factors set at 10% of K_I, the requirements for dislocation nucleation to occur before Griffith cleavage, again as estimated using the K^{eff} approximation, dropped to $\gamma_s/\gamma_{us} > 3.5$ (fcc) or 2.9 (bcc). If, provisionally, we also reduce these by the same 17% to correct for inadequacies of the K^{eff} approach, the limits become

$$\gamma_s/\gamma_{us} > 2.9 \text{ (fcc) or } 2.4 \text{ (bcc)}. \tag{1.81}$$

(The fcc numbers in each case also depend on γ_{sf}/γ_{us}, which has been taken as 1/3 in the above inequalities; $\nu = 0.3$ is used there too.)

If we tentatively accept the $\gamma_{us(EAM)}$ estimates as being close to correct, thus using the last column in Table 1.4 as estimates of γ_s/γ_{us}, then we come to the following conclusions for the fcc metals: All the fcc metals except Ir are incapable of cleaving, even if subjected to pure mode I loading. Ir would not cleave with the 10% shear mode loading discussed, but would behave in a ductile manner. Ni is moderately near the borderline and, as an indication that Ni may, plausibly, be thought of as a borderline material, in a brittle versus ductile sense, it is interesting to note that grain boundaries in Ni are rather easily rendered cleavable by segregation of S there and by the presence of H. If the true γ_{us} is close to the Frenkel estimate then, according to Table 1.4, both Ir and Ni would be cleavable under pure mode I, but Ni would be ductile with 10% shear mode loadings. Also, Pt, Al, and Cu are close to the borderline cleavable at low T if loaded in perfect mode I. This simplified discussion of fcc solids has assumed the same $\gamma_{us}/\mu_{slip}b$ in all material and also the same γ_{sf}/γ_{us}.

For the bcc metals, again first assume the $\gamma_{us(EAM)}$ is close to correct so that the last column of Table 1.4 gives γ_s/γ_{us}. Then the alkali metals, Li, Na and K are the standouts in terms of ductility, which is consistent with the general malleability of the alkali metals. The vanadium subgroup of the transition metals, in the order Nb, V and Ta, also stand out in Table 1.4. They fall below (marginally for Nb) the threshold for ductile crack tip response for pure mode I loading, but fit comfortably within the border for ductile response when mode I is accompanied by small loadings in the shear modes. Fe is predicted to be clearly cleavable, although it should likewise be ductilized by less than 10% shear loadings. By comparison, the chromium subgroup of transition metals, Cr, Mo and W, seem by our criterion to be irredeemably brittle, even with substantial shear mode loading.

If the Frenkel estimates of γ_{us} are, instead, somewhat closer to the mark, then the results of Table 1.4 still suggest that Li cannot be cleaved. The other alkali metals are slightly below the borderline, but are ductilized by modest shear mode loading, and Nb also would be ductilized by the 10% shear mode loading.

Diamond-cubic non-metals are also shown in Table 1.4 and γ_{us} has been extracted for them as for fcc metals, assuming that dislocations are generated by a partial route on $\{111\}$ planes, and assuming (quite questionably) the same scaling of γ_{us} with $\mu_{slip}b$. All of the diamond cubic solids are predicted to be cleavable by these considerations, for pure mode I loading, although Si is somewhat susceptible to ductilization by modest shear mode loading.

The discussion concerning the emission of dissociated dislocations given in Section 1.5, which properly treats the slip energy offset by stable stacking faults and the screening effect of the emitted partial, can be used to study dislocation emission in ordered intermetallics, in which there exist dislocations with a dissociated core in pairs, which are coupled by CSF or SISF surfaces in the $L1_2$ type intermetallics; Ni_3Al is an example. Complex paths are possible for dislocation nucleation in these materials. More details may be found in Sun et. al. (1991).

Since the rough γ_{us} estimates used in Table 1.4 scale directly with $\mu_{slip}b$ for a given crystal class, the characterization of crack tip response as brittle or ductile on the basis of the size of γ_s/γ_{us} shown in that table is equivalent to characterization on the basis of $\gamma_s/\mu_{slip}b$, much as advocated by Armstrong (1966) and Rice and Thomson (1974).

TABLE 1.5. Cleavage Versus Dislocation Nucleation; Anisotropic Formulation

Solid	β	$\gamma_s/\gamma_{us(EAM)}$	$\gamma_s/(0.83\beta)\gamma_{us(EAM)}$
fcc metals [(001) cracks growing along [$\bar{1}$10] with slip plane (1$\bar{1}$1), for nucleation of the pair of partials (1/6) [$\bar{1}$12] and (1/6) [$\bar{2}\bar{1}$1]]			
isotropic, $\nu = 0.3$	9.1	—	$\gamma_s/7.6\gamma_{us}$
Ag	11.1	16.1	1.74
Al	9.23	14.6	1.91
Au	11.0	20.1	2.20
Cu	11.8	15.1	1.54
Ir	10.85	4.8	0.537
Ni	12.0	10.7	1.07
Pb	10.8	21.1	2.35
Pt	9.33	14.3	1.85
bcc metals [(001) cracks growing along [010] with slip system (1/2)[11$\bar{1}$](011)]			
isotropic, $\nu = 0.3$	6.3	—	$\gamma_s/5.2\gamma_{us}$
Cr	5.83	2.1	0.432
Fe	8.77	4.1	0.559
K	14.16	7.4	0.627
Li	17.45	13.2	0.912
Mo	5.80	1.9	0.388
Na	16.33	7.9	0.584
Nb	4.88	5.6	1.38
Ta	7.34	4.7	0.777
V	5.68	5.1	1.07
W	6.38	2.1	0.395

Anisotropy considerations: We now extend the discussion to include anisotropic elastic effects. The quantity β was introduced in Eq. 1.61 and gives the bound, $\gamma_s/\gamma_{us} > \beta$, for dislocation nucleation to occur before Griffith cleavage. The expressions for β reported here are based on the K^{eff} concept. At the time of writing we have no idea of how significant the corrections, due to $\theta \neq 0$, are in the anisotropic case. We show β in Table 1.5 as it has been calculated from the elastic moduli of various fcc and bcc metals. For the fcc cases, it corresponds to nucleation of the second partial (last column of Table 1.1). A provisional guess, based on isotropic results as in Table 1.2 here, is that these may be 15% to 20% too high. We show γ_s/γ_{us}, estimated as $\gamma_s/\gamma_{us(EAM)}$ from the previous table, and show $\gamma_s/\beta\gamma_{us}$ with β provisionally replaced by 0.83 of β from the earlier columns (17% reduction, as in the isotropic case; Table 1.2). When the quantity $\gamma_s/\beta\gamma_{us}$ is greater than one, dislocation emission occurs prior to

Griffith crack extension, and when it is less than one, the opposite happens. The quantity $\gamma_s/\beta\gamma_{us}$ is tabulated in Table 1.5 for several bcc and fcc metals. The quantity β varies from 4.9 to 16.3 depending on the elastic anisotropy for bcc metals, from niobium having the lowest value to alkali metals having the highest. Therefore, the treatment of anisotropy in elasticity is important for bcc metals. In fact, it significantly changes conclusions. Nb, which was predicted to be borderline cleavable by the isotropic analysis, is now found to be ductile; Li, which was definitively ductile in the isotropic analysis is now borderline cleavable. On the other hand, β does not vary much for fcc metals, and Ir remains the standout as the cleavable fcc metal.

1.11 Extensions to Interfacial Failure

The results presented thus far may be generalized to cases where a crack lies on an interface been dissimilar materials. The case of joined isotropic solids has been worked out in detail by Beltz and Rice (1992a); a brief review of that development is given here.

Equations 1.72 may be generalized by making use of the interfacial crack tip field, in which stresses are given by

$$\sigma_{\alpha\beta} = \frac{1}{\sqrt{2\pi r}}\left[\mathrm{Re}(\mathrm{K}r^{i\varepsilon})\Sigma^{\mathrm{I}}_{\alpha\beta}(\theta) + \mathrm{Im}(\mathrm{K}r^{i\varepsilon})\Sigma^{\mathrm{II}}_{\alpha\beta}(\theta) + K_{\mathrm{III}}\Sigma^{\mathrm{III}}_{\alpha\beta}(\theta)\right] \quad (\alpha,\beta = r,\theta,z) \tag{1.82}$$

Only in-plane loadings are considered here. The functions $\Sigma_{\alpha\beta}(\theta)$ correspond to tractions across the interface at $\theta = 0$ of tensile, in-plane, and anti-plane shear type, so that

$$(\sigma_{\theta\theta} + i\sigma_{r\theta})_{\theta=0} = \frac{Kr^{i\varepsilon}}{\sqrt{2\pi r}}, \quad (\sigma_{z\theta})_{\theta=0} = \frac{K_{\mathrm{III}}}{\sqrt{2\pi r}}. \tag{1.83}$$

K is the complex stress intensity factor which characterizes the inherently coupled in-plane modes. The parameter ε is given by

$$\varepsilon = \frac{1}{2\pi}\ln\left[\frac{(3-4\nu_1)/\mu_1 + 1/\mu_2}{1/\mu_1 + (3-4\nu_2)/\mu_2}\right] \tag{1.84}$$

where μ and ν refer to the shear modulus and Poisson's ratio, respectively. Subscript 1 refers to the material on top, occupying $0 < \theta < \pi$, which is taken to be a metal (i.e., can sustain a dislocation-like process), and subscript 2 refers to a ceramic phase (i.e., no dislocation activity is assumed to occur). We have $\Sigma^{\mathrm{I}}_{\theta\theta}(0) = \Sigma^{\mathrm{II}}_{r\theta}(0) = 1$ and the full functions $\Sigma_{\alpha\beta}(\theta)$ are given by Rice, Suo, and Wang (1990) and can be extracted from discussions of the bimaterial elastic singular field (e.g., Rice (1988)).

The generalization of Eqs. 1.77 may now be written as

$$\begin{aligned}
&\tau\left(\Delta_r(r), \Delta_\theta(r)\right) \\
&\quad = \frac{1}{\sqrt{2\pi r}}\left\{\mathrm{Re}\left[\mathrm{K r}^{i\varepsilon}\right]\Sigma^{\mathrm{I}}_{r\theta}(\theta;\varepsilon) + \mathrm{Im}\left[\mathrm{K r}^{i\varepsilon}\right]\Sigma^{\mathrm{II}}_{r\theta}(\theta;\varepsilon)\right\} \\
&\quad - \frac{\mu_1}{2\pi(1-\nu_1)}\int_0^\infty g_{11}(r,\rho;\theta,\varepsilon)\frac{\partial\delta_r(\rho)}{\partial\rho}d\rho \\
&\quad - \frac{\mu_1}{2\pi(1-\nu_1)}\int_0^\infty g_{12}(r,\rho;\theta,\varepsilon)\frac{\partial\delta_\theta(\rho)}{\partial\rho}d\rho
\end{aligned} \tag{1.85a}$$

$$\begin{aligned}
&\sigma\left(\Delta_r(r), \Delta_\theta(r)\right) \\
&\quad = \frac{1}{\sqrt{2\pi r}}\left\{\mathrm{Re}\left[\mathrm{K r}^{i\varepsilon}\right]\Sigma^{\mathrm{I}}_{\theta\theta}(\theta;\varepsilon) + \mathrm{Im}\left[\mathrm{K r}^{i\varepsilon}\right]\Sigma^{\mathrm{II}}_{\theta\theta}(\theta;\varepsilon)\right\} \\
&\quad - \frac{\mu_1}{2\pi(1-\nu_1)}\int_0^\infty g_{21}(r,\rho;\theta,\varepsilon)\frac{\partial\delta_r(\rho)}{\partial\rho}d\rho \\
&\quad - \frac{\mu_1}{2\pi(1-\nu_1)}\int_0^\infty g_{22}(r,\rho;\theta,\varepsilon)\frac{\partial\delta_\theta(\rho)}{\partial\rho}d\rho
\end{aligned} \tag{1.85b}$$

The kernel functions g_{11}, g_{12}, g_{21}, and g_{22} are taken from the elasticity solution for a Volterra dislocation in the presence of an interfacial crack, and may be found in complex form (Suo, 1989).

Solutions to the pair of integral Eqs. 1.85 have been found using physical constants appropriate for copper bonded to sapphire (Beltz and Rice, 1992a) and iron bonded to titanium carbide (Beltz, 1991). As discussed by Rice, Suo, and Wang (1990), $r^{i\varepsilon}$ can be replaced by $b^{i\varepsilon}$, and the analysis is tenable when $Kb^{i\varepsilon}$ has a positive real part.

1.12 Experimental Observations

The actual observation of dislocation emission from crack tips has been achieved by the use of several experimental techniques. In work by Burns (1986), etch pit techniques were employed to observe edge dislocations on slip planes which emanated from a crack which had been cut parallel to the {110} planes in lithium fluoride. X-ray topography has been used by Michot and George (1986) to carry out similar observations in silicon. Possibly the most notable observations of dislocation emission is the T.E.M. work of Ohr (1985, 1986), which has the advantage that emission could be observed in-situ in several materials, including fcc and bcc metals with a high resolution. In these experiments, the critical applied stress intensity factor K_e to emit a dislocation was indirectly measured; they were in moderate agreement for several metals with the theoretical values of K_e as predicted by the Rice-Thomson model. More recently, Chiao and Clarke (1989) directly observed emitting dislocations in silicon and claimed reasonable agreement of the inferred K_e with Rice-Thomson modeling.

The first experimental evidence that the macroscopic behavior of an interface could be rationalized based on the competition between dislocation emission and cleavage was given by Wang and Anderson (1990), in their work on symmetric tilt bicrystals of copper. In this work, a directional effect on the toughness of the grain boundary in a Σ 9[110](2$\bar{2}$1) bicrystal was observed, in which two specimens were cut and notched along the boundary such that a crack would run in the opposite directions [$\bar{1}$14] and [1$\bar{1}\bar{4}$], respectively. The specimens were fatigued under a cyclic mode I loading of increasing amplitude. The specimen with the [1$\bar{1}\bar{4}$] cracking direction broke along the interface when the maximum normal stress reached σ = 28.1MPa, corresponding to $G \approx 28J/m^2$. An intergranular fracture surface with cleavage "tongues" was observed. The other specimen, with a cracking direction of [$\bar{1}$14], was loaded under identical conditions and eventually fractured at a normal stress of 76.7 MPa. The fracture surface contained large regions of ductile transgranular fracture and plastic tearing, and the G value, $> 210J/m^2$, was beyond the reliably measurable range for elastic fracture mechanics. The only difference between these two specimens was the cracking direction, hence it was concluded that the difference in ease with which dislocations could be nucleated at each crack tip was the cause of this behavior, as predicted nucleation loads are quite different for the two growth directions. Further, continuum plasticity analyses by Saeedvafa (1991) and Mohan et. al. (1991), suggested very little difference in the stress state ahead of the crack tip, for the two growth directions, and do not suggest a more macroscopic explanation of the experiments.

Most recently, Beltz and Wang (1992) have performed experiments on copper crystals bonded on the same {221} copper face to sapphire, to form a layered beam subjected to four-point bending (see Fig. 1.16). Again, a directional dependence of toughness was observed. In their experiment, the ductile direction was observed to be [1$\bar{1}\bar{4}$], the *opposite* of the ductile direction with the Wang-Anderson bicrystal specimen. This result was predicted by theory, however, and is elaborated on in Beltz and Rice (1992a) in terms of the Peierls-type nucleation model; it follows from different mode I/II mixture in the two specimens.

1.13 The Activation Energy for Dislocation Nucleation

Thus far, the analysis of dislocation nucleation rigorously holds true at zero Kelvin; i.e., thermal effects are neglected, except possibly through the weak temperature dependence of the elastic constants that enter the analysis. As discussed earlier in connection with Fig. 1.4, a saddle-point configuration

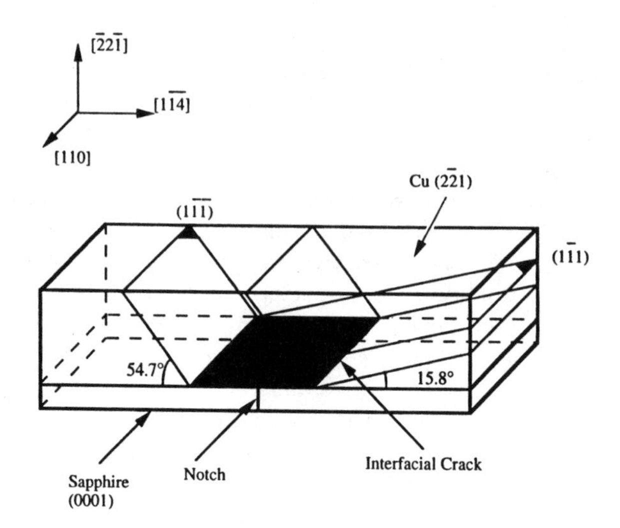

FIGURE 1.16. Diagram of specimen tested by Beltz and Wang (1992): a copper single crystal with {221} face bonded to sapphire; loaded in bending with crack tips along [110].

TABLE 1.6. Activation Energies

$\frac{G}{G_{\text{crit}}}$	$\frac{(1-\nu)\Delta_{\text{act}}}{\mu b^2}$	$\frac{\Delta E_{\text{Cu,partial}}}{kT_{\text{room}}}$	$\frac{\Delta E_{\text{Fe,fulldisl.}}}{kT_{\text{room}}}$
0.2	1.25×10^{-1}	29.4	239.6
0.3	9.05×10^{-2}	21.3	173.5
0.4	6.55×10^{-2}	15.4	125.5
0.5	4.62×10^{-2}	10.09	88.6
0.6	3.10×10^{-2}	7.29	59.4
0.7	1.90×10^{-2}	4.47	36.4
0.8	9.85×10^{-3}	2.31	18.9
0.9	3.32×10^{-3}	0.780	6.36
1.0	0	0	0

exists with 2D form corresponding to point C. The total energy corresponding to the system at C less the total energy at A would correspond to an activation energy; this energy could be thought of as the amount of energy due to thermal vibrations necessary to emit an incipient dislocation which is initially loaded below $G_{\rm crit}$. A two-dimensional simplification to the problem reduces to that of finding a *second* solution to Eq. 1.55, for a given applied load. To be realistic, the activation process would take place over a localized region, i.e. in the form of a dislocation loop that jumps out. At the time of writing, we have analyzed such solutions only for the case $\theta = 0$, $\phi = 0$, and using the Frenkel form of $\tau = f(\delta)$ and explicitly considering slip only, like in Sections 1.2 to 1.4 of the paper. In that case Eq. 1.55 corresponds to rendering stationary the energy functional (Rice, 1992)

$$U[\delta(r)] = U_0 + \int_0^\infty \Phi(\delta(r))dr + \int_0^\infty \frac{1}{2}s[\delta(r)]\delta(r)dr - \int_0^\infty \frac{K_{\rm II}}{\sqrt{2\pi r}}\delta(r)dr \tag{1.86}$$

with

$$s[\delta(r)] = \frac{\mu}{2\pi(1-\nu)}\int_0^\infty \sqrt{\frac{\rho}{r}}\frac{d\delta(\rho)/d\rho}{r-\rho}d\rho \tag{1.87}$$

Here $U[\delta(r)]$ is the energy of a slipped configuration per unit distance along the crack front. Thus, for $G < G_{\rm crit}$ ($= \gamma_{\rm us}$ in this case) and with $\delta_{\rm min}(r)$ and $\delta_{\rm sad}(r)$ representing $\delta(r)$ for the energy minimum and saddle-point solutions (with values of $\delta(0)$ corresponding respectively to points A and C in Fig. 1.4), we can calculate a 2D activation energy

$$\Delta U_{\rm act} = U[\delta_{\rm sad}(r)] - U[\delta_{\rm min}(r)] \tag{1.88}$$

Results are shown in Table 1.6. Also, we show by the dashed line in Fig. 1.11 the slip function $\delta_{\rm sad}(r)$ corresponding to $G = 0.9G_{\rm crit}$, a case for which $\delta_{\rm min}(r)$ is also shown.

The actual activation process is inherently three-dimensional, at least as regards the saddle point configuration. An asymptotic analysis is underway of this 3D phenomenon by Beltz and Rice (1992b). A very rough approximation to their result for the activation energy ΔE involves multiplying the two-dimensional activation energy $\Delta U_{\rm act}$ (an energy per unit dislocation length) by about five atomic spacings, which is a plausible length scale for the activation process.

Table 1.6 shows the results (from the two-dimensional analysis, assuming an activated dislocation length of 5b, $\Delta E \approx 5b\Delta U_{\rm act}$) for a partial dislocation in copper and a full dislocation in iron, with a coplanar slip plane and a mode II loading. The ΔE estimates are listed in units of kT as evaluated at room temperature. An elementary calculation of a "cutoff" $\Delta E/kT$ for spontaneous nucleation is discussed by Beltz and Rice (1992b) and uses the formula

$$v = n(c_{\text{shear}}/b)\exp(-\Delta E/kT) \tag{1.89}$$

where v is interpreted as the frequency of spontaneous nucleation events, n is taken as the number of nucleation sites in a typical span of crack front, taken here as 1 mm (i.e., $n = 1\text{mm}/5b$) and c_{shear} is the transverse shear wave speed, so that c_{shear}/b is an approximate attempt frequency. Here, c_{shear} is taken as 3 km/sec. Assuming that $v \approx 10^6$/sec describes spontaneous nucleation on a laboratory time scale, solution of Eq. 1.89 gives a borderline of $\Delta E/kT \approx 25$. Examination of Table 1.6 leads to the conclusion that thermal activation would be sufficient (at room temperature) to spontaneously emit a partial dislocation in copper at loadings of $G \approx (0.2-0.3)G_{\text{crit}}$ or greater, and a full dislocation in iron at loadings of $G \approx (0.7-0.8)G_{\text{crit}}$ or greater, where G_{crit} is the critical loading for dislocation nucleation without help from thermal activation. At $T = 2T_{\text{room}}$, these values for spontaneous nucleation would, e.g., change to approximately $0.1G_{\text{crit}}$ for Cu and $0.6G_{\text{crit}}$ for Fe.

1.14 Summary and Conclusions

A new analysis of dislocation nucleation from a crack tip is outlined based on the Peierls concept as applied to a slip plane emanating from the tip. An exact solution for the nucleation criterion is found using the J integral when the crack and slip plane coincide, at least within simplifying assumptions that consider only shear sliding between lattice planes, in forming a dislocation. The exact solution is also extended to the nucleation of dissociated dislocations, with complete results found for the nucleation of a pair of Shockley partials in fcc solids. For cases of greater interest, in which the slip and crack planes do not coincide ($\theta \neq 0$) but, rather, intersect along the crack tip, an approximate solutions for the nucleation criterion are given based on effective shear stress intensity factors along the slip plane, and exact solutions from numerical solution of appropriate integral equations are also discussed, including those which take fully into account the coupling between tension and shear across the slip plane.

The core width of the incipient dislocation at the threshold of instability is estimated to be about 3 times the corresponding width for an isolated dislocation in an otherwise perfect lattice, so that conditions seem favorable to use of the Peierls concept. Further, while previous treatments of nucleation have generally been based on elasticity solutions for fully formed dislocations located very near the crack tip, this analysis shows that maximum shear slippage at the tip is, at the moment of instability, only of the order of half that for a fully formed dislocation.

The results highlight a new solid state parameter γ_{us}, called the unstable stacking energy, which measures the resistance to dislocation nucleation at a crack tip. Critical stress intensity factors at nucleation scale with $\sqrt{\gamma_{us}}$. Here γ_{us} is the maximum energy, per unit area, encountered in the block-like shear of one half of a crystal relative to the other, along a slip plane in the direction of shear which forms a lattice dislocation. Also, some features of the 2D activated configuration (energy saddle point) have been derived for a crack tip loaded below the level for instantaneous nucleation.

There are, at present, only quite uncertain estimates of γ_{us}. The sheared atomic lattice geometry to which it corresponds is however, a relatively simple one, periodic in the two directions along the slip plane and involving simple block-like translation of atoms above and below. Thus, it is to be hoped that the parameter may be susceptible to quantum electronic calculation, and such work is encouraged (the same for stacking fault and anti-phase boundary energy terms, which also enter the nucleation criteria for dissociated dislocations).

Allowing for considerable uncertainties in γ_{us}, the evaluation of the competition over whether the condition for Griffith cleavage, or for dislocation generation and blunting, is met first at a crack tip leads to results that seem generally consistent with known brittle versus ductile response of fcc and bcc metals. The results also suggest that the outcome of this competition is often extremely sensitive to small amounts of mode II and mode III shear loading superposed on a basic mode I tensile loading; the shear loadings promote ductile response.

The new analysis of dislocation nucleation given here, like that formulated by Rice and Thomson (1974), is developed only for cases in which the crack tip lies in a slip plane. It has been noted (Argon, 1987; Dragone and Nix, 1988) that the maximally stressed slip plane is sometimes one which intersects the crack tip at a single point but does not contain it. There seems to be no simple way of extending the present approach to such cases.

Added Note: Our analysis of partial dislocation emission in §1.5 and §1.6 is incorrect in the following sense. As A.S. Argon and J. F. Knott indicated to us, the atomic geometry of the fcc slip plane forces partials to nucleate in an *ordered sequence*, not competitively. Our equations are correct if used first to evaluate the loads for nucleation of partial A (as in Eq. (1.26)) and then to evaluate the load for partial B (as in Eqs. (1.20) and (1.32)). For the crack geometry considered in §1.6 and §1.10, the proper sequence is first a partial with $\phi = \pm 60°$, and then the partial with $\phi = 0°$. Consequently, Eq. (1.67) should become $\gamma_s/\gamma_{us} > 11.8$, for emission before cleavage under mode I, and $\gamma_s/\gamma_{us} > 5.3$ under mode I plus 10% shear modes. The stacking fault energy γ_{sf} does not affect the nucleation of the first partial at $\theta = \pm 60°$ and the second, at $\phi = 0°$, follows spontaneously for the near-mode I conditions assumed. With the ~17% correction to the K^{eff} concept, the conditions for dislocation nucleation before cleavage in §1.10 change as follows: For a mode I load Eq. (1.80) should read $\gamma_s/\gamma_{us} > 9.8$ (fcc) or 5.2 (bcc), and for mode I plus 10% shear modes Eq. (1.81) should read $\gamma_s/\gamma_{us} >$ 4.4 (fcc) or 2.4 (bcc). These corrections do not affect our conclusions on brittle vs. ductile response for the fcc metals.

We are pleased to dedicate the paper to Professor Frank A. McClintock.

Acknowledgments: The group of studies reported here has been supported primarily by the Office of Naval Research, Mechanics Division (grant N00014-90-J-1379), and also by the National Science Foundation Materials Research Laboratory at Harvard (grant DMR-89-20490), and a University Research Initiative (subcontract P0AVB38639-0 from the University of California, Santa Barbara, based on ONR/DARPA contract N00014-86-K-0753). Some of the computations were carried out under NSF support at the Pittsburgh Supercomputing Center.

References

Anderson, P. M. (1986). Ductile and brittle crack tip response, Ph.D. Thesis, Div. of Applied Sciences, Harvard University, Cambridge, MA, USA.

Argon, A. S. (1987). Brittle to ductile transition in cleavage fracture, Acta Met., 35:185–196.

Armstrong, R. W. (1966). Cleavage crack propagation within crystals by the Griffith mechanism versus a dislocation mechanism, Mater. Sci. Eng., 1:251–256.

Ashcroft, N. W. and Mermin, N. D. (1976). Solid State Physics, Holt, Rinehart and Winston, New York.

Barnett, D. and Asaro, R. J. (1972). The fracture mechanics of slit-like cracks in anisotropic elastic media, J. Mech. Phys. Solids, 20: 353–366.

Beltz, G. E. (1991). Unpublished research, on the emission of dislocations on inclined slip planes in the Fe/TiC system.

Beltz, G. E. and Rice, J. R. (1991). Dislocation nucleation versus cleavage decohesion at crack tips, In Lowe, T. C., Rollett, A. D., Follansbee, P. S. and Daehn, G. S., editors, *Modeling the Deformation of Crystalline Solids*, TMS, pages 457–480.

Beltz, G. E., and Rice, J. R. (1992a). Dislocation nucleation at metal/ceramic interfaces, Acta Met., in press.

Beltz, G.E., and Rice, J. R. (1992b). Research in progress, on the 2D and 3D calculations of the activation energy for dislocation nucleation.

Beltz, G. E., and Wang, J.-S. (1992). Crack direction effects along copper/sapphire interfaces, Acta Met., in press.

Brandes, E. A. (1983). *Smithells Metals Reference Book*, 6th ed., Butterworths, London.

Burns, S. J. (1986). Crack tip dislocation nucleation observations in bulk specimens, Scripta Met., 20:1489–1494.

Cheung, K. (1990). Atomistic study of dislocation nucleation at a crack tip, Ph.D. Thesis, Dept. of Nuclear Engineering, MIT, Cambridge, MA, USA.

Cheung, K., Yip, S. and Argon, A. S. (1991). Activation analysis of dislocation nucleation from a crack tip in α-Fe, J. Appl. Phys., 69:2088–2096.

Chiao, Y.-H., and Clarke, D. R. (1989). Direct observation of dislocation emission from crack tips in silicon at high temperatures, Acta Met., 37:203–219.

Daw, M. S., and Baskes, M. I. (1984). Embedded-atom method: Derivation and application to impurities and other defects in metals, Phys. Rev. B, 29:6443–6453.

Dragone, T. L., and Nix, W. D. (1988). Crack tip stress fields and dislocation nucleation in anisotropic materials, Scripta Met., 22:431–435.

Duesbery, M. S., Michel, D. J., Kaxiras, E. and Joos, B. (1991). Molecular dynamics studies of defects in Si, In Bristowe, P. D., Epperson, J. E., Griffith, J. E. and Liliental-Weber, Z., editors, *Defects in Materials*, Materials Research Society, 209:125–130.

Eshelby, J. D. (1970). Energy relations and the energy-momentum tensor in continuum mechanics, In Kanninen, M. F., Adler, W. F., Rosenfield, A. R. and Jaffee, R. I., editors, *Inelastic Behavior of Solids*, McGraw-Hill, New York, pages 77–115.

Foiles, S. M., Baskes, M. I. and Daw, M. S. (1986). Embedded-atom-method functions for the fcc metals Cu, Ag, Au, Ni, Pd, Pt, and their alloys, Phys. Rev. B., 33:7983–7991.

Foiles, S. M., and Daw, M. S. (1987). Application of the embedded atom method to Ni^3Al, J. Mater. Res., 2:5–15.

Harrison, R. J., Spaepen, F., Voter, A. F. and Chen, A. F. (1990). Structure of grain boundaries in iron, In Olson, G. B., Azrin, M. and Wright, E. S., editors, *Innovations in Ultrahigh-Strength Steel Technology*, Plenum Press, pages 651–675.

Hirth, J. P., and Lothe, J. (1982). *Theory of Dislocations*, 2nd Edition, McGraw Hill, New York.

Hoagland, R. G., Daw, M. S., Foiles, S. M. and Baskes, M. I. (1990). An atomic model of crack tip deformation in aluminum using an embedded atom potential, J. Mater. Res., 5:313–324.

Kelly, A., Tyson, W. R. and Cottrell, A. H. (1967). Ductile and brittle crystals, Phil. Mag. 15, pages 567–586.

Lin, I.-H., and Thomson, R. (1986). Cleavage, dislocation emission, and shielding for cracks under general loading, Acta Met., 34:187–206.

Michot, G., and George, A. (1986). Dislocation emission from cracks — observations by x-ray topography in silicon, Scripta Met., 20:1495–1500.

Mohan, R., Ortiz, M and Shih, C. F. (1991). Crack-tip fields in ductile single crystals and bicrystals, In Lowe, T. C., Rollett, A. D., Follansbee, P. S. and Daehn, G. S., editors, *Modeling the Deformation of Crystalline Solids*, TMS, pages 481–498.

Nabarro, F. R. N. (1947). Dislocations in a simple cubic lattice, Proc. Phys. Soc., 59:256–272.

Ohr, S. M. (1985). An electron microscope study of crack tip deformation and its impact on the dislocation theory of fracture, Mat. Sci. and Engr., 72:1–35.

Ohr, S. M. (1986). Electron microscope studies of dislocation emission from cracks, Scripta Metall., 20:1501–1506.

Paxton, A. T., Gumbsch, P. and Methfessel, M. (1991). A quantum mechanical calculation of the theoretical strength of metals, Phil. Mag. Lett., 63:267–274.

Peierls, R. E. (1940). The size of a dislocation, Proc. Phys. Soc., 52:34–37.

Rice, J. R. (1968a). A path independent integral and the approximate analysis of strain concentration by notches and cracks, J. Appl. Mech., 35:379–386.

Rice, J. R. (1968b). Mathematical analysis in the mechanics of fracture, Ch. 3 of Liebowitz, H., editor, *Fracture: An Advanced Treatise* (vol. 2, *Mathematical Fundamentals*), Academic Press, NY, pages 191–311.

Rice, J. R. (1985). Conserved integrals and energetic forces, In Bilby, B. A., Miller, K. J. and Willis, J. R., *Fundamentals of Deformation and Fracture* (Eshelby Memorial Symposium), Cambridge University Press, pages 33–56.

Rice, J. R. (1987). Mechanics of brittle cracking of crystal lattices and interfaces, In Latanision, R. M. and Jones, R. H., editors, *Chemistry and Physics of Fracture*, Martinus Nijhoff Publishers, Dordrecht, pages 23–43.

Rice, J. R. (1988). Elastic fracture mechanics concepts for interfacial cracks, In J. Appl. Mech., 55:98–103.

Rice, J. R. (1992). Dislocation nucleation from a crack tip: an analysis based on the Peierls concept, to be published in J. Mech. Phys. Solids, 40:239–271.

Rice, J. R., Suo, Z. and Wang, J.-S. (1990). Mechanics and thermodynamics of brittle interfacial failure in bimaterial systems, In Rühle, M., Evans, A. G., Ashby, M. F. and Hirth, J. P., editors, *Metal-Ceramic Interfaces*, Pergamon Press, Oxford, pages 269–294.

Rice, J. R., and Thomson, R. M. (1974). Ductile *vs.* brittle behavior of crystals, Phil. Mag., 29:73–97.

Rice, J. R., and Wang, J.-S. (1989). Embrittlement of interfaces by solute segregation, Mat. Sci. and Engr., A107:23–40.

Saeedvafa, M. (1991). Orientation dependence of fracture in copper bicrystals with symmetric tilt boundaries, submitted to Mech. Mat.

Schoeck, G. (1991). Dislocation emission from crack tips, Phil. Mag., 63:111–120.

Stroh, A. H. (1958). Dislocations and cracks in anisotropic elasticity, Phil. Mag. 3:625–646.

Suo, Z. (1989). Mechanics of interface fracture, Ph.D. Thesis, Div. of Applied Sciences, Harvard University, Cambridge, MA, USA.

Sun, Y. (1991). Unpublished work, on EAM fits for α-Fe, Al, and Ni.

Sun, Y., Beltz, G. E. and Rice, J. R. (1992). Research in progress, on embedded atom models as a basis for estimating normal stress effects in dislocation nucleation.

Sun, Y., and Rice, J. R. (1992). Research in progress, on the anisotropic elastic formulation of dislocation nucleation.

Sun, Y., Rice, J. R. and Truskinovsky, L. (1991). Dislocation nucleation versus cleavage in Ni_3Al and Ni, In Johnson, L. A., Pope, D. T. and Stiegler, J. O., editors, *High-Temperature Ordered Intermetallic Alloys*, Materials Research Society, 213:243–248.

Tyson, W. R. (1975). Surface energies of solid metals, Canadian Metallurgical Quarterly, 14:307–314.

Vitek, V. (1968). Intrinsic stacking faults in body-centered cubic crystals, Phil. Mag., 18:773–786.

Vitek, V., Lejcek, L. and Bowen, D. K. (1972). On the factors controlling the structure of dislocation cores in bcc crystals, In Gehlen, P. C., Beeler, J. R. and Jaffee, R. I., editors, *Interatomic Potentials and Simulation of Lattice Defects*, Plenum Press, New York, pages 493–508.

Wang, J.-S., and Anderson, P. M. (1991). Fracture behavior of embrittled fcc metal bicrystals and its misorientation dependence, Acta Met., 39:779–789.

Weertman, J. (1981). Crack tip blunting by dislocation pair creation and separation, Phil. Mag., 43:1103–1123.

Willis, J. R. (1967). A comparison of the fracture criteria of Griffith and Barenblatt, J. Mech. Phys. of Solids, 15:151–162.

Yamaguchi, M., Vitek, V. and Pope, D. (1981). Planar faults in the $L1_2$ lattice, stability and structure, Phil. Mag., 43, 1027-1044.

2

Advances in Characterization of Elastic–Plastic Crack–Tip Fields

D. M. Parks

ABSTRACT We review basic features of recently-proposed two-parameter fracture mechanics descriptions of plane strain mode I elastic–plastic crack–tip fields. One of the parameters, the J-integral, essentially measures the scale of crack–tip deformation. Various choices have been proposed for a second crack–tip parameter characterizing the triaxiality of the crack–tip stress state. We discuss relative advantages and disadvantages of these second parameters in extending traditional fracture mechanics methodology. The additional resolution which a second parameter affords in describing the various levels of stress triaxiality encountered in laboratory specimens and structures suggests new approaches to the interpretation of experimental fracture data and to fracture mechanics applications in structures and components.

2.1 Introduction

Among McClintock's many distinguished contributions to the study of the mechanical behavior of materials, none surpasses the fact that he provided the principal impetus for bringing continuum plasticity theory to bear on problems of ductile fracture. His insights and analyses, at both the macroscopic and microstructural levels, have guided generations of researchers in the development of modern nonlinear fracture mechanics. At the risk of over-simplification, we will emphasize three major interrelated themes emerging from McClintock's research in ductile fracture.

First, is the notion that the kinetics of microstructural processes of fracture in a given material, e.g., ductile fracture by void growth (McClintock, 1967) can be considered as "driven" by continuum stress and deformation fields as averaged over larger dimensional scales. The rate of evolution of the ductile fracture process (per unit deformation) in general differs in material elements subjected to differing states of stress. In particular, at high triaxiality, void growth was shown to be essentially exponentially dependent on the triaxiality of stress.

Secondly, Hult and McClintock (1956) illustrated that under restricted conditions of limited-scale crack-tip plastic deformation, details of the local crack-tip elastic-plastic fields could be uniquely related to a single macroscopic parameter (e.g., the linear elastic stress intensity factor, K, or energy release rate, $\mathcal{G}$) scaling the intensity of crack-tip deformation. Thus, to the extent that local crack-tip fields can be characterized by a single parameter, and further, that fracture processes are driven by these fields, there exists a mechanistic rationale for constructing "single parameter" fracture mechanics correlations of crack extension.

Thirdly, using non-hardening plasticity theory, McClintock (1971) emphasized that at large-scale yielding, there is a wide range of plane strain crack-tip stress and deformation fields which depend critically on crack geometry and type of loading. Thus, to the extent that crack-tip fields of widely differing triaxiality drive fracture processes at differing rates, there is, in general, no mechanistic rationale for a unique "single parameter" fracture mechanics correlation of crack extension in large scale yielding.

The fracture mechanics community has had to deal with the far-reaching implications of this related set of observations. The introduction of the J-integral (Rice, 1967) and the asymptotic singular strain hardening crack-tip fields of Hutchinson (1968) and Rice and Rosengren (1968) (termed HRR) provided a basis for generalizing the concept of a dominant crack-tip parameter, and led to a substantial broadening of the range of loading magnitudes for which crack-tip fields were "well-described" over the fracture process zone size by a single parameter.

It remained apparent, however, that in general, no single parameter could uniquely correlate fully plastic fracture. Begley and Landes (1976) obtained dramatically higher slopes of experimental J vs. Δa "resistance" curves in fully plastic plane strain center-cracked plates under tension than in compact tension (CT) specimens loaded predominantly in bending. Hancock and Cowling (1980) found that the plane strain crack-tip opening displacement (δ_t) at the initiation of ductile tearing in an alloy similar to HY-80 ranged from $90\mu m$ in a deep, double-edged notch specimen to $\sim 900\mu m$ in a single-edge notch geometry subject to mid-ligament tension, both specimens being in the fully plastic regime at initiation. In these experiments, the geometries exhibiting higher measures of toughness were those whose fully plastic slip-line fields showed lower crack-tip triaxiality, consistent with McClintock's pioneering interpretations.

Fracture mechanics methodology is essentially the correlation of crack extension in two different cracked bodies (e.g., a laboratory specimen and an engineering structure) based on the similarity of their respective near crack-tip stress and deformation fields. Traditionally, the degree of similarity has been assessed in a one-parameter fashion, based on the strength of asymptotically-dominant singular crack-tip fields. However, the diverse range of crack-tip fields has made single-parameter assessment of the "similarity" of crack-tip fields somewhat elusive.

Crack configurations exhibiting the most triaxial of the plane strain fields generally lead to the lowest measure of toughness among the family. High levels of crack-tip stress triaxiality are associated with: (a) essentially all states of well–contained yielding; and (b) virtually all load levels in geometries involving predominant bending on uncracked ligaments sufficiently small compared to crack depth. Conversely, low levels of triaxiality occur in large-scale yielding and fully-plastic flow of single edge-cracked and center–cracked specimens under predominant tension loading, as well as in shallow edge–cracked specimens under bending.

In view of the technological importance of low toughness behavior in general and of the cracking behavior of large structures in small-scale yielding (SSY), attention has been primarily focused on conditions leading to higher crack-tip stress triaxiality and lower measures of toughness. McMeeking and Parks (1979) and Shih and German (1981) carefully examined the near-tip fields of detailed finite element solutions in plane strain tension and bending in order to quantify the evolving states of stress. They sought to develop guidelines regarding parametric limits on the magnitude of imposed crack-tip deformation level, relative to specimen dimensions such as crack length, remaining ligament, etc. (collectively denoted "ℓ"), such that "high" levels of crack-tip stress triaxiality, similar to those of SSY, were maintained at the crack-tip. McMeeking and Parks proposed that crack-tip stress triaxiality remained sufficiently "high" providing

$$J \leq \frac{\sigma_0 \ell}{\mu_{cr}}, \tag{2.1}$$

where σ_0 is tensile yield stress and μ_{cr} varied from ~ 25 for bending to ~ 200 for tension (Shih, 1985). For fixed specimen or structural geometry, when the imposed J–value exceeded these limits, stress triaxiality in the crack-tip region often decreased markedly. When stress triaxiality was "high" (as operationally defined by the satisfaction of Eq. 2.1), the crack-tip fields were taken to be "dominated", or adequately described, by a single parameter; e.g., the amplitude of the HRR singularity as measured by J.

A ductile fracture mechanics based on low toughness material data obtained from specimens of high triaxiality (e.g., CT specimens), when predictively applied to structural crack configurations of low triaxiality (e.g., a shallow surface flaw), is generally conservative in that the extent of cracking at any given load is over-predicted. However, the margin of conservatism is unknown.

There has recently been a surge of interest in analyzing, testing, and predicting crack growth under conditions of low crack-tip stress triaxiality. A primary impetus to this activity has been a set of efforts aimed at developing two-parameter descriptions of crack-tip fields. Retaining contact with traditional approaches, the first parameter reflects the scale of crack-tip deformation, as measured by J or δ_t. A second parameter is used to identify a particular member of a family of crack-tip fields of varying stress triaxiality. Monotonically increasing macroscopic loading of a given specimen or structure generates a trajectory in a two-parameter crack-tip loading space. In a fracture mechanics spirit, similar loading trajectories in a specimen and a component should cause similar crack extensions, while trajectories of significantly different triaxiality could lead to different measures of toughness appropriate to their respective conditions.

Three approaches to specifying families of mode I plane strain elastic-plastic crack-tip fields of varying triaxiality have been proposed. They are based on: (a) parametric descriptions correlated with the so-called T-stress, or second Williams (1957) coefficient in the expansion of isotropic linear elastic crack-tip fields (Betegón and Hancock, 1991; Al-Ani and Hancock, 1991; Du and Hancock, 1991); on (b) higher order asymptotic expansions of the crack-tip fields (Li and Wang, 1986; Sharma and Aravas, 1991); and on (c) approximate descriptions of large geometry change finite element solutions (O'Dowd and Shih, 1991a,b). Each of these approaches falls within the scope of two-parameter descriptions of crack-tip fields of varying triaxiality, and each method has certain relative advantages and disadvantages in usage.

In the following sections, we review major features of the three approaches to identification and utilization of a second crack-tip parameter, and discuss in some detail results which have been obtained with the T-stress approach.

2.2 Approaches to Two–Parameter Characterization

2.2.1 T-Stress Effects on Crack-Tip Plasticity

At load levels sufficiently small that crack-tip plasticity is well–contained, the loading imparted to the crack tip by the surrounding elastic material can be characterized by the two–term Williams (1957) expansion

$$\sigma_{\beta\gamma}(r,\theta) = \frac{K_I}{\sqrt{2\pi r}} f_{\beta\gamma}(\theta) + T\delta_{1\gamma}\delta_{1\beta}. \tag{2.2}$$

of the mode I linear elastic in–plane cartesian stress components (Greek subscripts range from one to two). Here cylindrical coordinates centered at the tip are (r, θ), $f_{\beta\gamma}(\theta)$ are the angular variations of the in–plane cartesian components of the elastic singular stress field, K_I is the stress intensity factor, and the "T" stress component represents a plane strain tension/compression stress parallel to the crack. The T-stress provides *the* rigorous "second" crack-tip parameter in linear elastic crack analysis, and it has recently been shown to be valuable in characterizing the stress triaxiality of elastic-plastic crack-tip fields.

The plane strain crack-tip fields resulting from the two-parameter loading of Eq 2.2 admit the length-dimensioned scaling parameter $(K_I/\sigma_0)^2$, where

σ_0 is tensile yield strength. The scale parameter can also be expressed in terms of the J–integral by using the identity $K_I^2 = JE/(1-\nu^2)$, where E is Young's modulus and ν is Poisson's ratio. In the case of contained yielding, a normalized form of a second crack-tip loading parameter is $\tau \equiv T/\sigma_0$. At large distances from the tip, the applied stress defined by Eq. 2.2 is dominated by the T-stress itself; in order that plastic response remain contained, it is necessary that $|T|$ be suitably small in comparison to σ_0. Adopting a Mises yield criterion, this requires $|\tau| < 1/\sqrt{1-\nu+\nu^2}$, or $|\tau| < 1.125$ when $\nu = 0.3$.

Consider the case of power law hardening materials which strain in uniaxial tension according to

$$\epsilon = \begin{cases} \sigma/E & \text{if } \sigma \leq \sigma_0 \\ \epsilon_0 \, (\sigma/\sigma_0)^n & \text{if } \sigma > \sigma_0 \end{cases}, \tag{2.3}$$

where $\epsilon_0 = \sigma_0/E$ and $n > 1$ is the stress exponent in the strain hardening expression. The plane strain crack-tip stress fields emerging from the two-term loadings Eq. 2.2, termed modified boundary layer (MBL) solutions, can be expressed as

$$\sigma_{ij}(r, \theta; J, \tau) = \sigma_{ij}(r/(J/\sigma_0), \theta; \tau) \equiv \sigma_{ij}^{MBL}. \tag{2.4}$$

For each value of n, there are also corresponding kinematical fields of displacement (u_i^{MBL}) and strain (ϵ_{ij}^{MBL}). These fields are fully described by the two load parameters J and τ. The special case $\tau = 0$ is the similarity solution of small-scale yielding: $\sigma_{ij}^{SSY} \equiv \sigma_{ij}^{MBL}(\tau = 0)$.

Figure 2.1 shows the near–tip crack opening stress profile, $\sigma_{22}(\theta = 0)$, for $n = 10$ and various values of τ. These results have been obtained using the small geometry change (SGC) assumption and a J_2 deformation theory of plasticity generalization of the uniaxial response of Eq. (2.3). Opening stress is normalized by σ_0 and distance ahead of the tip is normalized by J/σ_0. The case $\tau = 0$ (dark line) is the opening stress profile of SSY, while the open circles indicate the HRR field.

The stress profiles for different values of τ are roughly parallel to each other, suggesting that the deviation from SSY is essentially independent of normalized distance from the tip. Thus, the curves of Fig. 2.1 can be approximated well by the following fit to the crack opening stress profile:

$$\frac{\sigma_{22}^{MBL}(r; J, \tau)}{\sigma_0} = \frac{\sigma_{22}^{SSY}\,(r/\,(J/\sigma_0))}{\sigma_0} + A_n\,\tau + B_n\,\tau^2, + C_n\,\tau^3, \tag{2.5}$$

where A_n, B_n, and C_n are constants dependent on the strain hardening exponent n. This three-term polynomial fit was suggested by Wang (1991a) as a slight modification of the two-term $(C_n \equiv 0)$ fit initially proposed by Betegón and Hancock (1991). The results in Fig. 2.1 are well–fit by the constants $(A_n, B_n, C_n) = (0.617, -0.565, 0.123)$ for $n = 10$, as illustrated

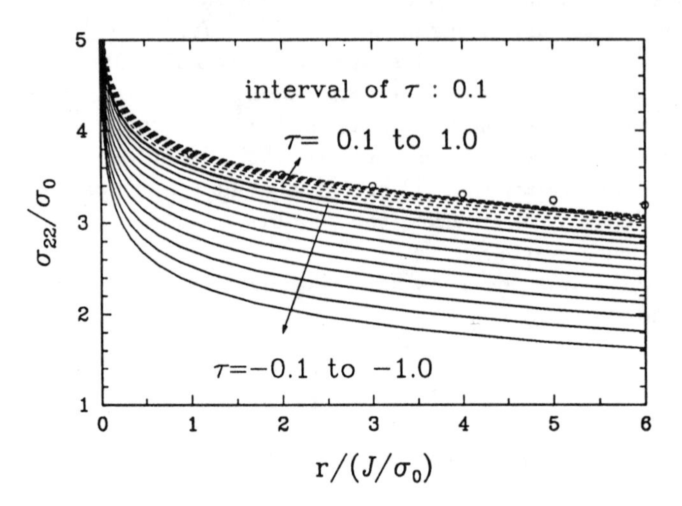

FIGURE 2.1. Normalized crack opening stress distribution in plane strain MBL solutions for hardening exponent $n = 10$, for various values of $\tau = T/\sigma_0$. The circles are the HRR field, while the dark line is that of small-scale yielding, $\tau = 0$.

in Fig. 2.2. Similar conclusions regarding the essential relation between τ and local crack-tip stress in the MBL loadings were also reached by Bilby, et al., (1986) in large geometry change (LGC) solutions of non-hardening materials and by O'Dowd and Shih (1991a,b) in LGC solutions of power law hardening materials.

The remarkable range of crack opening stress profiles exhibited by the τ-family of loadings is primarily associated with variations in the level of hydrostatic (mean) stress, $\sigma_m = \sigma_{kk}/3$. This point is most clearly evident in the non-hardening MBL finite element solutions of Du and Hancock (1991). They showed that the limiting Prandtl field, consisting of constant state regions on the crack flanks connected by centered fans of angular extent $\pi/2$ to a constant state region ahead of the tip, was only obtained for sufficiently positive values of τ. In this special case of yielding extending from the crack plane back to the flanks, the crack-tip hydrostatic stress can be computed using slip-line theory as $\sigma_m/\sigma_0 = (1+\pi)/\sqrt{3} \doteq 2.39$ (Mises yield). In contrast, when $\tau \leq 0$, an elastic zone of angular extent $\geq \pi/4$ emerges from the crack flanks, cutting into the extent of the centered fan within which hydrostatic stress builds up. The more negative τ becomes, the greater is the reduction in fan extent and crack-tip stress triaxiality. Computational results for the circumferential variation of near-tip stress triaxiality with τ are shown in Fig. 2.3 for the case of strain hardening exponent $n = 10$.

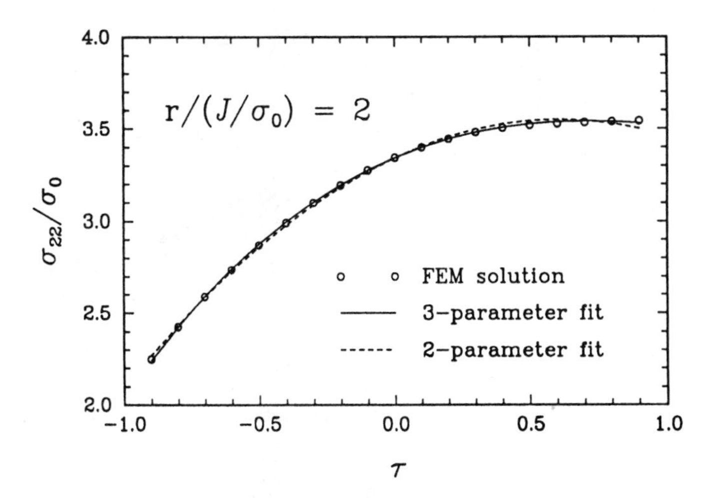

FIGURE 2.2. Comparison of two-term and three-term polynomial fits, Eq. 2.5, to the normalized crack opening stress a distance $2J/\sigma_0$ ahead of the crack tip in plane strain MBL solutions for $n = 10$.

The MBL fields generated by the remote elastic loading of Eq. 2.2 possess distinct kinematic fields, depending upon τ. The dependence of the size and shape of the plastic zone on τ was given originally by Larsson and Carlsson (1973) and Rice (1974). The maximum radius of the plastic zone size can be written as $r_p = \Lambda(\tau)\ (K_I/\sigma_0)^2$, where the minimum value of the factor Λ is 0.1, occurring for $\tau \doteq 0.3$. Under SSY conditions ($\tau = 0$), $\Lambda = 0.15$. For $|\tau| > 0.5$, the plastic zone size grows rapidly, and, Λ reaches 1.5 at $\tau = 1.0$ and -0.85. The angular orientation of the lobe of $r_p^{\max}$ rotates from near $\theta = 135^o$ at $\tau = 1.0$ to $\theta = 45^o$ for $\tau = -1.0$.

In the near-tip region, there is a corresponding rotation and scaling of the angular distribution of equivalent plastic strain distribution, ϵ^p, as shown in Fig. 2.4. Strictly, the small geometry change solutions shown in Figs. 2.3–2.4 are sampled too close to the tip ($r = 1.22J/\sigma_0$) to be unaffected by the neglected large geometry change, but they nonetheless represent the τ-effects on the near-tip stress and deformation fields.

On the other hand, the relation between crack-tip opening displacement, δ_t, and J/σ_0 is much less sensitive to τ. Under MBL conditions

$$\delta_t = d_n(\tau;\, n, \epsilon_0)\, \frac{J}{\sigma_0}. \tag{2.6}$$

Here, δ_t is the $\pm 45^o$ intercept definition of CTOD suggested by Rice, and d_n depends on strain hardening exponent n, the yield strain ϵ_0, and τ. For

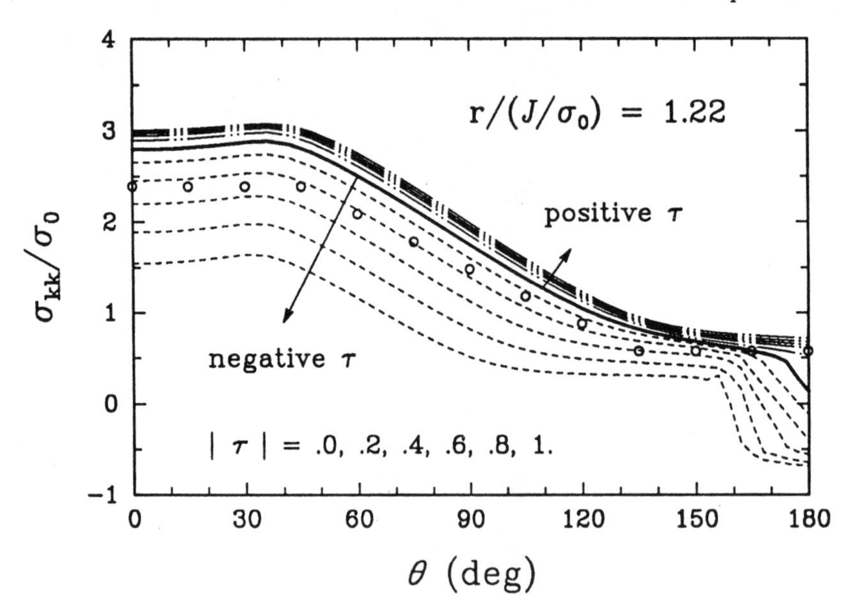

FIGURE 2.3. Angular variation of near-tip normalized hydrostatic stress for various values of τ in plane strain MBL solutions with $n = 10$.

$n = 10$ and $\epsilon_0 = 0.0025$, Wang's (1991b) large geometry change flow theory plasticity calculations gave $d_n = 0.59$ at $\tau = 0$. Shih (1981) manipulated the small geometry change HRR displacement fields to estimate d_n; for the case considered, this estimate is 0.51. Figure 2.5 shows that the value 0.59 is essentially the minimum value of d_n over the range of $-0.9 \leq \tau \leq 0.9$, but the maximum value for d_n over this range was only 30% greater, occurring at $\tau = -0.9$. Thus, for all practical purposes, the MBL relation between J and δ_t for moderate strain hardening is essentially independent of τ.

The key feature of Figs. 2.1–2.3 and Eq. 2.5 is that negative T–stress ($\tau < 0$) is associated with substantial reduction in crack tip stress triaxiality (compared to SSY), while positive T-stress results in only modest elevation of triaxiality above SSY. Consider, for example, a crack geometry in which the correlation between nominal stress, σ^{nom} (> 0), and T-stress is negative; that is

$$T = \sigma^{\text{nom}}\,\hat{t}, \tag{2.7}$$

where, for a given crack geometry, the dimensionless function $\hat{t} < 0$ depends on loading type and on dimensionless ratios of geometric lengths. Such crack configurations will display decreasing crack-tip triaxiality with increasing load, since as load increases, T and τ become progressively more negative. Indeed, Al–Ani and Hancock (1991) examined the complete near tip crack opening stress profiles in plane strain edge–cracked geometries of various depths subject to remote tension or bending loads ranging from SSY

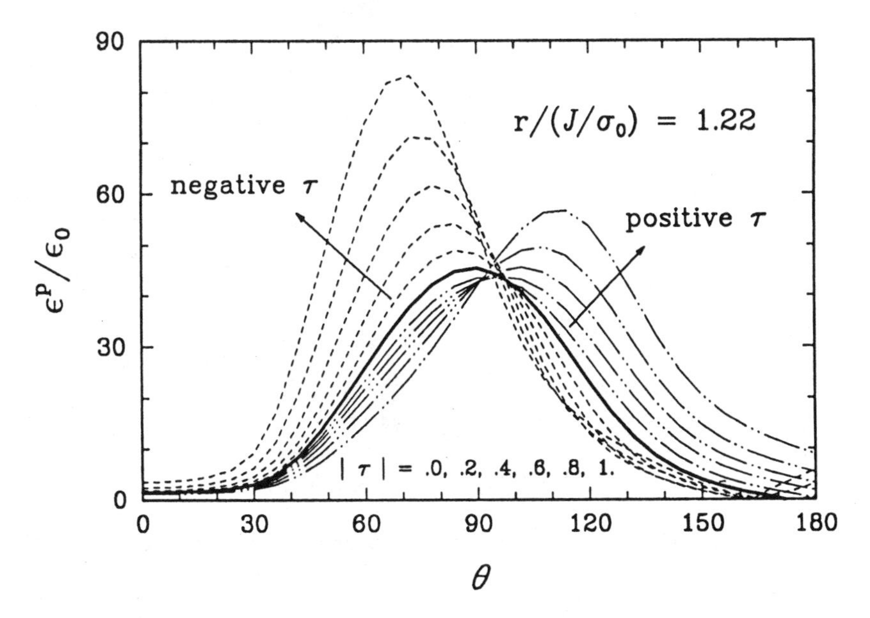

FIGURE 2.4. Angular variation of near-tip normalized equivalent plastic strain for various values of τ in plane strain MBL solutions with $n = 10$.

through fully plastic conditions. At the lowest loadings, local fields were very close to those of SSY. At higher load levels, the opening stress profiles deviated from SSY in a manner identical to that predicted by the MBL analysis of Eq. 2.5, based on the geometry– and load amplitude–dependent J and on elastically–computed T–stress values (i.e., from Eq. 2.7) of the respective problems. Deviations from SSY were especially prominent in shallow-cracked bend bars having crack depth to width ratios $a/w < 0.3$, corresponding to the range for which $\hat{t} < 0$. In contrast, in bend specimens having $a/w > 0.3$, $\hat{t} > 0$, and the elevations of stress profile above SSY were relatively small.

The main theoretical objection to the T-stress as a broadly applicable correlator of elastic-plastic crack-tip stress triaxiality is that its derivation relies on the assumption of a surrounding linear elastic region characterized by the two-term Williams expansion of Eq. 2.2. At fully plastic conditions of any geometry, no such zone exists.

The nominal stress at limit load can be written as $\sigma_{\text{limit}}^{\text{nom}} = \sigma_0 \, \hat{g}$, where $\hat{g}$ also depends on geometry. Thus τ can be expressed as $\tau = \Sigma \, \hat{\tau}$, where $\hat{\tau} = \hat{t}\hat{g}$ and $\Sigma \equiv \sigma^{\text{nom}}/\sigma_{\text{limit}}^{\text{nom}}$ is load/limit load. In any crack problem, there is a critical value of load level, Σ_{cr}, beyond which the theoretical basis for T vanishes, and a corresponding limit, $\tau_{cr} = \Sigma_{cr} \, \hat{\tau}$. However, as in assessing limits of J–dominance (McMeeking and Parks, 1979), there is no clear, objective criterion for precisely establishing theoretical limits of applicability of τ-based correlations of crack-tip triaxiality. Satisfactory

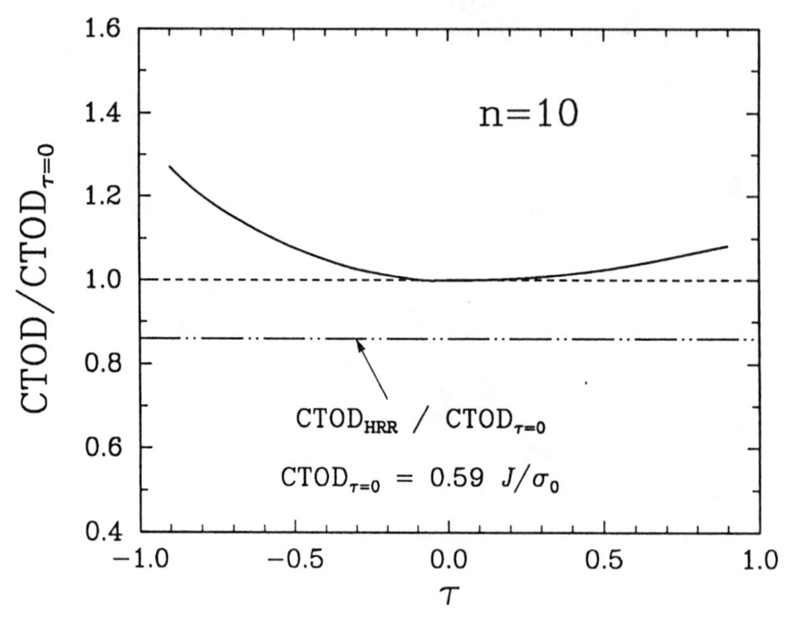

FIGURE 2.5. Normalized crack tip opening displacement ($\pm 45^o$ intercept definition) in large geometry change solutions of plane strain MBL loading.

answers in both cases depend on the context in which the results are to be interpreted and used, the trade-offs between costs and benefits associated with implementation of more sophisticated approaches, etc.

In this spirit, then, Betegón and Hancock (1991), Al-Ani and Hancock (1991), and Betegón, et al. (1991) have shown that in practice, the formal application of the elastically-calculated, τ-based methodology continues to provide extremely good estimates of crack-tip stress triaxiality even beyond limit load ($\Sigma > 1$) in a wide variety of plane strain and axisymmetric crack geometries and loadings. This claim is also broadly consistent with results of O'Dowd and Shih (1991b), even though their interpretation of their results seems excessively pessimistic on this point.

Parks and co-workers (Parks, 1991; Wang, 1991a) have demonstrated that the varying stress triaxiality along surface-cracked plates can be accurately predicted up to and beyond the limit load by using τ as a constraint parameter. Using detailed continuum finite element solutions, Wang (1991a) analyzed wide plates containing semi-elliptical part-through surface cracks of aspect ratio $a/c = 0.24$, subject to remote tension and to remote bending. The plates had thickness t, total width $2b$, and total length $2h$. The surface cracks had maximum penetration a and total surface length $2c$. Two relative crack depths were investigated: $a/t = 0.15$ ("shallow") and $a/t = 0.6$ ("deep"). Figure 2.6 shows one–quarter of the structural geometry. In the post-processing of the data obtained near the crack front,

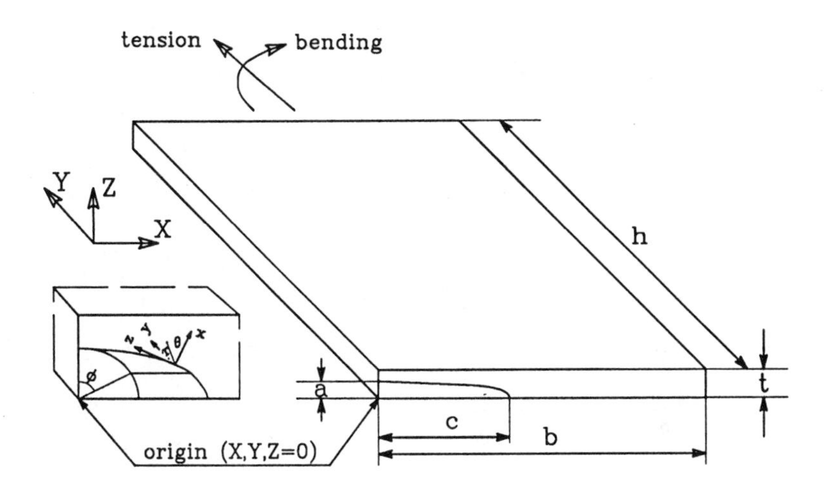

FIGURE 2.6. Schematic view of one–fourth of a surface–cracked plate subject to remote tension or bending, showing global coordinates (X, Y, Z) and local coordinates (x, y, z) along a semi–elliptical crack front. Crack front position parameter ϕ is also indicated.

local coordinates (x,y,z), indicated in the figure, were used. In addition, the parametric angle ϕ locating positions along the semi–elliptical crack front given by $(X/c)^2 + (Z/a)^2 = 1$ is shown, where $X = c \sin \phi$ and $Z = a \cos \phi$. The local z–axis is tangent to the crack front, and the local y–axis, which coincides with the global Y–axis, is normal to the crack plane. In the local coordinate system, cylindrical coordinates (r, θ) are given by $r = \sqrt{x^2 + y^2}$ and $\theta = \arctan(y/x)$. The geometrical ratios chosen for the plate were $b/t = 8$ and $h/t = 16$. The constitutive model was the same small geometry change J_2 deformation theory generalization of Eq 2.3 used in the MBL analyses; the strain hardening exponent chosen was $n = 10$.

Using virtual crack extension/domain integral methods (Parks, 1977; Li, Shih, and Needleman, 1985), local values of J were determined at various positions, ϕ, along the crack front at each value of Σ. For the tension loadings, Σ was defined as σ^{∞}/σ_0, where the total tensile load remotely applied to the specimen is $\sigma^{\infty} 2tb$. For the bending load, the nominal bending stress is $\sigma^{\text{nom}} = 6M/bt^2$, where M is the total applied bending moment. The limit moment of the uncracked section is $M_{\text{limit}} = \sigma_0 bt^2/4$, so $\Sigma \equiv M/M_{\text{limit}} = 1.5\sigma^{\text{nom}}/\sigma_0$. In comparing the applied loads with respective limit loads, values of limit load for the remote uncracked section have been used. Since the actual limit loads for the cracked structures must be less than estimated (especially so in the deep crack case, where the crack

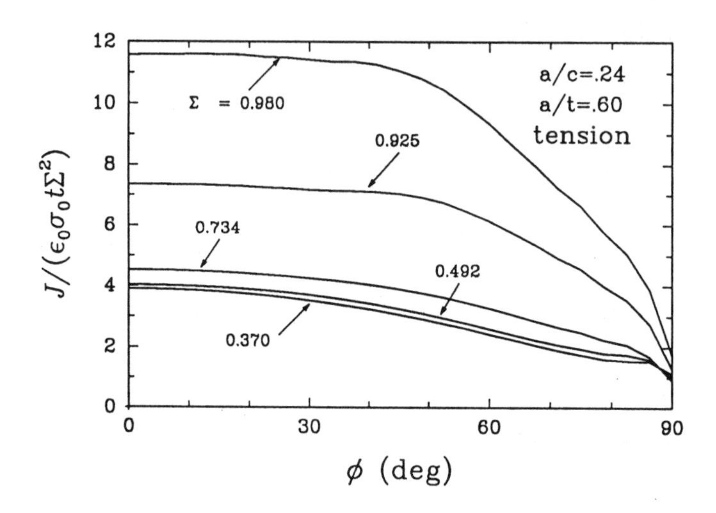

FIGURE 2.7. Normalized J distribution along the surface crack front in a deeply-cracked plate under remote tension.

plane area, $\pi ac/2$, is 14.7% of the uncracked area, $2bt$), the actual fractions of limit load must exceed those reported here.

Figures 2.7 and 2.8 show the distributions of normalized J-value along the crack front for the deep-cracked geometry under tension and bending, respectively. At low loads ($\Sigma \leq 0.5$), the J distribution is essentially as elastically calculated. The maximum J in the deep bend specimen occurs near $\phi = 60^o$, since the center portion of the crack front extends beyond the uncracked neutral axis. Normalized J profiles for the shallow crack under both tension and bending resemble Figure 2.7 in that the peak J occurs at $\phi = 0$ and monotonically decreases with increasing ϕ at all load levels.

The crack opening stress component, σ_{yy}, computed directly ahead of the crack front ($\theta = 0$) is normalized by σ_0 and plotted in Figs. 2.9–2.12 versus distance ahead of the crack front (local coordinate x in Fig. 2.6) in the range $r/(J/\sigma_0) \leq 6$; also shown in each figure is the crack opening profile of small-scale yielding. Figure 2.9 shows results from the deep crack under tension for the crack front positions $\phi = 0^o$, 30°, and 67.5^o at loads ranging from $0.370 \leq \Sigma \leq 0.98$. Figure 2.10 shows the deep crack under bending, while Figs. 2.11 and 2.12 show corresponding results for the shallow crack under tension and bending, respectively. Details of the stress profiles are omitted near the free surface ($\phi > 67.5^o$) due to inadequate mesh refinement in this region; in view of the high aspect ratio of the cracks ($a/c = 0.24$), only a small fraction of the crack front is actually omitted. Clearly, a wide range of local crack-tip stress triaxiality is evidenced in these solutions.

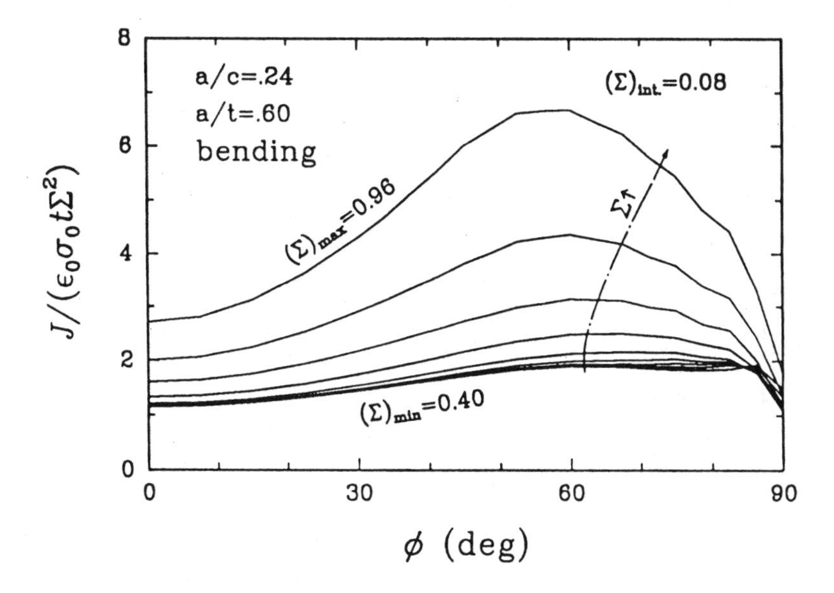

FIGURE 2.8. Normalized J distribution along the surface crack front in a deeply-cracked plate under remote bending.

In an effort to rationalize these dependencies, Wang (1991a) determined the T-stress distribution along the respective crack fronts using both continuum and line-spring (Wang and Parks, 1991) methods. Figure 2.13 shows local T-stress along the crack front for $\nu = 0.3$, normalized by nominal tensile and bending stress, for the four cases studied. The deep crack in bending has positive T-stress along most of its length, in contrast to the negative T-stress profiles of the other three configurations.

Wang (1991a) then showed that the entire set of crack opening stress profiles in Figs. 2.9–2.12 could be fit by the MBL fields Eq. 2.4, by making use of the local $\tau(\phi) = T(\phi)/\sigma_0 = \hat{t}(\phi)\,\sigma^{\rm nom}/\sigma_0$, plus the local $J(\phi)$. With the conventions used for Σ and $\sigma^{\rm nom}$, in tension, $\tau(\phi) = \Sigma\,\hat{t}(\phi)$, while under bending, $\tau(\phi) = 1.5\Sigma\,\hat{t}(\phi)$. Thus, for tension, the predicted crack opening profile is

$$\begin{aligned} \sigma_{22}(r, \theta = 0,\, \phi;\, J^{\rm local},\, \Sigma) \quad &= \quad \sigma_{22}^{SSY}\left(r/\left(J^{\rm local}(\phi)/\sigma_0\right)\right) \\ &\quad + \sigma_0 \cdot \left[A_n \hat{t}\,\Sigma + B_n\,\left(\hat{t}\,\Sigma\right)^2 + C_n\,\left(\hat{t}\,\Sigma\right)^3\right]. \end{aligned} \tag{2.8}$$

For convenience, the comparison is made at the particular distance $r = 2J/\sigma_0$. A composite of the predictions for crack opening stress at all locations ϕ along all four crack fronts (shallow/deep; tension/bending) for loads ranging up to limit load is shown in Fig. 2.14. The solid curve is the

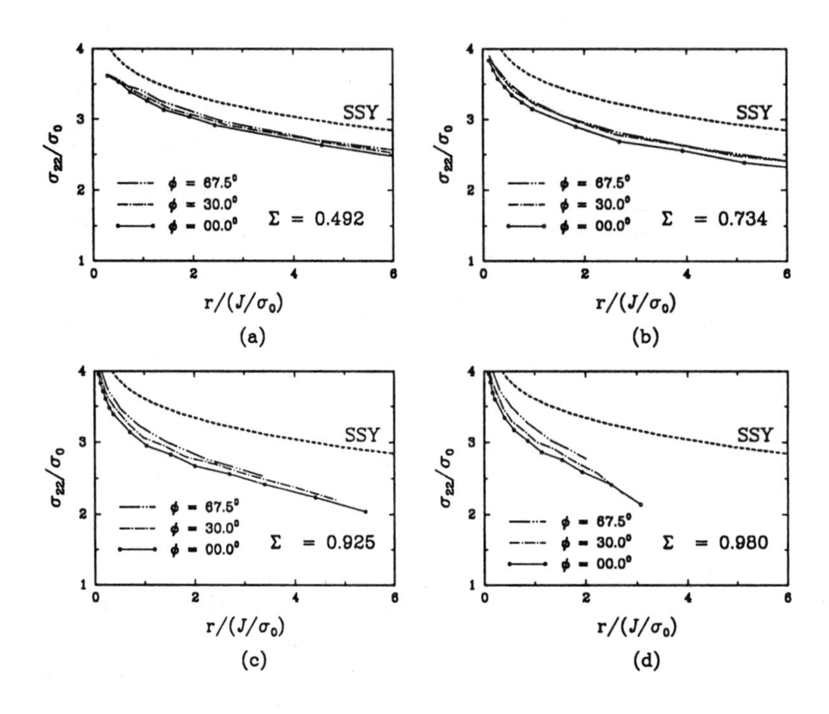

FIGURE 2.9. Normalized crack opening stress profiles in deeply-cracked surface-flawed plates under remote tension at various load levels and crack locations.

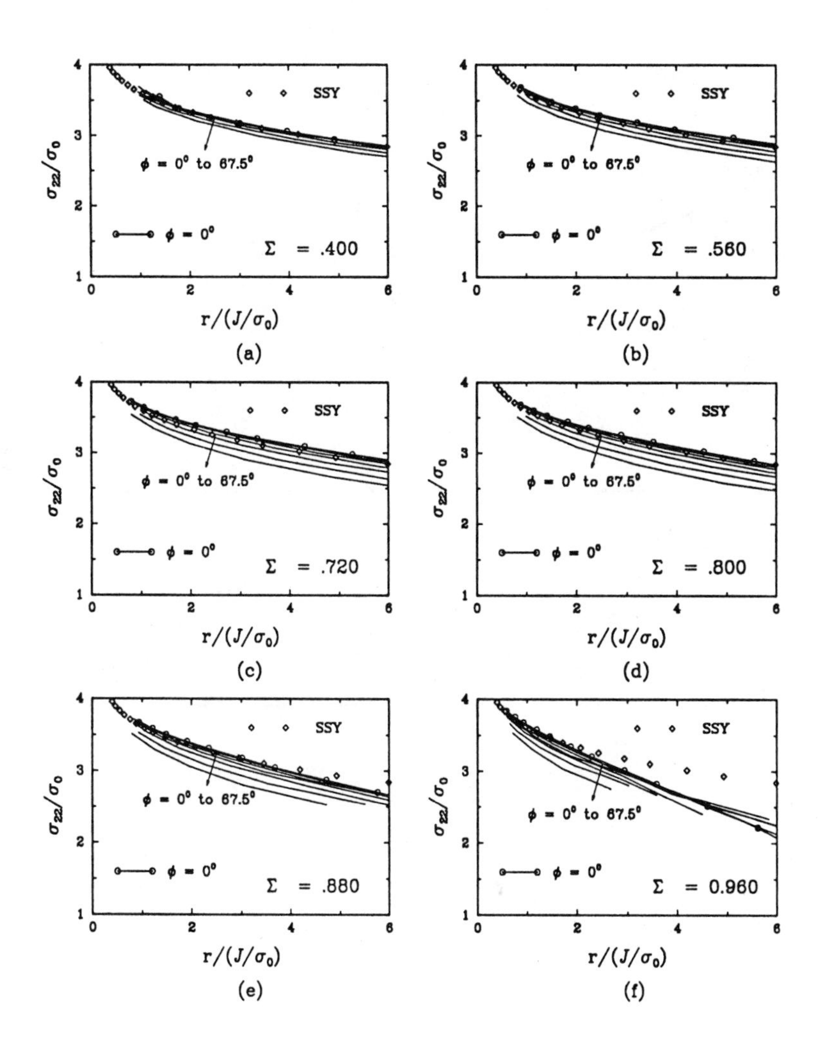

FIGURE 2.10. Normalized crack opening stress profiles in deeply-cracked surface-flawed plates under remote bending at various load levels and crack locations.

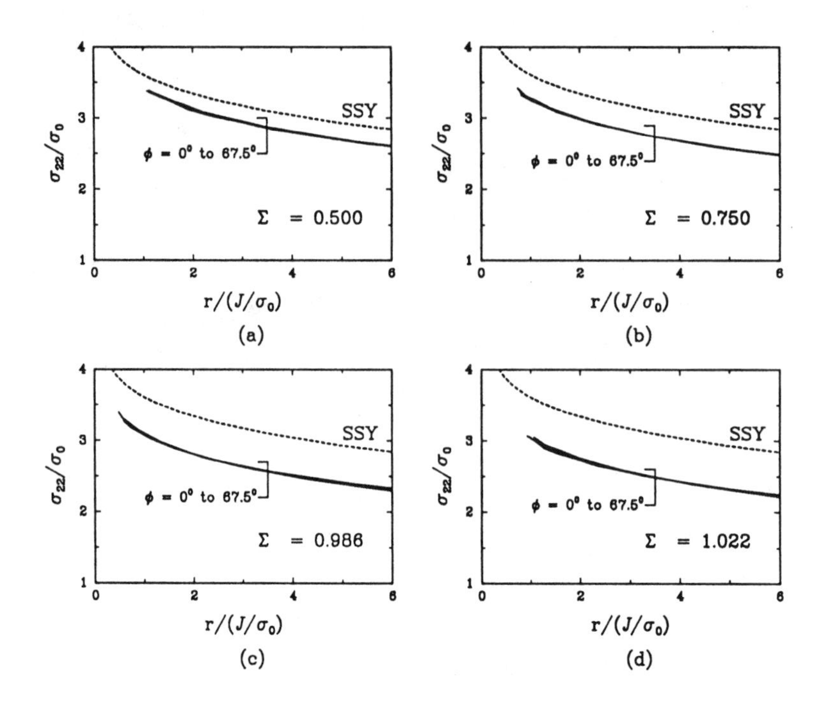

FIGURE 2.11. Normalized crack opening stress profiles in shallow-cracked surface-flawed plates under remote tension at various load levels and crack locations.

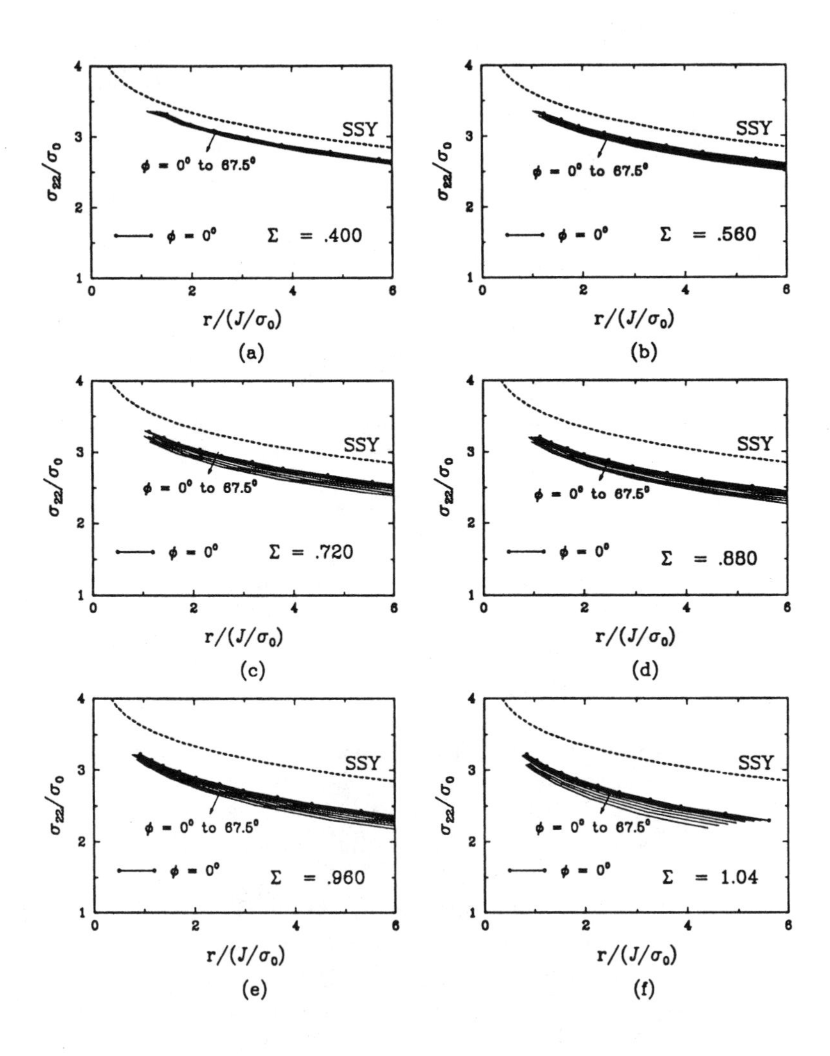

FIGURE 2.12. Normalized crack opening stress profiles in shallow-cracked surface-flawed plates under remote bending at various load levels and crack locations.

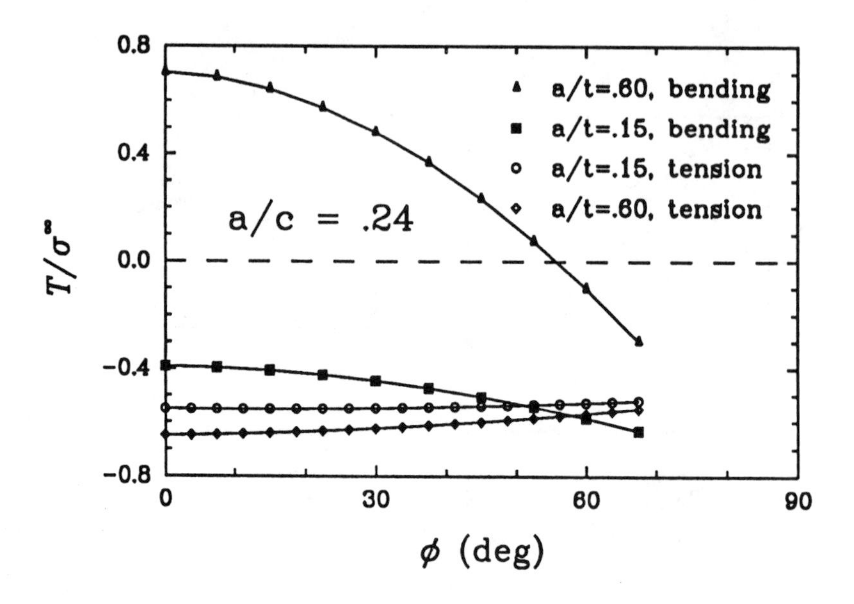

FIGURE 2.13. Normalized elastic T-stress variation along crack fronts in deep and shallow-cracked surface-flawed plates under remote tension and bending. The normalizing stress, σ^{∞}, is remote tensile stress and remote bending stress, respectively.

MBL prediction, and the dotted line represents 95% of the predicted stress (note zero offset on ordinate scale). Up to the load levels shown, the stress is within five percent of the prediction, this despite the fact that in some cases the stress has decreased from SSY values by upwards of 20%.

The diamonds marking the data furthest from the prediction are for the deeply-cracked bend specimen, and merely reflect the fact that at high load, the distance $2J/\sigma_0$ ahead of the crack front is a significant fraction of the distance to the neutral stress axis of the remaining ligament. The globally negative stress gradient of plastic bending is impinging to the crack tip, resulting in lower stress than predicted. (See also discussion following Eq. 2.23, below). At the highest load shown, the midplane J value in the deep-cracked bend specimen is $J \doteq \sigma_0(t-a)/15$. The minimum value of the relative size parameter μ_{cr} in Eq. 2.1 which has been proposed in order that single-parameter fracture mechanics be applicable is $\mu_{cr} = 25$; no less stringent restriction should apply for any two-parameter fracture mechanics methodology.

In summary, the τ-family of crack-tip fields defined by the modified boundary layer formulation describes a wide range of crack-tip stress triaxiality. There is a rigorous relationship between τ and explicitly local measures of crack-tip stress triaxiality such as the crack opening stress profile for loads up through moderate-scale yielding. For such loads, the param-

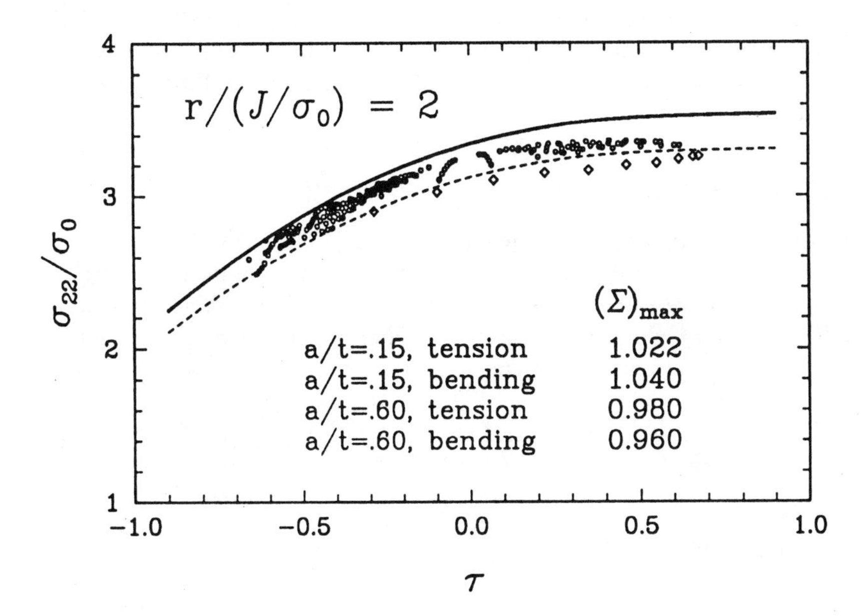

FIGURE 2.14. Composite prediction of normalized crack opening stress at $r = 2J/\sigma_0$ at all crack front locations ($\phi \leq 67.5^o$) for loads ranging to limit load. The dark line is the plane strain MBL relation, and the dashed line is 95% of the MBL crack opening stress level. The abscissa for each data point is based on elastically-calculated T-stress for the respective problems, at the ϕ and load level. At low loads, all data points are near the SSY solution ($\tau = 0$). Data points marked with diamonds are the deep-cracked bend at the highest loads, where the uncracked ligament is only $\sim 15J/\sigma_0$ in extent.

eter τ, as elastically calculated based on load level, plus the parameter J, as calculated based on the actual elastic-plastic deformation field, rigorously and accurately describe the local crack-tip stress and deformation. At higher loads, near and beyond limit load, the rigor of the correlation diminishes, although excellent qualitative and good quantitative predictions of crack-tip stress triaxiality have been made at load levels in this range for a wide variety of plane strain, axisymmetric, and three-dimensional crack configurations.

2.2.2 Higher-Order Asymptotics

Li and Wang (1986) and Sharma and Aravas (1991) have performed higher order asymptotic solutions for power law hardening materials using small geometry change and a "baseline" solution precisely matching the dominant HRR solution, which is scaled by J. Sharma and Aravas attempted to construct a solution in the form

$$\frac{\sigma_{ij}(r,\theta)}{\sigma_0} = \left(\frac{J}{\epsilon_0\sigma_0 I_n r}\right)^{\frac{1}{1+n}} \tilde{\sigma}_{ij}(\theta;\, n) + K^{(1)}\, r^t\, \tilde{\sigma}_{ij}^{(1)}(\theta;\, n); \quad (2.9)$$

$$\frac{\epsilon_{ij}(r,\theta)}{\epsilon_0} = \left(\frac{J}{\epsilon_0\sigma_0 I_n r}\right)^{\frac{n}{n+1}} \tilde{\epsilon}_{ij}(\theta;\, n) + \left(\frac{J}{\epsilon_0\sigma_0 I_n r}\right)^{\frac{n-1}{n+1}} K^{(1)}\, r^t\, \tilde{\epsilon}_{ij}^{(1)}(\theta;\, n); \quad (2.10)$$

$$\frac{u_i(r,\theta) - u_i^0}{\epsilon_0} = \left(\frac{J}{\epsilon_0\sigma_0 I_n r}\right)^{\frac{n}{n+1}} r\, \tilde{u}_i(\theta;\, n) + \left(\frac{J}{\epsilon_0\sigma_0 I_n r}\right)^{\frac{n-1}{n+1}} K^{(1)}\, r^{t+1}\, \tilde{u}_i^{(1)}(\theta;\, n), \quad (2.11)$$

where: I_n is a normalizing constant depending on n; $\tilde{\sigma}_{ij}$, $\tilde{\epsilon}_{ij}$, and $\tilde{u}_i$ are the angular distributions of the dominant HRR stress, strain, and displacement fields, respectively; u_i^0 is a rigid body crack-tip translation; $t > -1/(n+1)$ is the exponent of the radial dependence in the second-order term in the expansion of the stress field; $\tilde{\sigma}_{ij}^{(1)}$ and $\tilde{\epsilon}_{ij}^{(1)}$ are the normalized angular variations of the second term expansions for stress and strain, respectively; and $K^{(1)}$ is the amplitude of the second term. The first terms on the right sides of Eqs. 2.9–2.11, respectively, are precisely those of the HRR solution, and, as noted above, the magnitude of these fields is scaled by J. The exponent t depends on n, and its value is determined by posing an eigenvalue problem involving equilibrium, compatibility, and the power law deformation theory constitutive model. In plane strain, the eigenvalue lies in the range $0.05 \le t(n) \le 0.07$ for $5 \le n \le 20$. With t known, the second-order eigenfields $\tilde{\sigma}_{ij}^{(1)}$, $\tilde{\epsilon}_{ij}^{(1)}$, and $\tilde{u}_i^{(1)}$ can be determined. The functional form shown in Eqs 2.9–2.11 applies only for cases of realistically low strain hardening

($n > 3.2$), in which cases the amplitude parameter $K^{(1)}$ is formally arbitrary, and as such, can be considered as a second independent parameter describing the crack tip stress field. (For unrealistically high levels of strain hardening ($n < 3.2$), the second term in the series expansion is fully determined by J alone, and different functional forms apply; for details, see Sharma and Aravas (1991)). In a closely related formulation, Li and Wang (1986) obtained a similar structure for a two-term series expansion of the crack-tip fields in power law materials. They matched the two-term representation of crack opening stress to the center-cracked plate finite element solutions of Shih and German (1981) to track the evolution (with increasing loading) of the numerical value of the second (free) coefficient.

A major problem in applying this approach, however, is that apparently the only way to determine the second amplitude parameter in a given crack problem is by (a) calculating the crack-tip scale J, (b) obtaining an accurate description of the complete crack-tip fields over distances of order J/σ_0, then (c) "subtracting" the J–based HRR field from the complete field in order to quantify the residual with the parameter $K^{(1)}$. This elaborate local procedure for evaluating the second parameter describing the crack-tip field is in stark contrast to the relative ease with which the first parameter, J, can be evaluated by more "global" means such as virtual crack extension (Parks, 1977) or domain integral formulations (Li, Shih, and Needleman, 1985). Evidently, it is necessary, when using this approach, to have a complete, accurate, and highly-detailed description (numerical) of the crack-tip field before the two–parameter characterization can be specified. In application, such an approach must be considered descriptive rather than predictive. Moreover, the *sine qua non* of the parameterization, the detailed crack-tip solution, involves considerable computational effort, and may not always be easily obtained for engineering applications.

A second drawback which this approach shares with all SGC solutions is that the zone of dominance of a two-term asymptotic expansion may sometimes (especially, low strain hardening, $n >> 1$) be small compared to physically relevant length scales of the fracture process zone. In particular, since ductile fracture by hole growth is inherently a large deformation process, it is necessary that the description of the crack-tip field be a faithful representation of the small strain fields exterior to the blunting zone, $r >\sim 3\delta_t$ (McMeeking, 1977). At small but finite distances from the tip, a higher order expansion (three or more terms) may be required to accurately describe the fields surrounding the fracture process zone.

Thus, while the mathematical basis of higher-order asymptotic expansions provides a rigorous framework for developing two-parameter descriptions of crack-tip fields, the approach is insufficiently robust to provide a viable basis for two-parameter fracture mechanics applications.

2.2.3 Descriptions of Large Strain Numerical Solutions

In an approach related to both of those previously-noted, O'Dowd and Shih (1991a,b) performed very detailed large deformation finite element studies in the manner of McMeeking (1977) and McMeeking and Parks (1979). They systematically investigated the differences between the computed fields and a J–scaled reference solution in an annular region surrounding the tip which extends between $\sim J/\sigma_0 < r < 5J/\sigma_0$.

Following Aravas and Sharma (1991) and Li and Wang (1986), O'Dowd and Shih motivate their interpretation of their numerical results by suggesting a series representation of the stress field in a small-strain annulus surrounding the finitely-blunting crack-tip. They propose an expression which is formally similar to Eq. 2.9:

$$\frac{\sigma_{ij}}{\sigma_0} = \frac{\sigma_{ij}^{HRR}}{\sigma_0} + Q\left(\frac{r}{J/\sigma_0}\right)^q \hat{\sigma}_{ij}, \tag{2.12}$$

where the dimensionless functions $\hat{\sigma}_{ij}$ depend on n and θ. The functional forms of Eq. 2.9 and Eq. 2.12 agree if $t = q$, if $\hat{\sigma}_{ij} = \tilde{\sigma}_{ij}^{(1)}$, and if $K^{(1)}(J/\sigma_0)^q = Q$. However, O'Dowd and Shih intend to use Eq. 2.12 as a description of the LGC crack-tip field in the region $J/\sigma_0 < r < 5J/\sigma_0$, while Sharma and Aravas demonstrate that the form Eq. 2.9 is a rigorous two-term eigen-expansion of the SGC solution as r asymptotically approaches zero. In contrast, no justification other than dimensional consistency is offered in support of Eq. 2.12, so this form, like the crack opening stress profile Eq. 2.5, is a useful "curve-fit" only to the extent that it matches essential features of the crack-tip fields emerging from detailed numerical computations.

Following Betegón and Hancock (1991), O'Dowd and Shih generated a particular set of LGC solutions using the T-stress in the MBL loading of Eq. 2.2. Based on examination of the computed stress fields for these loadings in the annular region described above, they proposed a further simplification to the fields Eq. 2.12 as

$$\frac{\sigma_{ij}}{\sigma_0} = \left(\frac{J}{\epsilon_0 \sigma_0 I_n r}\right)^{\frac{1}{1+n}} \tilde{\sigma}_{ij} + Q\delta_{ij}, \tag{2.13}$$

where δ_{ij} is the Kronecker delta. That is, they effectively set $q = 0$ and take the perturbation stress fields (from the HRR reference value) as purely hydrostatic, at least in the angular range $|\theta| <\sim 90^o$. In this form, the physical basis of the second crack tip parameter (Q) becomes clear: it explicitly measures the deviation in crack tip triaxiality from a particular reference value of triaxiality. In the context of the MBL loadings, the parameters τ and Q are in rigorous one-to-one relation. A comparison between the crack opening stress profiles of Eq. 2.5 and 2.13 shows that

$$Q = Q_0 + A_n \tau + B_n \tau^2 + C_n \tau^3, \tag{2.14}$$

where $Q_0 = \sigma_{22}^{SSY} - \sigma_{22}^{HRR}$ represents the difference between HRR and SSY crack opening stress at a reference distance from the tip (e.g., $r = 2J/\sigma_0$). If, instead, the J-based reference field introduced by O'Dowd and Shih had been chosen as that of SSY, with a modified definition of the triaxiality parameter as $\tilde{Q}$ according to

$$\frac{\sigma_{ij}}{\sigma_0} = \frac{\sigma_{ij}^{SSY}}{\sigma_0} + \tilde{Q}\delta_{ij}, \tag{2.15}$$

then $\tilde{Q} = A_n \tau + B_n \tau^2 + C_n \tau^3$ precisely.

O'Dowd and Shih used the MBL loading to "generate" the crack-tip fields to which the approximate two-parameter (J, Q) description Eq. 2.13 was fit, and the limitations on admissible values of $|\tau|$ similarly limit the range of Q-fields which can be generated.

O'Dowd and Shih (1991b) applied the same data reduction technique to the detailed crack tip fields of complete numerical solutions in various plane-strain crack geometries at loads ranging from SSY through fully plastic conditions. They report that the local fields were always well-described by Eq. 2.13, for some instantaneous value of Q, dependent on load magnitude, crack geometry, and loading type, although they note that the precise value to assign to Q depends somewhat on the location $(r/(J/\sigma_0), \theta)$ under investigation and on which stress component is being fit, especially in large scale yielding.

As in the approach of Li and Wang, a requisite feature for strict application of this two–parameter description is first obtaining a complete numerical description of the crack tip which is accurate over distances of order blunted crack opening displacement, then subtracting off a reference one–parameter stress measure in order to scale the differential with Q. The O'Dowd–Shih approach addresses the open question of neglected large geometry change in the Li–Wang method, but at the extra cost of a requiring a more involved numerical formulation. Incorporation of LGC is not an essential feature in assessing varying crack-tip triaxiality, since careful J–dominance interpretations of SGC solutions (Shih and German, 1981) provided results fully consistent with those of LGC solutions (McMeeking and Parks, 1979).

Thus, the proposed Q-family, generated by MBL loadings, describes LCG crack-tip fields of substantially different stress triaxiality. Strictly, Q should be determined as scaling the difference (at the scale of crack tip opening displacement) between a complete local stress field and a reference field (HRR). The local focus of Q assures that it can be operationally evaluated from finite element solutions at all load levels.

2.3 Discussion of Two-Parameter Characterizations

Of the three approaches to two-parameter characterization of crack tip fields outlined in the previous section, currently only those based on normalized T-stress (τ) and on the local triaxiality perturbation (Q) seem likely to prove sufficiently adaptable to serve as a basis for extending fracture mechanics methodology. These two approaches and parameters have much in common, but there are differences which can have important practical implications. In this section, we critically discuss relative advantages and disadvantages of the two methods.

Ideally, a second parameter characterizing crack-tip stress triaxiality should possess the following features:

1. Its value should be readily calculable (with only modest effort) as a function of loading and geometry for a broad range of laboratory and structural crack configurations (global).

2. It should admit a close correspondence with physically–based aspects of the crack-tip stress field (local).

3. It should be applicable in an accurate and predictive manner over a broad range of loading and geometries (robust).

4. It should provide means for quantitative assessment of fracture toughness under conditions leading to a broad range of stress triaxiality, based on a limited number of laboratory experiments (useful!).

In evaluating the parameters τ and Q in the context of this four-fold desideratum, we should note that there is currently only limited experimental data with which to confront item number four. This issue will be briefly addressed in a following section. On the other hand, the extensive theoretical and computational studies which have been noted provide considerable opportunity to critically discuss the first three issues.

2.3.1 Parameter Calculation

There are several methods in the literature (Larsson and Carlsson, 1973; Leevers and Radon, 1982; Kfouri, 1986; Sham, 1991) available for calculating the T-stress in two-dimensional planar elastic crack configurations, and hence for obtaining the calibration function $\hat{t}$ of Eq. 2.7. Larsson and Carlsson (1973) used a very refined mesh, and computed T from the local stress field, assuming that K_I was independently available; i.e., $T = \sigma_{11}(r,\theta) - K_I f_{11}(\theta)/\sqrt{2\pi r}$. Sham (1991) developed a method for calculating generalized weight functions in a given geometry, so that T can be obtained through quadrature of applied loading distributions. Kfouri

(1986) implemented a theorem of Eshelby for evaluating T in terms of the difference in J-integral of two solutions, with and without a superimposed crack-tip point load in the crack advance direction. However, only the "brute force" method of Larsson and Carlsson is directly applicable to axisymmetry or three-dimensional bodies.

Recently Nakamura and Parks (1991) extended the Eshelby method to three-dimensional applications through the introduction of a domain interaction integral. The 3-D extension of the mode I elastic crack-tip fields Eq. 2.1 can be expressed as

$$\sigma_{33} = \frac{K_I}{\sqrt{2\pi r}} 2\nu \cos\frac{\theta}{2} + E\epsilon_{33} + \nu T, \tag{2.16}$$

where ϵ_{33} is the (bounded) strain tangent to the crack front. (There can also be a non-zero second order stress component T_{13} in 3-D applications; we do not consider this point further here as its presence does not affect the primary results.)

The 3-D stress field corresponding to a plane strain line load f per unit length along the crack, applied in the crack extension direction (x_1), is

$$\sigma^L_{\beta\gamma} = \frac{f}{\pi r} l_{\beta\gamma}(\theta), \tag{2.17}$$

with $l_{11} = \cos^3\theta$, $l_{22} = \cos\theta\sin^2\theta$, $l_{12} = \cos^2\theta\sin\theta$, and $\sigma^L_{33} = \nu(\sigma^L_{11} + \sigma^L_{22})$. The strains and displacements corresponding to the 3-D crack stress field and to the crack line-load can be readily calculated.

Consider, as in Fig. 2.15, a location s on a 3-D crack front, with local crack extension direction cosines $\mu_i(s)$, where crack front parameters are $K_I(s)$ and $T(s)$. Imagine the line load $f_i = f\,\mu_i(s)$ also being applied, and consider the interaction integral

$$I(s) = \lim_{\Gamma\to 0} \mu_k(s) \int_{\Gamma(s)} \left[\sigma_{ij}\epsilon^L_{ij} n_k - n_j\left(\sigma_{ij}\frac{\partial u^L_i}{\partial x_k} + \sigma^L_{ij}\frac{\partial u_i}{\partial x_k}\right)\right] d\Gamma. \tag{2.18}$$

Here the superscript 'L' denotes the line-load solution. The path $\Gamma(s)$ surrounds the crack front at s and lies in the plane orthogonal to the crack front tangent. This integral is path-independent in the limit, and evaluation on a limiting circle of radius $r \to 0$ provides

$$I(s) = \frac{f}{E}\left[T(s)(1-\nu^2) - \nu E\epsilon_{33}(s)\right]. \tag{2.19}$$

The line interaction integral Eq. 2.18 can be transformed into an equivalent domain (volume) interaction integral suitable for finite element evaluation (Nakamura and Parks, 1989; 1991). In plane strain, $\epsilon_{33} = 0$ and the method corresponds with that used by Kfouri (1986). In 3-D applications, $\epsilon_{33}(s)$ of the actual crack solution is determined directly from nodal displacements,

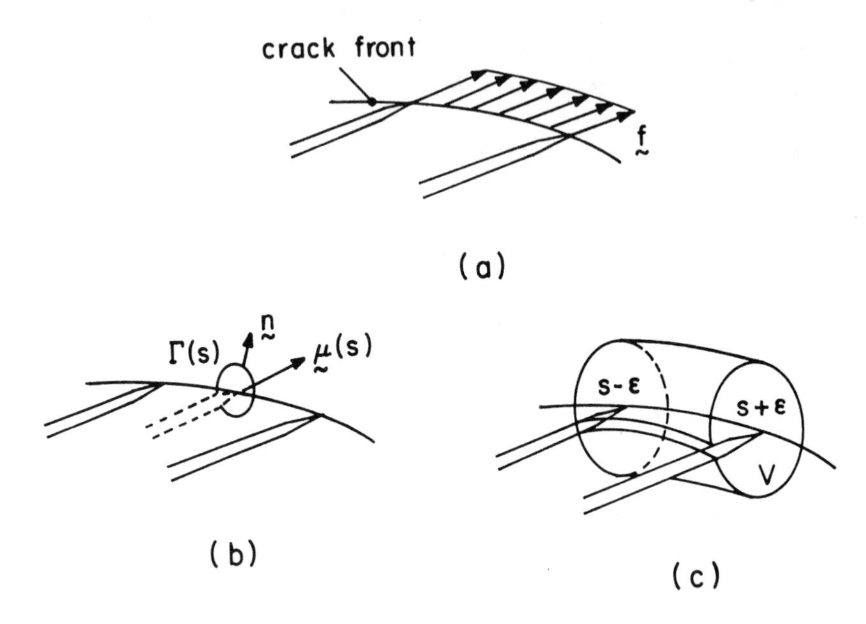

FIGURE 2.15. (a) Line-load applied in the direction of crack advance along the crack front. (b) Crack tip contour Γ in the plane locally perpendicular to the crack front, where s represents arc-length along the crack front. (c) Volume V enclosing the crack front segment where domain integral $\bar{I}$ is evaluated.

and thus the variation $T(s)$ can be obtained. Nakamura and Parks (1991) have applied this method to straight through–crack geometries and to surface cracked plates. Its numerical implementation is quite comparable to domain integral evaluations of J (Li, Shih, and Needleman, 1985).

Wang and Parks (1991) obtained approximate estimates of the T–stress distribution in a wide range of surface–cracked plates under tension and bending by using the line–spring model. The procedure consisted of extracting the membrane force per unit cut length, $N(X)$, and the bending moment per unit length, $M(X)$, from a linear elastic line–spring finite element solution. The standard line–spring model (Rice and Levy, 1972) estimates the local stress intensity factor as

$$K_I(X) \doteq N(X)\, k_N(a(X),\, t) + M(X)\, k_M(a(X),\, t), \tag{2.20}$$

where k_N and k_M are stress intensity calibration functions for a single edge crack geometry of crack depth a and width t subject to tension and bending, respectively. Wang and Parks analogously estimated

$$T(X) \doteq N(X)\, t_N(a(X),\, t) + M(X)\, t_M(a(X),\, t), \tag{2.21}$$

where t_N and t_M are T–stress calibration functions for the single edge crack in tension and bending (Sham, 1991). Their results were in excellent agreement with T-estimates from highly detailed 3-D finite element solutions, as well as with the surface crack T-distribution obtained by domain interaction integral methods (Nakamura and Parks, 1991).

In summary, the parameter T can be related to loading and geometry for arbitrary crack problems with only a modest computational effort.

In contrast, however, the detailed computational procedures required to obtain accurate values of Q by differencing the local stress fields (analogous to Larsson and Carlsson's approach for calculating T) represents a much higher level of effort. In 3-D applications, the required computational resources and data preparation and reduction make direct implementation of this approach all but prohibitive. O'Dowd and Shih (1991b) recognized this limitation of Q-based approaches to crack-tip stress triaxiality, and proposed several alternatives for approximate (indirect) evaluation of Q as a function of geometry and loading which would not pose such onerous computational burdens. First and foremost, they note that in small- to moderate-scale yielding, Q is isomorphic to τ, so that an accurate assessment of T from methods like those noted above would likewise provide precise and rigorous values of Q through, e.g., Eq. 2.14.

O'Dowd and Shih (1991b) noted that Q could be estimated under fully plastic (limit load) conditions in non-hardening (and lightly strain hardening) materials by using the known crack tip triaxialities of the corresponding slipline fields, as tabulated, for example, by McClintock (1971). Thus $Q^{LIM} \doteq (\sigma_m^{LIM} - \sigma_m^{HRR})/\sigma_0$, where σ_m is the mean hydrostatic stress of the respective slipline and HRR stress fields. This estimation procedure would seem to be limited to plane strain geometries where precise slipline-based estimates of crack tip stress triaxiality are available, but problematic in axisymmetric or truly 3-D crack geometries.

Finally, O'Dowd and Shih (1991b) also outlined a procedure for interpolating estimates of Q based on T-correlations at small-scale yielding (elastic) and an apparently ad hoc assumption for the scaling behavior of crack-tip stress fields in SGC pure power law (fully plastic) solutions. This suggestion was motivated by the analogous interpolation procedure for estimation of J in the simplified engineering analysis of fracture (Kumar, German, and Shih, 1981). No results were given, so the efficacy of such an estimation/interpolation procedure for Q remains undetermined.

In summary, while T can be readily calculated for arbitrary elastic crack configurations, equally simple and direct methods for calculating Q do not exist. Indirect estimates of Q which are readily achieved involve the τ–Q relation of MBL, Eq. 2.14, as well as some procedure(s) to explicitly account for crack tip triaxiality under fully plastic conditions. Implicit in the creation of fully-plastic "corrections" to Q (as otherwise calculated based solely on elastic load/T and MBL-based τ/Q correlations) is the notion of unacceptably large errors in (uncorrected) predictions of crack tip stress triaxiality, a contention which we believe is debatable.

2.3.2 Local Definition of Stress Triaxiality

The parameters τ and Q serve as essentially identical measures of crack tip stress triaxiality in small- and moderate-scale yielding, so no meaningful differentiation between them exists.

O'Dowd and Shih (1991b) emphasize that "the Q-family of fields can exist over the entire range of plastic yielding and does not depend on the existence of the elastic field" (of Eq. 2.2), while "Of course, T has no relevance under fully yielded conditions." Such broad statements should not be uncritically endorsed. First, the presumption of the existence of the Q-family can lead to uncritical data reduction procedures which are sure to "find" a value for Q. For example, a sampling of the crack opening stress profile (σ_{22} at $\theta = 0$) at a specified distance (e.g., $r = 2J/\sigma_0$) "automatically" provides Q as

$$\begin{aligned} Q &\equiv \frac{\sigma_{22}(r = 2J/\sigma_0,\, \theta = 0) - \sigma_{22}^{HRR}(r = 2J/\sigma_0,\, \theta = 0)}{\sigma_0} \\ &= \frac{\sigma_{22}(r = 2J/\sigma_0,\, \theta = 0)}{\sigma_0} - \left(\frac{1}{2\epsilon_0 I_n}\right)^{\frac{1}{n+1}} \tilde{\sigma}_{22}(\theta = 0,\, n), \end{aligned} \quad (2.22)$$

providing the Q-family of Eq. 2.13 is indeed applicable over the range $J/\sigma_0 \leq r \leq 5J/\sigma_0$. However, mere "point matching" of this sort is a much weaker condition than the more stringent requirements of "field matching" which the methodology leading to Eq. 2.13 espouses. If the Q-family of crack-tip fields is generated under MBL loadings, and if it is also to describe the near-tip fields in largely yielded bodies, then, for example, the slopes of the crack opening stress profile should also match at the selected point:

$$\left.\frac{\partial\, \sigma_{22}}{\partial\, (r/(J/\sigma_0))}\right|_{Q,\, r=2J/\sigma_0} = \left.\frac{\partial\, \sigma_{22}^{MBL}}{\partial\, (r/(J/\sigma_0))}\right|_{Q,\, r=2J/\sigma_0}. \quad (2.23)$$

In fully-plastic bending of deeply-cracked geometries, the strongly negative gradient of the global bending stress distribution on the ligament can impinge so near the crack tip (Al-Ani and Hancock, 1991; O'Dowd and Shih, 1991b), that Eq. 2.22 is not satisfied. In such an event, it is not clear to what extent even "Q" truly characterizes the crack-tip fracture process zone.

We have also seen that τ-based predictions of local crack opening stress depart less than 5% from actual values at loads up to limit load.

2.3.3 Robustness of Correlation with Crack Tip Stress Triaxiality

When the τ-formalism is used to estimate crack tip stress triaxiality in the fully-plastic range, certain approximations are involved. An important question is "how serious are the errors in estimates of crack tip stress triaxiality which are introduced by these approximations?" O'Dowd and Shih have chosen to measure this "error" by assessing how closely the elastically-predicted value of Q (using, e.g., Eq. 2.7 and Eq. 2.14 is to their explicitly local measure, Q_{local}, as determined, e.g., by Eq. 2.22. This basis for comparison is perversely biased: Q is a measure of the perturbation in the stress field from a reference stress field (HRR) of inherently high stress triaxiality, so small errors in predictions of overall stress triaxiality are manifest as large errors in the perturbation parameter (Q). Moreover, the micromechanisms of fracture by hole growth or by cleavage are sensitive to the triaxiality of the *total stress state*, and not to the triaxiality of the perturbation stress field! Thus it is appropriate to examine the quantitative consequences of errors in predicted crack tip stress triaxiality.

Suppose that the Q-based description of Eq. 2.13 is precise, with amplitude Q_{local}. Further, suppose that an implementation of the τ-based (or other approximation) method provides an estimate of local stress perturbation given by

$$Q_{est} = Q_{local} + \delta Q, \tag{2.24}$$

where δQ represents the "error" in local estimate of stress triaxiality. O'Dowd and Shih (1991b) claim that errors of order $|\delta Q| \leq 0.5$ can be obtained at loads within the limits of $25J/\sigma_0 < \ell$, although in most cases $|\delta Q| < 0.3$ or less. Wang (1991a) found that $|\delta Q| < 0.05$ for all locations along a variety of surface crack fronts, at loads up to limit load.

What are the consequences of such errors on micromechanical predictions of ductile cavity growth? Rice and Tracey (1969) considered the growth of an isolated spherical cavity of radius R embedded in a rigid/perfectly-plastic matrix material having tensile flow strength σ_0, undergoing remote equivalent strain-rate $\dot{\epsilon}^p$ and subjected to remote hydrostatic (mean) tensile stress $\sigma_m = \sigma_{kk}/3$. Letting $V = 4\pi R^3/3$ be cavity volume and $\dot{V}$ be rate of increase of cavity volume, the Rice-Tracey results for high stress triaxiality can be expressed as

$$\frac{\dot{V}}{\dot{\epsilon}^p V} = D \exp\left(\frac{3\sigma_m}{2\sigma_0}\right). \tag{2.25}$$

Rice and Tracey (1969) suggested that the constant $D = 0.85$, a result consistent with the findings of Budiansky, et al. (1982), but recent computations by Huang, et al. (1991) show that this result is too low by a factor of near 1.5; a more accurate value is $D = 1.275$. In any event, based on Eq. 2.13, the term $3\sigma_m/2\sigma_0$ can be estimated at any particular distance from the crack as

$$\frac{3\sigma_m}{2\sigma_0} = \frac{3\sigma_m^{HRR}}{2\sigma_0} + \frac{3Q}{2}. \tag{2.26}$$

Now, on combining Eqs. 2.24, 2.25, and 2.26, the ratio of estimated void volume increase, $\dot{V}_{est}$, to "actual" growth rate, $\dot{V}$, is

$$\frac{\dot{V}_{est}}{\dot{V}} = \exp\left(\frac{3}{2}\delta Q\right) \tag{2.27}$$

When $\delta Q = 0.1$, the predicted local void volume growth rate is 16% too high; for $\delta Q = 0.2$, the corresponding error is 35%. These differences in estimates of local cavity growth rate are not negligible, but in view of the considerable idealization in the model, experimental scatter, etc., neither should they be considered unacceptable. It can reasonably be concluded that small "errors" in predictions of crack tip triaxiality should not cause the methodology to be subject to a priori disqualification from further consideration.

A similar conclusion may be drawn, even if the micromechanism of fracture is cleavage. Ritchie, Knott, and Rice (1971) showed that the essential feature of cleavage crack initiation at a macroscopic crack tip involved achieving a critical level of tensile stress over a critical material length scale. Potential sites for nucleation of cleavage microcracks in ferritic steel are often coarse grain boundary carbide particles which are essentially randomly distributed. Accordingly, subsequent analyses (Lin, Evans, and Ritchie, 1986a,b, 1987; Evans, 1983; Beremin, 1983) have emphasized statistical aspects of cleavage initiation.

Under certain conditions, the kinetics of cleavage crack nucleation can be treated by extreme value statistics. Let N be the number of particles per unit volume, and f be the fraction of these particles eligible to participate in the cleavage process. Assuming the particle strength distribution is $g(S)$ (the number of particles per unit volume having strength S), then the probability of failure of a representative material element of volume δV is

$$\delta P_f = 1 - \exp\left(-\delta V \int^{\sigma} g(S)\, dS\right). \tag{2.28}$$

The total failure probability, P_f, is the integral of the elemental failure probability over the volume. A widely-used expression for the strength distribution is the three-parameter Weibull (1938, 1939) distribution

$$\int_{\sigma_u}^{\sigma} g(S)\, dS = fN\left(\frac{\sigma - \sigma_u}{S_0}\right)^m, \tag{2.29}$$

where m is the Weibull shape factor, σ_u is the lowest (threshold) stress at which cleavage can be initiated, and S_0 is a scaling parameter with dimensions of stress. In applications, the stress measure σ is taken as the maximum principal stress.

Now again consider a material element ahead of the crack tip where crack opening stress is $\sigma = \sigma_{22}^{HRR} + \sigma_0 Q_{local}$. Again assuming that the estimated value of Q contains error δQ, a first-order expansion of $(\sigma - \sigma_u)^m$ about the actual value based on Q_{local} provides

$$(1 - \delta P_f)_{est} = (1 - \delta P_f)_{local} \cdot \exp\left(\frac{m\, \sigma_0\, \delta Q}{\sigma - \sigma_u}\right). \tag{2.30}$$

Assuming a nominal value of crack tip stress minus threshold as $2\sigma_0$ (e.g., $\sigma = 3\sigma_0$; $\sigma_u = \sigma_0$), and a Weibull exponent of $m = 2$ provides the ratio of estimated to "actual" local survival probabilities as $\exp(\delta Q)$. For $\delta Q = 0.3$, this ratio is only 1.35. Clearly, the local errors would grow with increasing m and with decreasing $(\sigma - \sigma_u)/\sigma_0$, but again, in view of the extremes of idealization and simplification involved in the fracture model, the sensitivity of the methodology to "error" does not seem to be unacceptably large.

Two simplified models of local fracture processes which might occur in crack tip fields are relatively insensitive to errors in estimates of the Q parameter, providing the errors are sufficiently small. The essential reason for this insensitivity is that Q (or τ) essentially measures the triaxiality of the perturbation in the crack tip stress, relative to a reference solution (HRR or SSY) of inherently high stress triaxiality.

2.4 Two-Parameter Analysis of Experimental Data

The existence of a rational, tractable two-parameter formalism for accurately predicting the history of crack-tip deformation (J) and stress triaxiality (τ or Q) brings fracture mechanics to the brink of a new era. What types of experimentation and interpretation will be needed to achieve the benefits which this enhanced analytical resolution seems to promise? Answers to this question are only beginning to emerge. Here we briefly note two classes of experimental investigations which indicate the promise of quantitatively unifying toughness measures in specimens of differing geometry and stress triaxiality.

2.4.1 Triaxiality Effects on Cleavage Initiation

One application of two-parameter fracture mechanics is the interpretation of "specimen geometry effects" on cleavage fracture initiation toughness in structural steels. To the extent that brittle cleavage initiating within the crack-tip plastic zone can be considered an "event," the macroscopic conditions for fracture should appear as a locus in a (J, τ) plane.

Betegón and Hancock (1990) interpreted the variation of cleavage initiation toughness with bend specimen crack depth in terms of such a locus. The shallowest cracks, those with $a/w < 0.3$, have $T < 0$. The shortest of these cracks have the most negative T (at fixed nominal bending stress), and they gave the highest J-values at fracture. For specimens having $a/w > 0.3$, the $T > 0$. Since positive T-stress elevates local stress triaxiality only slightly, in comparison to the SSY result, the locus of J-value at fracture is quite flat (independent of τ) for these relative crack depths.

Betegón's experimental data was analyzed by Wang (1991b) in terms of weakest link statistics, using LGC crack-tip stress fields under τ-based MBL loadings. Wang integrated the elemental failure probability, δP_f, over the intersection of the plastic zone and the region in which maximum principal stress exceeded σ_u, including parameter studies of effects of strain hardening exponent n, Weibull exponent, m, and ratio of threshold stress to yield stress, σ_u/σ_0. Extending the dimensional analysis of Lin, Evans, and Ritchie, Wang showed that in planar specimens of uniform thickness subject to crack tip fields of the MBL loadings, the locus of constant failure probability (e.g., $P_f = 0.5$) could be expressed as the following locus in a (J, τ) plane:

$$J_{Ic}(\tau) = J_{Ic}\big|_{\tau=0} \cdot H(\tau, n, m, \sigma_u/\sigma_0). \tag{2.31}$$

Here $J_{Ic}(\tau)$ is the critical value of J at a given value of τ (for a particular value of P_f), and the scaling toughness, $J_{Ic}\big|_{\tau=0}$ is the corresponding level of toughness under SSY conditions; hence, the latter value could also be directly related to the toughness measure K_{Ic}. The dimensionless function H depends on the argument list shown. However, for fixed values of the material properties, n, m, and σ_u/σ_0, the shape of the curve as a function of the geometry-dependent stress triaxiality parameter τ is indeed similar to our expectations and to the experimental data of Betegón. Figure 2.16, from Wang's thesis, shows the sensitivity of H to threshold stress at low strain hardening ($n = 10$) and Weibull shape parameter $m = 2$. A similar trend, but with reduced sensitivity, applies when m is varied over the range $0.5 \leq m \leq 3$. Cases of higher strain hardening ($n = 7, 5$) showed less overall sensitivity to τ. Figure 2.17 shows Betegón's experimental data, plotted in a J vs. τ locus, and Fig. 2.18 shows Wang's curve fit to the data. Obviously, this fitting procedure is not meant to be a rigorous means for determining the strict applicability of weakest link statistics to this problem, much less of determining constants of the three-parameter Weibull distribution. Rather, it is intended primarily as an exercise illustrating the potential inherent in constraint-sensitive fracture mechanics.

In an approach similar to that of Wang (1991b), O'Dowd and Shih (1991b) have used the simple model of Ritchie, Knott, and Rice (1972) to illustrate the effect of varying crack tip triaxiality on macroscopic cleavage fracture toughness. The condition of achieving a critical level of tensile stress, σ_c, at a critical microstructural distance r_c ahead of the crack (which was assumed to lie within the (J, Q) annulus in which stress is well-described by Eq. 2.12) was shown to lead to the following prediction of the dependence of (deterministic) mode I plane strain cleavage initiation toughness, J_c:

$$\frac{J_c}{J_c\,|_{T=0}} = \left(\frac{\frac{\sigma_c}{\sigma_0} - Q}{\frac{\sigma_c}{\sigma_0} - Q_{T=0}} \right)^{n+1}. \tag{2.32}$$

This functional form predicts trends similar to the statistically-based

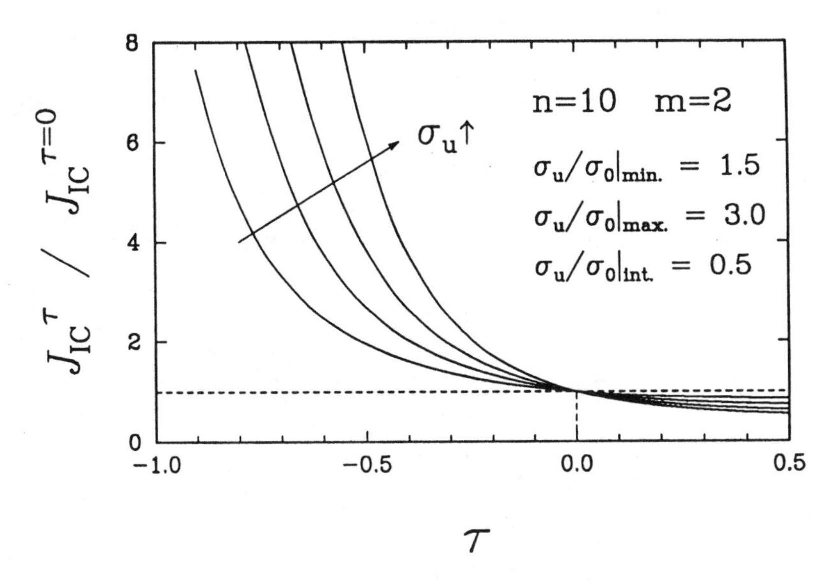

FIGURE 2.16. Predicted variation of constant probability of normalized J-scaled cleavage fracture toughness with τ based on large geometry change MBL stress fields and local three-parameter Weibull statistics for a range of normalized threshold stress levels (Wang, 1991b).

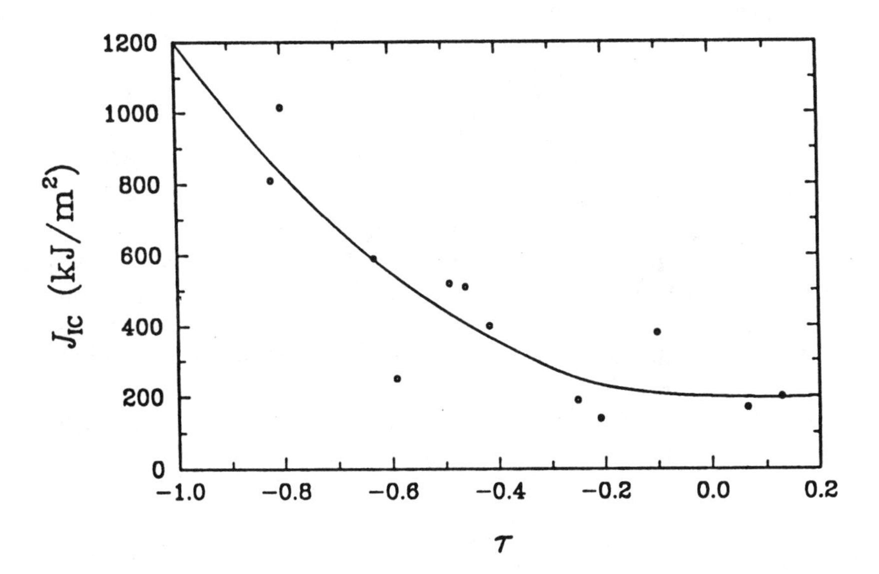

FIGURE 2.17. Cleavage fracture toughness vs. τ, as obtained by Betegón and Hancock (1990) in tests of an as-quenched steel using shallow and deep single edge crack bend specimens. Solid line is a curve-fit to the data (Wang, 1991b).

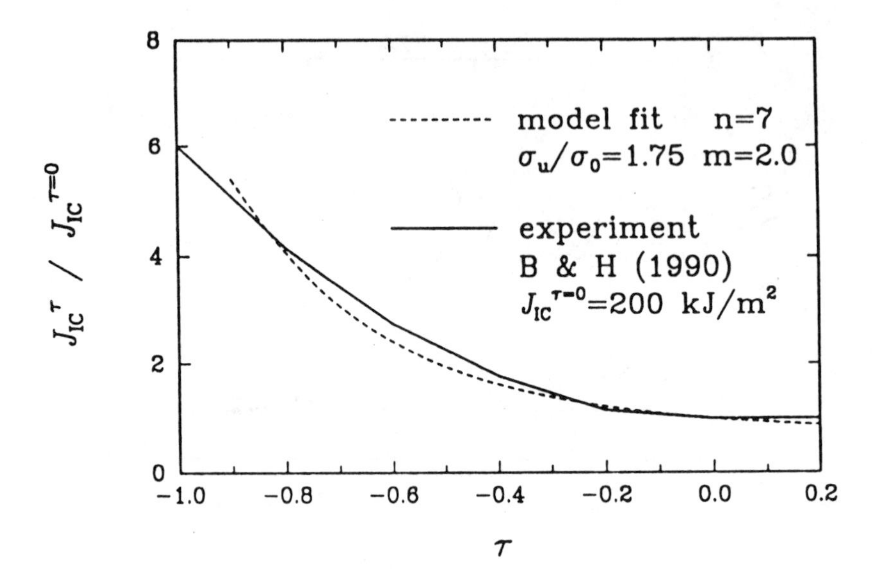

FIGURE 2.18. Comparison of variation in normalized cleavage toughness data (Betegón and Hancock, 1990) with τ against fitted functional form of predicted toughness variation with τ (Wang, 1991b).

Eq. 2.31 and to the data of Figs. 2.17–2.18. The sensitivity of the predicted fracture toughness to errors in Q parallels that of Eq. 2.30.

The data of Betegón (1990), which was used by Betegón and Hancock (1990), was obtained from bend specimens of constant size and varying crack depth, a. From this data alone, it is not clear how (if) the (J, τ) approach to cleavage initiation correlates specimen geometry effects over specimens of fundamentally different geometry. This issue has recently been addressed by Sumpter (1991), who performed a series of tests on a 1960's era mild steel, using bend specimens of varying crack depth and a center-cracked tension specimen of relative crack size $2a/w = 0.7$. This specimen was chosen to have essentially the same elastically-calculated τ value at limit load as the shallowest of the bend bars. These data, shown in Fig. 2.19, exhibit a good deal of scatter, but there is clearly at tendency for the (J, τ) rationalization to bring the data from the different specimen geometries into correspondence. More work of this sort seems promising.

2.4.2 Triaxiality Effects on Ductile Tearing

Since ductile fracture by hole nucleation, growth, and coalescence is a process (rather than an "event"), there could be "loading history effects" on fracture. That is, the loading trajectory through the (J, τ) loading space could affect "critical" conditions. The simplification of a (unique) critical

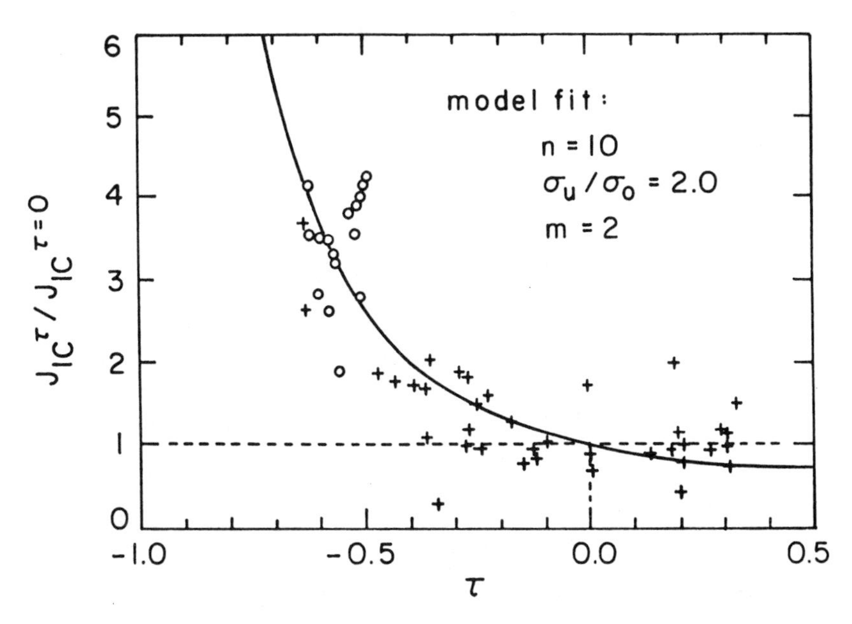

FIGURE 2.19. Comparison of variation in normalized cleavage toughness with τ in various specimen geometries (Sumpter, 1991) and fitted functional form of predicted toughness variation with τ (Wang, 1991b). Open circles are from CCP specimens and +'s are from bend and CT specimens.

fracture locus may not be obtained. Hancock, et al. (1991) have examined the early slopes of resistance curves in a number of through–crack and surface–cracked specimens of A710 steel. There is a clear correlation of the steepness of the resistance curve with τ: the steepest slopes are obtained in specimens having the most negative τ–values. Figure 2.20 shows the loci of J vs. τ corresponding to values of ductile crack extension of $\Delta a = 0$, 200, and 400μm. In this material, the "initiation" value of J corresponding to negligible crack extension is relatively insensitive to stress triaxiality, but the resistance to ductile tearing (treated here as a quasi-stationary crack) shows an expected trend. Again, the ranking of the stress triaxiality (τ) over the range of bend, compact tension, center-cracked tension, and surface-cracked tension specimens of varying crack aspect ratio is in good correspondence with the overall trend of the data.

More experimental data sets of this type, illustrating the sensitivity of a given material to crack tip stress triaxiality, should be collected. Ultimately, a limited number of laboratory tests which span a relevant range of triaxiality (for the application under consideration) should suffice to quantify margins of structural integrity with respect to fracture, even in low constraint geometries currently beyond the scope of one-parameter fracture mechanics methodology.

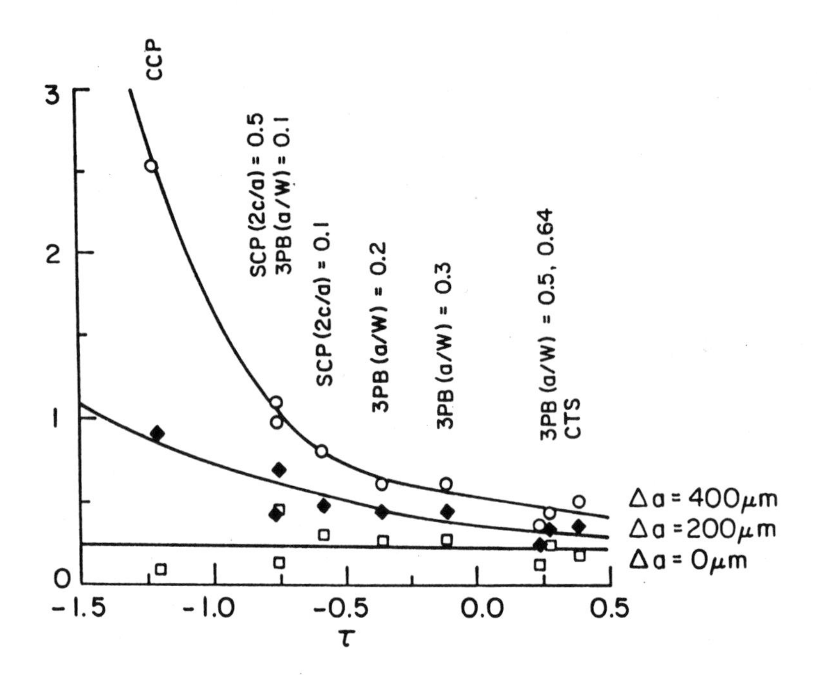

FIGURE 2.20. Variation in J in various specimen geometries corresponding to ductile crack growth of 0, 200, and 400 μm, plotted versus $\tau = T/\sigma_0$ (Hancock, Reuter, and Parks, 1991). The initiation toughness is relatively insensitive to crack tip stress constraint in this A710 steel, but the slope of the initial portion of the resistance curve steepens with the reduced triaxiality corresponding to negative τ.

Acknowledgments: This work was supported by the Office of Basic Energy Sciences, Department of Energy, under Grant # DE-FG02-85ER13331. Computations were performed on an Alliant FX-8 computer obtained under D.A.R.P.A. Grant # N00014-86-K-0768 and on the MIT Cray-2. The ABAQUS finite element program was made available under academic license from Hibbitt, Karlsson, and Sorensen, Inc., Pawtucket, RI. I am pleased to acknowledge useful interactions and discussions with Drs. Y. Wang and C. Betegón and Professors J. W. Hancock and T. Nakamura.

REFERENCES

Al-Ani, A. M. and Hancock, J. W. (1991). J. Mech. Phys. Solids, 39:23-43.

Begley, J. A. and Landes, J. D. (1976). Int. J. Fract., 12:764-766.

Beremin, F. M. (1983). Met. Trans. A, 14A:2277-2287.

Betegón, C. (1990). Ph. D. Thesis, University of Oviedo, Spain.

Betegón, C. and Hancock, J. W. (1990). in ECF 8 *Fracture Behaviour and Design of Materials and Structures, Volume II*, Ed. D. Firraro, Engineering Materials Advisory Services, Ltd., Warly, U.K., pages 999-1002.

Betegón, C. and Hancock, J. W. (1991). J. Appl. Mech., 58:104-110.

Betegón, C., Parks, D. M., Hancock, J. W., and Wang, Y. (1991). manuscript in preparation.

Bilby, B. A., Cardew, G. E., Goldthorpe, M. R., and Howard, I. C. (1986). in *Size Effects in Fracture*, Institution of Mechanical Engineers, London, pages 37-46.

Budiansky, B., Hutchinson, J. W., and Slutsky, S. (1982). in *Mechanics of Solids, The Rodney Hill 60th Anniversary Volume*, Eds. H. G. Hopkins and M. J. Sewell, page 13. Pergamon Press, Oxford.

Du, Z.-Z. and Hancock, J. W. (1991). J. Mech. Phys. Solids, 39:555-567.

Evans, A. G. (1983). Met. Trans. A, 14A:1349-1355.

Hancock, J. W. and Cowling, M. J. (1980). Metal Science, August-September 1980:293-304.

Hancock, J. W., Reuter, W. G., and Parks, D. M. (1991). to appear, Proceedings ASTM Symposium on Crack-Tip Constraint, Indianapolis, May, 1991.

Huang, Y., Hutchinson, J. W., and Tvergaard, V. (1991). J. Mech. Phys. Solids, 39:223–241.

Hult, J. A. and McClintock, F. A. (1956). in *Proceedings 9th International Congress for Applied Mechanics*, Brussels, pages 51–58.

Hutchinson, J. W. (1968). J. Mech. Phys. Solids, 16:13–31.

Kfouri, A. P. (1986). Int. J. Fract., 30:301–315.

Kumar, V., German, M. D., and Shih, C. F. (1981). *An Engineering Approach for Elastic-Plastic Fracture Analysis*, EPRI Topical Report NP-1931, EPRI, Palo Alto, CA.

Larsson, S. G. and Carlsson, A. J. (1973). J. Mech. Phys. Solids, 21:263–278.

Leevers, P. S. and Radon, J. C. (1982). Int. J. Fract., 19:311–325.

Li, F. Z., Shih, C. F., and Needleman, A. (1985). Engr. Fract. Mech., 21:405–421.

Li, Y. C. and Wang, T. Z. (1986). Scientia Sinica (Series A), 29:941–955.

Lin, T., Evans, A. G., and Ritchie, R. O. (1986a). J. Mech. Phys. Solids, 34:477–497.

Lin, T., Evans, A. G., and Ritchie, R. O. (1986b). Acta Met., 34:2205–2216.

Lin, T., Evans, A. G., and Ritchie, R. O. (1987). Met. Trans. A, 18A:614–651.

McClintock, F. A. (1967). J. Appl. Mech., 35:363–371.

McClintock, F. A. (1971). in *Fracture: an Advanced Treatise, Vol. III*, ed. H. Leibowitz, Academic Press, New York, pages 47–225.

McClintock, F. A. (1974). in *Fracture Mechanics of Ceramics, Volume 1*, Eds. R. C. Bradt, D. D. H. Hasselman, and F. F. Lange, Plenum Press, NY, pages 99–113.

McMeeking, R. M. (1977). J. Mech. Phys. Solids, 25:357–381.

McMeeking, R. M. and Parks, D. M. (1979). in *Elastic-Plastic Fracture, ASTM STP 668*, Eds. J. D. Landes, et al., ASTM, Phil., pages 175–194.

Nakamura, T. and Parks, D. M. (1989). Int. J. Solids Structures, 25:1411–1426.

Nakamura, T. and Parks, D. M. (1991). MIT Report, July, 1991. To appear, Int. J. Solids Structures.

O'Dowd, N. P., and Shih, C, F. (1991a). J. Mech. Phys. Solids, Vol. 39, No. 8, pp 989-1015.

O'Dowd, N. P., and Shih, C. F. (1991b). Brown University Report, August, 1991. to appear, J. Mech. Phys. Solids.

Parks, D. M. (1977). Comp. Meth. Appl. Mech. Engr., 12:353–364.

Parks, D. M. (1991). in *Defect Assessment in Components — Fundamentals and Applications*, ESIS/EGF9, Eds. J. G. Blauel and K.-H. Schwalbe, Mechanical Engineering Publications, London, pages 205–231.

Rice, J. R. (1967). J. Appl. Mech., 35:379–386.

Rice, J. R. (1974). J. Mech. Phys. Solids, 22:17–26.

Rice, J. R. and Levy, N. (1972). J. Appl. Mech., 39:185–192.

Rice, J. R. and Rosengren, G. F. (1968). J. Mech. Phys. Solids, 16:1–12.

Rice, J. R. and Tracey, D. M. (1969). J. Mech. Phys. Solids, 19:201.

Ritchie, R. O., Knott, J. F., and Rice, J. R. (1973). J. Mech. Phys. Solids, 21:395–410.

Sham, T. L. (1991). Int. J. Fract., 48:81–102.

Sharma, S. M. and Aravas, N. A. (1991). J. Mech. Phys. Solids, 39:8.

Shih, C. F. (1981). J. Mech. Phys. Solids, 29:305–326.

Shih, C. F. (1985). Int. J. Fracture, 29:73–84.

Shih, C. F. and German, M. D. (1981). Int. J. Fract., 17:27–43.

Sumpter, J. D. (1991). to appear, Proceedings ASTM Symposium on Crack-Tip Constraint, Indianapolis, May, 1991.

Wang, Y. (1991a). to appear, Proceedings ASTM Symposium on Crack-Tip Constraint, Indianapolis, May 1991.

Wang, Y. (1991b). Ph. D. Thesis, Department of Mechanical Engineering, MIT, May, 1991.

Wang, Y. and Parks, D. M. (1991). MIT Report, to appear, Int. J. Fracture.

Weibull, W. (1938). Ingenioersvetenskapakad, Handl., 151:45.

Weibull, W. (1939). Ingenioersvetenskapakad, Handl., 153:55.

Williams, M. L. (1957). J. Appl. Mech., 24:111–114.

3

Constraint and Stress State Effects in Ductile Fracture

J. W. Hancock

ABSTRACT The mechanics and mechanisms of ductile fracture are reviewed, emphasizing the effect of multi-axial states of stress on the mechanisms of hole nucleation, growth, and coalescence. This provides a basis for a discussion of crack extension by ductile failure mechanisms, with particular reference to ways in which the constraint can be quantified in a two parameter constraint based fracture mechanics methodology.

3.1 Introduction

Engineering structures are usually required to have a global elastic response. This is achieved by design processes which attempt to restrict the nominal stresses in relation to the yield stress. However at sites of local stress concentration, the stress system frequently violates the yield criterion, indicating the occurrence of local plasticity.

Plasticity has a vital role in ensuring structural integrity. Yield and subsequent plastic flow limit the high stresses that would otherwise occur in a perfectly elastic material but replace the stress concentration with a strain concentration. To guarantee that local failure does not occur in the plastic strain concentration, it is necessary to ensure that the material has an appropriate level of ductility. However the stress systems in which plasticity and failure occurs are rarely simple, and this leads to a need to understand material behavior in general three dimensional states of stress and strain.

A severe example of this problem occurs at cracks. Elastic analysis indicates the existence of a stress singularity, which is transformed into a strain singularity by the finite deformation plasticity at the crack tip. Failure processes in such crack tip fields are largely paralleled by failure processes that occur in simpler deformation fields where the importance of microstructural variables on failure processes are well recognized. However in both cases, deformation analysis is conveniently handled at a continuum level, when the detailed micro-structural features are ignored except in the way that they affect the constitutive equations.

The study of deformation and fracture may thus occur at different size scales, with the motive of scientific curiosity or a practical need to use and develop materials. It is hoped that these motives are related, since materials can be used most effectively when the detailed mechanisms by which they deform and break are understood. Similarly the practical use of materials provides the most stringent test of how well the fundamental mechanisms are understood. In this context the present Chapter develops the theme of failure in multi-axial states of stress in an incomplete and prejudiced manner. The work attempts to review some aspects of a theme developed earlier by McClintock, while freely admitting "an incomplete mastery of his precepts."

3.2 Plasticity

The foundations of continuum plasticity derive largely from the work of Hill (1950), Prager and Hodge (1951), and Drucker (1960). Yield is presumed to occur in an isotropic solid at a combination of stresses which define a yield criterion, and which can be expressed graphically as a surface when plotted in stress space. Discussion is limited to the Mises yield criterion which can be expressed, either in terms of the stresses, σ_{ij}:

$$\bar{\sigma} = \left\{ \frac{1}{2}\left[(\sigma_{11}-\sigma_{22})^2 + (\sigma_{22}-\sigma_{33})^2 + (\sigma_{33}-\sigma_{11})^2\right] + 3\sigma_{12}^2 + 3\sigma_{23}^2 + 3\sigma_{31}^2 \right\}^{1/2} \tag{3.1}$$

or more compactly in terms of the stress deviators, s_{ij}[1]:

$$s_{ij} = \sigma_{ij} - \sigma_{kk}/3 \tag{3.2}$$

$$\bar{\sigma} = \sqrt{\frac{3}{2} s_{ij} s_{ij}}. \tag{3.3}$$

Yield occurs when the equivalent stress $\bar{\sigma}$ reaches the yield stress in tension, Y, i.e.,

$$\bar{\sigma} = Y. \tag{3.4}$$

Alternatively the yield criterion can be expressed in terms of the yield stress in shear, k

$$\bar{\sigma} = \sqrt{3}k. \tag{3.5}$$

Assuming that isotropy is maintained under plastic deformation, the flow stress increases as a function of the equivalent plastic strain, defined by integrating the plastic strain increments, de_{ij}^p through the deformation

[1]Here and throughout this chapter, repeated indices imply summation of terms.

history:

$$d\bar{e}^p = \left\{\frac{2}{9}\left[(de^p_{11} - de^p_{22})^2 + (de^p_{22} - de^p_{33})^2 + (de^p_{33} - de^p_{11})^2\right] + \frac{1}{3}\left[d\gamma^p_{12}{}^2 + d\gamma^p_{23}{}^2 + d\gamma^p_{31}{}^2\right]\right\}^{1/2} \tag{3.6}$$

$$\bar{e}^p = \int d\bar{e}^p. \tag{3.7}$$

In classical plasticity the relationship between the equivalent plastic strain and the equivalent stress is independent of the stress or strain state, and as such can be determined from a single uni-axial tensile test. Yield and plastic flow are independent of the hydrostatic or mean stress, σ_m:

$$\sigma_m = \sigma_{kk}/3 = (\sigma_1 + \sigma_2 + \sigma_3)/3. \tag{3.8}$$

The mean and equivalent stress can however be combined into a single non-dimensional parameter, which defines the triaxiality of the stress state $(\sigma_m/\bar{\sigma})$. At yield in pure shear, $(\sigma_1 = k = -\sigma_2, \sigma_3 = 0)$, the triaxiality is zero, while in perfect hydrostatic tension or compression $(\sigma_1 = \sigma_2 = \sigma_3 = \sigma_m)$ it is infinite.

To simplify the analysis it is convenient to write the stress-strain relationship in a form such as:

$$\bar{\sigma} = \sigma_0(\bar{e}^p)^n, \tag{3.9}$$

where σ_0 and the strain hardening exponent n are material properties. A further simplification can be affected by neglecting elastic deformation in comparison to the plastic strains. The material is thus assumed to be elastically rigid, with an infinite Young's modulus. This simplification used in conjunction with an assumption that the material does not strain harden, defines a response, described as rigid, perfectly plastic. The assumptions of rigid perfectly plastic deformation under plane strain conditions allow a particularly useful way of visualizing and analyzing plastic deformation through plane strain slip line fields. The method has been thoroughly described by Hill (1950) and Johnson, Sowerby and Venter (1982), and notably exploited by McClintock(1969) to elucidate the role of plasticity in fracture. Incompressibility allows the stress-strain relations to define the out of plane stress σ_3 in plane strain as:

$$\sigma_3 = (\sigma_1 + \sigma_2) = \sigma_m = \sigma_{kk}/3. \tag{3.10}$$

The yield criterion limits the difference of the in-plane stresses (σ_1, σ_2) to $2k$

$$\sigma_1 - \sigma_2 = 2k. \tag{3.11}$$

This allows the three principal stresses to be written in terms of the mean stress, σ_m

$$\sigma_1 = \sigma_m + k \tag{3.12a}$$

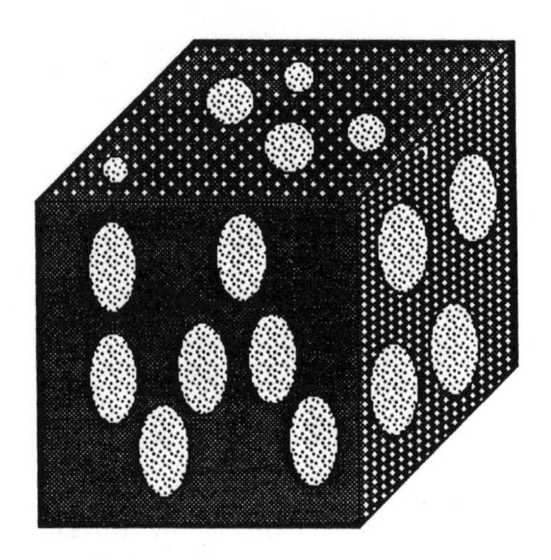

FIGURE 3.1. A porous solid.

$$\sigma_2 = \sigma_m - k \tag{3.12b}$$

$$\sigma_3 = \sigma_m. \tag{3.12c}$$

At a free surface one of these stresses (say σ_2) must be zero and this defines the local stress state as either plane strain tension or compression,

$$\sigma_1 = \pm 2k, \sigma_2 = 0, \sigma_3 = \sigma_m = k. \tag{3.13}$$

In analyzing plane strain perfectly plastic deformation, it is convenient to work in co-ordinates (α, β) which correspond to the directions of maximum shear stress. In a body undergoing homogeneous deformation the axes are straight, but in general stress gradients allow the axes (α, β) to be curvilinear, but always orthogonal. Referred to such axes, or slip lines the equilibrium equations adopt a simple form. The Hencky equations, as they are known, give the change in mean normal stress in terms of the rotation of a slip line direction:

$$d\sigma_m = 2k\, d\phi \quad \text{on an } \alpha \text{ line} \tag{3.14a}$$

$$d\sigma_m = -2k\, d\phi \quad \text{on a } \beta \text{ line.} \tag{3.14b}$$

A given slip line field can then be analyzed by following the rotation of the slip lines from free surface where the stress state and is known and the slip lines necessarily are inclined at $\pi/4$ to the surface.

Classical plasticity is based on the incompressibility of plastic strains. However during plastic deformation it is observed experimentally that voids nucleate at second phase particles and grow in the strain field, and thus the porous material can be regarded as a composite of matrix and voids as illustrated in Figure 3.1. Such a material is capable of exhibiting plastic dilation due to void growth, or shrinkage. The plastic behavior of such a material was initially described in a qualitative sense by Berg (1970), and quantitatively by Gurson (1977) in terms of a yield criterion based on the volume fraction of voids, f, and the hydrostatic stress, $\Sigma_{kk}/3$, on the porous aggregate. In order to distinguish between properties of the matrix and that of the porous aggregate it is now convenient to use capital letters to denote stresses applied to the porous aggregate and small letters for the properties of the matrix. Thus $\bar{\Sigma}$ is now the equivalent or Mises stress on the porous aggregate while $\bar{\sigma}$ is the Mises stress of the matrix alone. In the light of an upper bound analysis of spherical and cylindrically symmetric cells of a non-hardening material, Gurson (1977) proposed a yield criterion of the form

$$\phi = (\bar{\Sigma}/\bar{\sigma}) - g(\Sigma_{kk}, f) = 0 \tag{3.15}$$

with

$$g(\Sigma_{kk}, f) = 1 + q_2 f^2 - 2 f q_1 \cosh\left(\frac{\Sigma_{kk}}{2\bar{\sigma}}\right) \tag{3.16}$$

and

$$q_1 = q_2 = 1. \tag{3.17}$$

Setting $f = 0$ recovers the Mises yield criterion for an incompressible material, where,

$$\bar{\Sigma} = \bar{\sigma} \tag{3.18}$$

Similarly setting $\Sigma_{kk} = 0$ shows that in stress states such as pure shear in which the voids do not increase in volume, the flow stress of the material is simply reduced by the volume fraction of voids

$$\bar{\Sigma} = \bar{\sigma}(1 - f). \tag{3.19}$$

Gurson's (1977) expression has undergone several empirical modifications based on adjusting the parameters q_1 and q_2 while retaining the identity

$$q_1^2 = q_2. \tag{3.20}$$

This has the effect of essentially increasing the volume fraction of voids in Gurson's analysis to $(q_1 f)$, giving the shear behavior as,

$$\bar{\Sigma} = \bar{\sigma}(1 - q_1 f). \tag{3.21}$$

Tvergaard (1982) has suggested $q_1 = 1.5$, which gave improved agreement with a numerical power law hardening matrix, while experimental evidence suggests q_1 to be in the range 1.5 to 2.4. For an isotropic material normality of the matrix deformation implies normality of the aggregate, through an argument advanced by Bishop and Hill (1951) and discussed by Berg (1970). The dilating plastic material thus follows an associated flow rule in which the pressure dependence matches the dilation rate. Using the Gurson form of the yield surface, the dilation strain becomes,

$$\frac{\partial E_{kk}}{\partial \bar{E}} = f \sinh\left(\frac{\Sigma_{kk}}{2\bar{\sigma}}\right) / (\bar{\Sigma}/\bar{\sigma}). \tag{3.22}$$

As the volume fraction of voids tends to zero and $(\bar{\Sigma}/\bar{\sigma})$ tends to one, the dilation rate may be interpreted in terms of the growth rate of a single isolated void, assuming that it remains basically spherical. This is a good approximation at high triaxialities as discussed by Berg (1970) and Budiansky, Hutchinson and Slutsky (1981) and gives

$$\frac{\partial R}{\partial \bar{e}} = \frac{R}{3} \sinh\left(\frac{\Sigma_{kk}}{2\bar{\sigma}}\right). \tag{3.23}$$

This makes contact with the void growth analyses of McClintock (1968) and Rice and Tracey (1969) for the growth of a single isolated void in an infinite block of material. Gurson's dilating plastic model thus allows complete analysis of deformation while incorporating the development of porosity as an internal variable. Further refinements have been added to the basic concept including stress controlled void nucleation, which causes a loss of normality in the flow rule as discussed by Saje Pan and Needleman (1982).

3.3 Failure Mechanisms

3.3.1 INTRODUCTION

Stress state appears to have been first formally included in the failure of metals by Ludwik (1923) and Haigh (1923). Ludwik did not differentiate between the various failure modes, but postulated that failure should occur at a critical value of the maximum principal stress. An account of the classical origins of failure mechanics has been given by Orowan (1945). The role of the stress state was however most clearly recognized in Bridgman's (1952) study of the effect of external pressure on the tensile ductility of a range of metals. A salient result arising from Bridgman's work is that the ductility, as a measure of the extent of plastic strain, is markedly increased by superimposing a hydrostatic compression, to the extent that material which previously failed could be made to neck to a point. More recently Zok et al. (1988) have demonstrated the effect of hydrostatic pressure on the ductility of metal matrix composites, while the effect on polymer matrix composites was demonstrated by Parry and Wronski (1985, 1986).

The effect of hydrostatic compression, in enabling high levels of ductility, is of course fundamental to the large plastic strains utilized in forming processes such as rolling or extrusion. However in structural applications failure usually occurs in stress states with a tensile mean stress. Historically, failure under multi-axial tensile stress states has been studied by examining the deformation and failure of tubes under combinations of internal pressure, torsion and tension. Although such specimens were essential to the understanding of yielding in multi-axial stress states (Lode 1926), their contribution towards understanding the failure modes of ductile materials has been limited. Failure frequently occurs in within some localized deformation mode in which the stress state differs from that of the homogeneous deformation field which precedes it, and the anisotropy of the material from which the tubes are made further complicates simple interpretations of the data. A more productive line of experimental work originated from M.I.T. where Neimark (1959), under the supervision of McClintock, tested and analyzed notched tensile bars, making use of Bridgman's solution for the stress and strain distribution in a naturally necking bar.

3.3.2 Naturally Necking Bars

The basis of Bridgman's (1952) solution for the deformation of a naturally necking bar is the assumption that the state of strain is simply uniaxial tension across the neck. In fully plastic deformation the true strain at all points on the minimum cross section of the neck is given by

$$de_z^p = d\bar{e}^p = -2de_\theta^p = -2de_r^p \tag{3.24}$$

$$\bar{e}^p - 2\log_e\left(\frac{a_0}{a}\right). \tag{3.25}$$

An assumption about the trajectories of the maximum principal stress then allowed the equilibrium equations to be integrated to give the stress state. On the minimum section the result is expressed in terms of the internal and external radii of the neck, a and R, and the radial co-ordinate, r, from the center of the bar, as illustrated in Figure 3.2.

$$\sigma_z = \bar{\sigma}\left[1 + \log_e\left(\frac{a^2 + 2aR - r^2}{2aR}\right)\right] \tag{3.26}$$

$$\sigma_r = \sigma_\theta = \bar{\sigma}\left[\log_e\left(\frac{a^2 + 2aR - r^2}{2aR}\right)\right] \tag{3.27}$$

$$\frac{\sigma_m}{\bar{\sigma}} = \left[\frac{1}{3} + \log_e\left(\frac{a^2 + 2aR - r^2}{2aR}\right)\right] \tag{3.28}$$

Thus although the strain state comprises simple uniaxial tension there is a stress gradient such that there is an elevation of hydrostatic tensile stress field at the center of the bar. This simply depends on the ratio of the internal and external radii of the neck (a/R). A range of experiments show

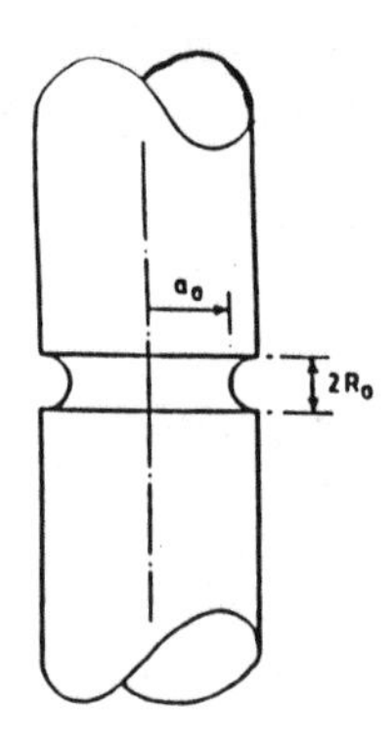

Axisymmetric Specimens

Notch	a_0	R_0
A	3·8mm	3 · 8mm
D	3·8mm	1 · 27mm
Unnotched	3·8mm	∞

The dimensions of the notched axisymmetric specimens.

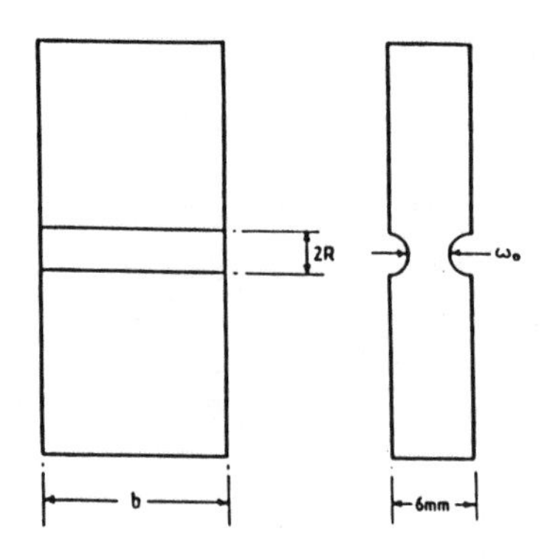

Dimensions of Plane Strain Notches

	Notch	ω_0	b	R
(mm)	A	3	25·4	1 · 5
	D	3	25 · 4	0 · 5

The dimensions of the notched plane strain specimens.

FIGURE 3.2. Axisymmetric and plane strain notched bars.

that after the onset of necking the ratio (a/R) is well approximated by the additional plastic strain after necking, $\bar{\epsilon}^p_n$.

$$\frac{a}{R} = \bar{\epsilon}^p - \bar{\epsilon}^p_n \;\; (\bar{\epsilon}^p > \bar{\epsilon}^p_n) \tag{3.29}$$

$$\frac{a}{R} = 0 \;\; (\bar{\epsilon}^p \leq \bar{\epsilon}^p_n) . \tag{3.30}$$

If the material follows a Ludwig power hardening relation the necking strain given by Considere's criterion is simply equal to the strain hardening exponent, n, and is typically of the order of 0.1 to 0.2. This allows the triaxiality of the center of a naturally necking bar to be followed as a function of deformation history. The problem of a naturally necking round bar has subsequently been examined by Needleman (1972), Norris, Moran, Scudder and Quinones (1978) using finite element techniques. These confirm the basic correctness of Bridgman's analysis and show the results to be remarkably accurate. Detailed numerical analyses generally show that the strain develops in accord with Bridgman's analysis, but differ slightly on the amplification of triaxiality within the neck.

The problem of the formation of a natural neck under plane strain conditions is fundamentally different, and has been discussed by Onat and Prager (1954), Cowper and Onat (1962), McMeeking and Rice (1975), Burke and Nix (1979) and Tvergaard, Needleman and Lo (1981). In perfect plasticity the homogeneous deformation before the onset of necking comprises a set of straight slip lines inclined at $\pm 45°$ to the tensile axis, giving the simple stress field:

$$\sigma_1 = 2k, \sigma_2 = 0, \sigma_3 = \sigma_m = k \tag{3.31}$$

and a triaxiality

$$\frac{\sigma_m}{\bar{\sigma}} = \frac{1}{\sqrt{3}} . \tag{3.32}$$

Given that the specimen is long enough for the ends of the bar not to inhibit necking, localized deformation can occur by sliding off on one of the slip lines, or on alternate slip lines. In the absence of strain hardening the necking mode is not unique, as sliding could occur on any slip line. The infinite strain concentration associated with such a discontinuous displacement field is possible in the absence of strain hardening, but is inhibited by strain hardening. The sharp concentration of plastic flow into a thin band is more appropriately discussed as a localization problem. However before the occurrence of such sharp localizations the development of a diffuse natural neck in plane strain has been discussed by Cowper and Onat (1962), who have given solutions for the distribution of strain along the axis and the amplitude of the neck.

Plane strain notched bars have been analysed and tested less than their axi-symmetric counterparts. The main deterrent is the absence of simple analytic expressions for the plastic strain distribution across the neck, although numerical solutions have been given by Hancock and Brown (1983).

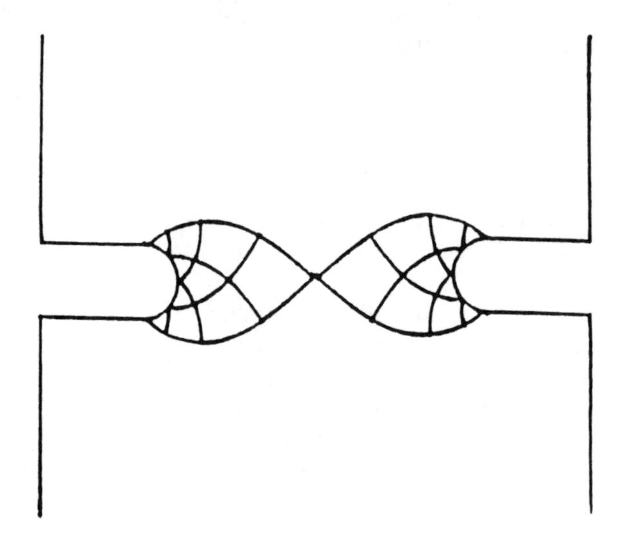

FIGURE 3.3. Deformation of a plane strain notched tensile specimen represented by interlocking log-spirals.

However the form of the stress distribution is reasonably represented by a slip line field consisting of interlocking log-spirals as shown in Figure 3.3 for ($a/R \leq 0.5$), indicating severely triaxial stress states in the center of the specimen. The ability to use such specimens to determine failure conditions in multi-axial states of stress is however offset by the fact that the strain concentration moves to the root of the notch as the notch profile becomes sharper. The effect occurs in both axisymmetric and plane strain notched bars, but is probably more pronounced in the latter. This produces a competition between failure in high strains and low triaxialities at the base of the notch, and low strains and severe stress states at the center of the specimen.

3.3.3 Experimental Tests on Notched Bars

Scots may be forgiven for feeling that the first attempt to use notched bars to elucidate failure in multi-axial states of stress originated with Kirkaldy in 1860. However the systematic use of such specimens clearly derives from the work of Neimark under the supervision of McClintock at MIT in 1959. Hancock and Mackenzie (1976) and Mackenzie Hancock and Brown (1977) and Beremin (1983) used such specimens to measure the ductility of a range of structural materials in multi-axial states of stress. The strains were measured through the change in diameter of notched bars, initially using the Bridgman solution and later finite element solutions (Hancock

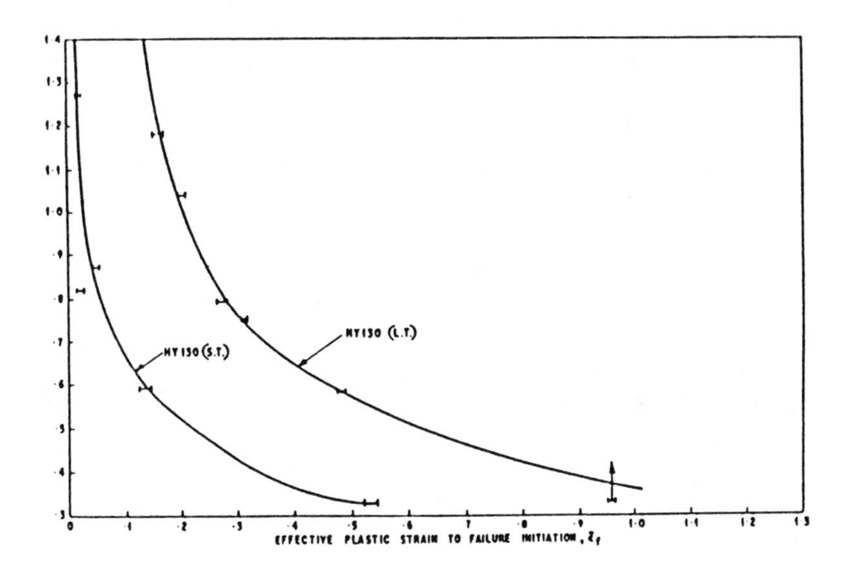

FIGURE 3.4. Failure locus for two orientations (LT and ST) of a quenched and tempered steel denoted HY130, showing the equivalent plastic strain to failure initiation as a function of triaxiality after Hancock and Mackenzie (1977).

and Brown 1983). Hancock and Mackenzie (1976) expressed the results in the form of a failure locus, such as that shown in Figure 3.4, where the strain to initiate a distinct crack by void coalescence in the center of the specimen is given as a function of the triaxiality, $(\sigma_m/\bar{\sigma})$. The important conclusion is that ductility decreases markedly with triaxiality. The effect is particularly pronounced for steels tested in the through thickness direction (denoted ST in Figure 3.4) of a rolled plate. In these circumstances Hancock and Mackenzie (1977) showed that the strain to initiate failure of an unnotched tensile specimen of a quenched and tempered steel designated HY130 was 55% while at a triaxiality $(\sigma_m/\bar{\sigma})$ of 1.25 the equivalent plastic strain to failure was 1–2%. Such steels show strong tendencies to delamination failure under welded joints where severely triaxial residual stress fields are to be found, and the necessary ductility to prevent structural failure is not well quantified by uni-axial tests.

In general wrought structural steels show a broad tendency for the ductility to decrease exponentially with stress state as indicated by the form of the void growth equations of McClintock (1968) and Rice and Tracey (1969). Metallurgical investigations show that the ductility of these steels was controlled by the growth of voids from weakly bonded inclusions, from which voids nucleated at very small strains. The ductility of such materials is thus largely controlled by void growth. The anisotropy of these steels arises from the non-random distribution of inclusions, which are found in

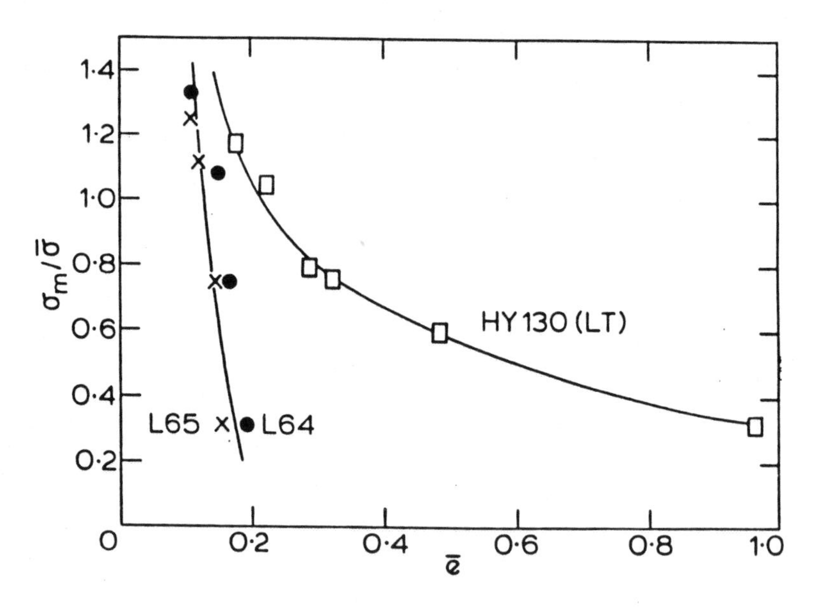

FIGURE 3.5. Failure locus for two aluminium alloys designated L64 and L65 which exhibit nucleation controlled ductility compared to the void growth controlled ductility of HY130 after Hancock and Cowling (1980).

plate-like colonies in the rolling plane. Void coalescence occurs first with the colonies of inclusions leading to highly eccentric voids. When tested in the short transverse direction, the major axis of these voids are oriented perpendicular to the maximum principal stress, and the voids grow very much more rapidly than when tested in the rolling direction. This leads to marked fracture anisotropy of wrought steels. Final linkage of the voids frequently occurred by void nucleation and coalescence of a smaller generation of carbides further complicating the fracture process. It would thus be naive to expect a detailed agreement between void growth calculations base on isolated spherical voids to give a correct analysis of the failure process. However in such materials ductility is clearly a function of the triaxiality which controls void growth.

The exponential decrease in ductility with triaxiality is however not a universal feature of the results, and one which has been perhaps over emphasized. Figure 3.5 shows the failure locus for two aluminum alloys designated L64 and L65. The ductility of these alloys is weakly dependent on stress state and almost occurs at a critical strain. It is thought that in these materials the controlling step is void nucleation. On nucleation there are enough voids to coalesce, with minimal additional void growth, giving a weak dependency of ductility on stress state nucleation. Stress controlled void nucleation is linearly dependent on triaxiality, as discussed

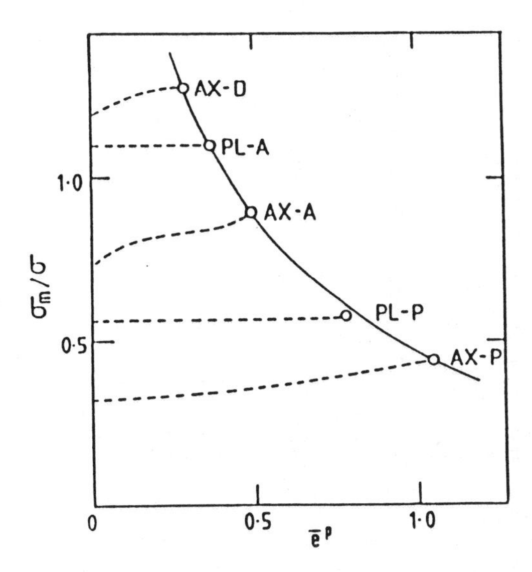

FIGURE 3.6. Failure locus for plane and axisymmetric states of strain for a model material known as Swedish Iron, which contains a 1 volume percent of iron oxide inclusions. Axisymmetric specimens are prefixed by AX and plane strain specimens by P.

in Section 3.4.1, while strain controlled nucleation is of course independent of stress state. The weak, or non-existent dependence of nucleation controlled ductility, thus contrasts strongly the exponential dependence of void growth controlled failure.

Plane strain bars have been tested experimentally by Clausing (1970), who reported that the ductility of plane strain tension specimens was less than that of axisymmetric specimens of the same material. Clausing's interpretation of these results was that the ductility in plane strain was less than that in axisymmetric states of strain, and his original experiments were widely given that interpretation. However the results are ambiguous, even if the problem of deformation history associated with necking is neglected. In comparing the ductility of plane strain and axisymmetric specimens, it is necessary to recognize that both the stress and the strain state have changed.

Hancock and Brown (1983) attempted to resolve this issue by testing plane strain and axisymmetric notched bars with the same profile. The stress states were calculated by elastic-plastic finite element stress analysis featuring plastic incompressibility. On this basis the strain to initiate failure in plane and axisymmetric states of strain were closely similar and fell on the same failure locus, as illustrated for a simple model material called Swedish Iron in Figure 3.6. This basically indicates insensitivity to the effect

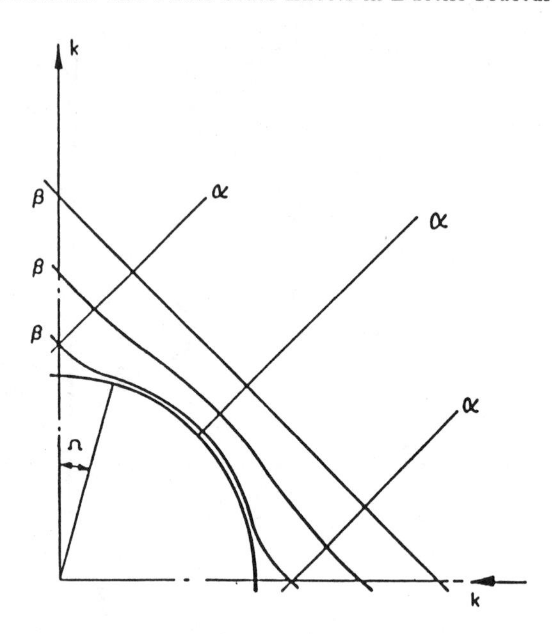

FIGURE 3.7. Slip line field for a rigid cylindrical inclusion.

of strain state in the materials tested, given the cautionary note that failure loci as presented do not represent the effect of stress history. Subsequently, Needleman and Tvergaard (1984) re-analyzed the experiments of Hancock and Brown (1983) on a plain carbon steel, using a modification of the Gurson model for the flow of porous plastic solids in an attempt to represent the plastic dilation associated with void growth. A variety of arbitrary void nucleation criteria were used, and an unrealistically high void volume fraction. However in the steels tested the effect is at best clearly very weak or non-existent. In addition the analysis severely exaggerates the effect of plastic dilation by overestimating the volume fraction of particles which nucleate voids by an order of magnitude.

The necessary interpretation is that the void coalescence process in these materials occurs within local statistical inhomogeneities, such as those originally discussed by McClintock (1967). On the basis that the void nucleating particles are randomly dispersed, then typical cells inside tensile specimens are expected to contain approximately ten times the average volume fraction of voids. The local stress and strain state in such inhomogeneities does not correspond to that of the remote field. As an example consider a material with an average porosity of 1%, the statistics indicate that a reasonably sized tensile specimen will contain a local inhomogeneity where the porosity is 10%. The application of plane strain conditions in the remote field does guarantee identical strain conditions within inhomogeneity. In fact it is known that if plane strain conditions are applied remote to a single void (being the limiting case of an inhomogeneity with $f = 1$) in an incompressible material, then the void grows in the plane strain direction, producing local out of plane displacements. The point is that the identity of the remote strain state is not generally maintained inside the inhomogeneity. As a consequence strain state effects, as measured in the remote field, must not be expected in "dirty" materials containing significant volume fractions of hole nucleating inclusions, in which crack formation is dominated by inhomogeneities. However in clean materials in which such inhomogeneities do not exist, or failure occurs over distances large compared to the inhomogeneity, then strain state effects must be expected, and are indeed observed as shown in the experiments of Anand and Spitzig (1980).

3.4 Ductile Failure Mechanisms

3.4.1 Void Nucleation

The key processes leading to ductile failure were first identified by Tipper (1949), as void nucleation, growth, and coalescence. The first stage in the process is the nucleation of voids by decohesion of the particle-matrix interface or by cracking of the particle itself. The general problem of the stress and strain fields for ellipsoidal elastic inclusions has been presented by Eshelby (1957). The solutions for spherical elastic or cylindrical inclusions may be derived as special cases, although it is usually more convenient to retrieve the necessary results from the earlier work of Goodier (1933). In the particular case of a rigid spherical inclusion the maximum interfacial stress occurs on the interface in the loading direction, and for a cylindrical inclusion has a magnitude:

$$\sigma_{rr\,\max} = 1.5\sigma \tag{3.33}$$

while for a spherical inclusion:

$$\sigma_{rr\,\max} = 2\sigma. \tag{3.34}$$

The solution for general multi-axial loadings can be obtained as by superimposing uniaxial solutions. For example, the solution for shear is obtained by adding the stress field for uniaxial tension of magnitude $+\tau$ to that of a uniaxial compression of magnitude $-\tau$ rotated by $\pi/2$. The maximum interfacial shear stress for a rigid cylindrical inclusion then becomes:

$$\sigma_{rr\,\max} = 2\tau. \tag{3.35}$$

In terms of the equivalent stress this may be written as

$$\sigma_{rr\,\max} = \frac{2\bar{\sigma}}{\sqrt{3}}. \tag{3.36}$$

Since the superposition principle is valid for linear materials a hydrostatic stress σ_m can be superimposed to give the general result

$$\sigma_{rr\,\max} = \frac{2\overline{\sigma}}{\sqrt{3}} + \sigma_m. \tag{3.37}$$

Nucleation has been studied at two distinct size scales. For particles whose size is comparable to the dislocation or slip band spacing, the local conditions are most appropriately discussed in terms of dislocation mechanisms, such as those reviewed by Goods and Brown (1979). Alternatively if the particles are large compared to the slip band spacing the stresses and strains near the particle matrix interface are most conveniently described in terms of continuum plasticity. On this basis numerical solutions for cylindrical inclusions have been given by Argon, Im and Safoglu (1975), and for spherical particles by Thomson and Hancock (1984).

Numerical solutions may be used as a basis for discussing the form of a possible slip line field for a rigid cylindrical inclusion embedded in a perfectly plastic matrix. Given the small strain nature of a slip line field, the solution for a given loading, will also apply to a loading of the reverse sign. That is, within the restrictions of small strain theory, the field for a tensile problem is admissible for the equivalent compression problem, with the sign of the stresses and displacements reversed. Interest is restricted initially to a remote stress state of pure shear ($\sigma_1 = -\sigma_2 = k, \sigma_{m\infty} = 0$), recalling that for an incompressible material a pure hydrostatic stress may be superimposed later. The symmetry of the problem implies that slip lines that meet the inclusion at multiples of $\pi/4$ must remain straight from the remote field to the interface At these sites the stress state at the particle matrix interface is the same as that of the remote matrix.

$$\sigma_{m(r=a)} = \sigma_{m(r=\infty)} \text{ at } \Omega = \pi/4. \tag{3.38}$$

At the interface between a plastically deforming matrix and a rigid particle the slip lines must meet the interface radially and tangentially. This ensures that the compatibility requirement that $e_{\theta\theta} = 0$ is met at the plastically deforming part of the interface. The slip line tangential to the interface must break away from the interface at some angle ω to cross the remote tensile axis at $\pi/4$ to ensure that intersecting slip lines from adjacent quadrants remain orthogonal, as illustrated in Figure 3.7. This leaves a small rigid cap on the poles and equator of the cylindrical inclusion, whose extent is described by the angle Ω_c. This angle is not determined at present, but has limiting values 0 and $\pi/4$. Application of the Hencky equations gives the maximum stress across the plastically deforming sector of the particle matrix interface as

$$\sigma_{rr} = 2k(\pi/4 - \Omega). \tag{3.39}$$

In the case of $\Omega_c = \pi/4$ the slip lines form a square surrounding the inclusion, leaving only one point on the interface plastic. The maximum stress, however occurs when $\Omega_c \rightarrow 0$ when the maximum stress is

$$\sigma_{rr} \rightarrow \pi k/2. \tag{3.40}$$

This is very close to the numerical result of Argon, Im and Safoglu (1975) who found that

$$\sigma_{rr} = 1.5k. \tag{3.41}$$

The minor difference between the two results corresponding to the fact that the maximum stress is displaced by a small angle from the pole of the particle in the numerical solutions.

Argon, Im and Safoglu (1975) have used notched tensile specimens to determine the local failure conditions at the particle-matrix decohesion. The technique involved determining the point on the specimen axis at which void nucleation was just starting, and using the local stresses as boundary conditions for the inclusion problem. Interface decohesion was interpreted in terms of a critical radial stress. Using similar specimens Thomson and Hancock (1984) determined the strain that was necessary to grow metallurgically observed voids from the particle size to the size at which they were observed on the minimum cross section. This then allowed the nucleation strain and the interfacial stresses to be estimated. A feature of all experiments on void nucleation is the inherent scatter in the interfacial strengths, such that nucleation occurs over a range of strains. This makes tests of interfacial decohesion theories difficult, although noteworthy advances have been made in this respect by Needleman (1990).

3.4.2 Void Coalescence and Flow Localization

Void Growth and Coalescence

The first quantitative analysis of ductile failure by void growth was presented by McClintock (1968) using an approximate method of McClintock and Rhee (1962) to interpolate between the linear viscous solution of Berg (1970) and non-hardening solutions. Later this problem was addressed by Rice and Tracey (1969), Budiansky, Slutsky and Hutchinson (1982), and Huang (1991) using variational methods. The analyses emphasize the strong dependence of the void growth rate on the triaxiality of the stress state and the non-linear behavior of the matrix. Thus at high triaxialities Rice and Tracey (1969) give the growth of the void radius R as:

$$\frac{\partial R}{\partial \bar{e}} = 0.28 R \sinh\left(\frac{\Sigma_{kk}}{2\bar{\sigma}}\right) \tag{3.42}$$

This behavior is reflected in the experimental observations which show the strong dependence of the ductility of structural metals on triaxiality reported by Hancock and Mackenzie (1977). Simple calculations of the strain to cause void coalescence, using the behavior of a single void to approximate the behavior of an array of voids, naturally show the correct stress state dependence but generally overestimate the strains to cause failure. This discrepancy arises largely because such calculations neglect void interaction which become a strong effect as the distance between neighboring voids decreases.

The size scale of the void interaction and subsequent fracture processes are vital to the way that they are modelled. For crack formation which occurs by necking of the ligaments between individual adjacent voids, the finite geometry changes of discrete cavities are clearly important. This problem has been addressed by Needleman (1972) who has analysed the growth of a bi-periodic array of cylindrical voids in plane strain. The results of a similar analysis by Hancock (1987) are shown in Figure 3.8a,b for a material with a strain hardening exponent of 0.2 and a non-hardening material, following J_2 flow theory. Strain concentrates into the ligament between the voids after the maximum load on the ligament is reached, and the subsequent increase in the transverse void growth rate leads to necking of the ligament. In periodic arrays of voids the traction maximum on the ligament corresponds to a traction maximum in the remote field, however for the void cluster necking of the ligament occurs while the traction in the remote field is still increasing. For the non-hardening matrix both correspond to a geometric condition in which the height of the void in the tensile direction is approximately equal to the void ligament. Such a simple geometric criterion for void coalescence was originally proposed on semi-empirical grounds by Brown and Embury (1973), and would correspond to a local volume fraction of the order of 0.15. The basis of this suggestion is that this allows the formation of 45° slip lines between rectangular holes as idealized in Figure 3.9, although clearly more constrained modes of deformation are likely to be possible, as discussed by Thomason (1971) and Nagpal, McClintock, Berg and Subudhi (1973).

The non-hardening plane strain analysis of isolated void pairs and periodic arrays agree with the criterion proposed by Brown and Embury (1973) in that void coalescence occurs very close to a geometry in which the void height equals the ligament. However strain hardening has a strong effect, requiring increased void growth, and increasing the height to ligament ratio. This is in accord with the idea that strain hardening increases the ductility of metals both by decreasing the void growth rate, but more importantly by delaying void interactions which lead to coalescence.

The effect of void volume fraction on ductility is most clearly seen in the experiments of Edelson and Baldwin (1962) which indicate that ductility is basically proportional to $(1 - f)/f$. In the experiments of Thomson and Hancock (1984), calculations of the void growth from a 0.01 volume fraction of nucleating particles, suggest that at the formation of a distinct crack the average porosity was calculated to be 3.8±1.4%. This is markedly less than the 0.15 required by Brown and Embury's simple criterion, however in the statistical inhomogeneities the porosity is clearly of a comparable order. It is also noteworthy that there is a consistent trend for the calculated

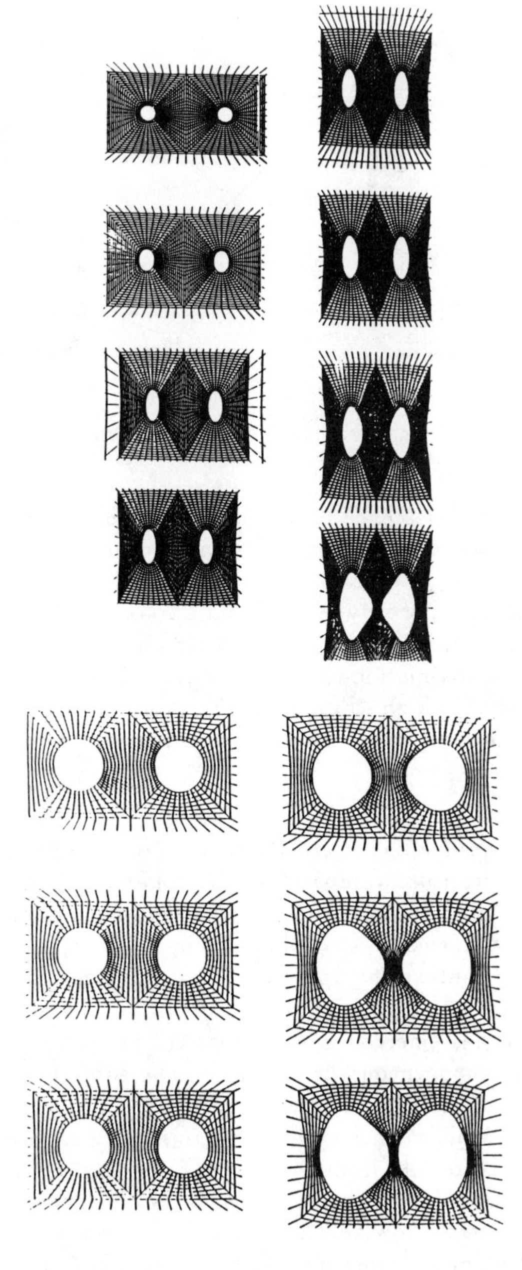

FIGURE 3.8. (a) Coalescence of a pair of voids in plane strain deformation, strain hardening exponent, $n = 0.2$. (b) Coalescence of a pair of voids in non-hardening plane strain deformation.

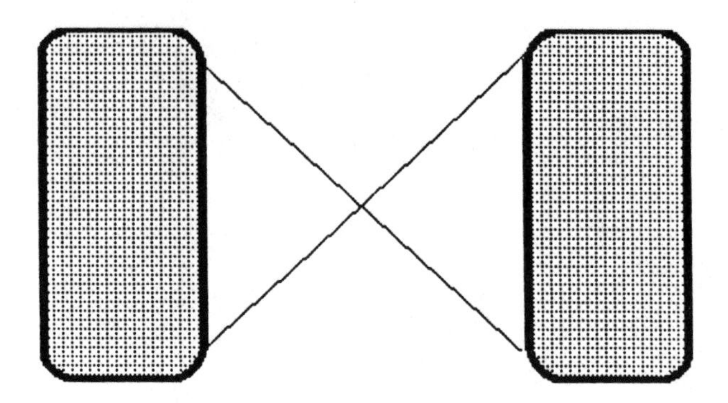

FIGURE 3.9. Void coalescence criterion of Brown and Embury (1973).

porosity at void coalescence to decrease with increasing triaxiality. Thus, in a sharply notched specimen coalescence occurred at a porosity of 3%, and a triaxiality $\sigma_m/\bar{\sigma}$ of 1.28. In the unnotched tensile specimens coalescence occurred at a porosity of 2.6% and a triaxiality of 0.45. Intermediate notches gave intermediate porosities. A simple geometric criterion is thus a useful first approximation to void coalescence but the detailed conditions are clearly such that high strain hardening rates and low triaxialities inhibit coalescence, such that more void growth is required before coalescence.

Localization

If void coalescence leading to crack formation occurs over a size scale which is large in all directions compared to the void spacing and radius, then it seems entirely appropriate to model the coalescence process with a dilating plastic continuum, such as that discussed by Gurson (1977) and Tvergaard (1982). In this context the approach is to search for conditions in which flow can localize into a thin band.

Localization has emerged as an important factor in plasticity, limiting the extent of homogeneous deformation. The subject has been reviewed by Rice (1976), Needleman and Tvergaard (1984), and Hutchinson (1989). Two types or localization may be distinguished. First a diffuse geometric instability which results from the geometry change associated with finite deformation. Secondly a sharp localization, which occurs as a bifurcation from a homogeneous flow field due to a change in the nature of the equations of continuing deformation. Although this is formally distinct from fracture, the additional remote displacements to cause fracture once flow has localized into a thin band are negligible, so that ductility is limited by the conditions for localization. Alternatively in materials without voids, localization can occur by dislocation processes, indeed slip on a crystal plane can be regarded in this way, as discussed by Asaro and Rice (1977).

Two conditions for localization arise; firstly the need for compatibility between the localized band and the bulk material, and a strain hardening requirement to ensure equilibrium of the two parts. The equilibrium requirement has been widely identified as that of a traction maximum. However detailed calculations by Rudnicki and Rice (1975) show that the process is sensitive to the exact form of the constitutive equations. A range of constitutive laws have been discussed in the literature including, isotropic, and kinematic hardening, vertex yield relations, including dislocation slip models and dilating plasticity. Localization in all cases is highly sensitive to the type of material model that is used. Hutchinson and Tvergaard (1981) have pointed out that shear localizations inherently involve non-proportional loading due to material rotations within the band, and as such are very sensitive to the curvature of the yield surface. The low curvature of isotropically hardening yield surfaces associated with incompressible flow virtually excludes the possibility of localization at realistic strains. Rotation of the stress system in localization inherently leads to non-proportional loading, making yield vertex and non-normality effects particularly important. Non-normality may arise in dislocation slip processes when slip is affected both by the critical resolved shear stress and stress components normal to the slip plane.

Kinematic compatibility requires continuity of certain components of the displacement field across the boundaries of the localized band. These conditions are automatically satisfied when the boundaries are non-deforming surfaces, such that the normal strains are zero. These conditions are satisfied for states of incremental plane strain. For example in a plane strain slip line field the slip lines, being lines of zero extension, fulfill the kinematic conditions for localization, and many slip line fields exhibit displacement discontinuities. This leads to a general feature of all shear localization arguments that the plane strain deformation is expected to be less stable to flow localization than axisymmetric states of strain. In fact the analysis of Rudnicki and Rice (1975) shows that for the Prandtl-Reuss flow rule localization is associated with negative strain hardening rate except for states of plane strain when the rate is zero, so that axisymmetric tension or compression is more resistant to localization than plane strain.

The plane strain tension results of Clausing (1970) and Hancock and Brown (1983) can be discussed in terms of the yield vertex analysis of Needleman and Rice (1978). In this case the ratio of homogeneous strain in plane strain to axisymmetric deformation is $2\sqrt{[n/(1+3n)]}$ which is 0.71 for a strain hardening exponent $n = 0.2$. This is a sensible ratio for many structural materials although flow localization controlled by yield vertices does not admit a role for the nucleation and growth of voids which are experimentally observed to control the ductile failure processes in these materials, and the absolute values of the localization strain are generally too large.

Localization models admit the possibility of hole growth through dilating plastic stress-strain relations. This approach has been pursued by Yamamoto (1978) who pointed out that in the case of axisymmetric deformation that localization could only be expected at unrealistically large strains. This can be rationalized by recognizing the role of statistical inhomogeneities, modelled by Yamamoto (1978) as a thin planar band extending to the limits of the body. The procedure is to consider an inhomogeneity in the form of a thin band where the porosity is higher than average and then examine the possibility of localization at every possible band orientation in a search for the most favorable angle. Although localization can occur in a band normal to the tensile axis of a uni-axial tension specimen, it is generally favoured at some inclined angle, giving a significant shear character to the localization. Thus, in Yamamoto's calculations, an initial void volume fraction of 1% and a local porosity of 6% result in localization in plane strain tension at a strain of 0.28, and in axisymmetric tension at a strain of 0.75 This is broadly comparable to the experimental results shown in Figure 3.6 for a material with one volume percent of inclusions which nucleate at strains around 25%. Inhomogeneities in the form of parallel bands of inclusions do exist in wrought steels and are a major cause of their anisotropic failure characteristics, as discussed by Hancock and Mackenzie (1977), however the unbounded planar inhomogeneities envisaged by Yamamoto (1978) were not strictly intended to represent features of the metallurgical microstructure, but were rather a formal device for representing statistical inhomogeneities at any orientation, such that localization occurs at the most favorable angle. In fact the inhomogeneities are better represented as small volumes of enhanced porosity contained in material of average porosity. Such inhomogeneities admit internal stress and strain gradients, while being kinematically contained by the surrounding material. An approach to this problem has been presented by Ohno and Hutchinson (1984) who considered the role of disc shaped inhomogeneities in which the porosity varies in a sinusoidal manner from the maximum enhanced value inside the imperfection to the average value at the edges.

It seems impossible to deny the existence of statistical inhomogeneities or the fact that void coalescence must occur first in such regions. However the significance of such inhomogeneities is arguable. In the author's opinion, inhomogeneities are significant, and the open question is condition under which a cluster of coalesced voids starts to behave in crack like manner rather than as a single large void. This introduces a size scale into the problem, such size scales being absent in localization models, which are simply pointwise criteria. Crack like behavior thus requires the inhomogeneity to produce a strain and triaxiality concentration over a distance which encompasses the void spacing in the average material, thus causing an enhanced void growth rate leading to propagation of void coalescence in a crack like manner.

3.5 Crack Tip Constraint

3.5.1 INTRODUCTION

The mechanisms of failure have been discussed thus far with particular reference to multi-axial states of stress. This is of particular relevance to the problem of failure at the stress and strain concentrations associated with cracks. There are essentially two approaches to this problem. The direct approach essentially attempts to apply specific failure criteria to the local crack tip stress and strain fields, recognizing both that multi-axial states of stress, and severe stress and strain gradients are involved. The second approach is to attempt to finesse the problem (Parks 1991). This involves finding parameters which characterize crack tip deformation with sufficient accuracy, that can be used to characterize fracture. The critical test of both approaches is the ability to measure the toughness of one specimen and predict the failure of a specimen or component of a completely different geometry. The two approaches are of course not exclusive as local failure criteria are often used to predict the critical value of a parameter which characterizes fracture toughness.

The characterization of crack tip stress and strain fields is fundamental to fracture mechanics. In classical linear elastic fracture mechanics, an appropriate parameter is the stress intensity factor K, whose significance can be seen by expanding the stress field in cylindrical co-ordinates (r, θ) about the crack tip, following the work of Williams (1957).

$$\sigma_{ij} = A_{ij}(\theta)r^{-1/2} + B_{ij}(\theta) + C_{ij}(\theta)r^{1/2} + \qquad (3.43)$$

For a central Griffith crack of length $2a$ lying on the 1 axis of a set of cartesian axes, and subject to a remote uniaxial tension in the 2 direction, the stresses directly ahead of the crack ($\theta = 0$) can be expressed in the form:

$$\frac{\sigma_{22}}{\sigma} = \frac{1}{\sqrt{2}}\left(\frac{r}{a}\right)^{-1/2} - \frac{5}{32\sqrt{2}}\left(\frac{r}{a}\right)^{3/2} + \frac{7}{128\sqrt{2}}\left(\frac{r}{a}\right)^{5/2} + [0]\left(\frac{r}{a}\right)^{7/2} \qquad (3.44a)$$

$$\frac{\sigma_{11}}{\sigma} = \frac{\sigma_{22}}{\sigma} - 1 \qquad (3.44b)$$

The first term in all such expansions is singular at the crack tip, where the remaining terms are either finite or zero. Elastic fracture mechanics is thus based on the premise that fracture processes which occur close to the crack tip are only determined by the first term in the expansion, allowing the asymptotic elastic stress field of a Griffith crack to be expressed in the particular form

$$\sigma_{ij} = \frac{\sigma\sqrt{\pi a}}{\sqrt{2\pi r}} \qquad (3.45)$$

Such results are however most usefully generalized through the stress intensity factor K.

$$\sigma_{ij} = \frac{K}{\sqrt{2\pi r}} \tag{3.46}$$

3.5.2 Small Scale Yielding

The application of linear elastic fracture mechanics is subject to size limitations (ASTM (1983)) intended to ensure that plasticity is restricted to a local perturbation of the elastic field. Within this context, Larsson and Carlsson (1975) have demonstrated that the second term in the Williams series has a significant effect on the shape and size of the plastic zone which develops at the crack tip. In the notation of Rice (1974), the second term in the expansion is denoted the T stress and can be regarded as a stress parallel to the crack flanks.

$$\begin{bmatrix} \sigma_{11} & \sigma_{12} \\ \sigma_{21} & \sigma_{22} \end{bmatrix} = \frac{K}{\sqrt{2\pi r}} \begin{bmatrix} f_{11}(\theta) & f_{12}(\theta) \\ f_{21}(\theta) & f_{22}(\theta) \end{bmatrix} + \begin{bmatrix} T & 0 \\ 0 & 0 \end{bmatrix} \tag{3.47}$$

For the example of the Griffith crack, T is simply equal to $-\sigma$. For a bi-axially loaded crack T is zero. The T stress has now been tabulated for a wide range of geometries, in which the results are either expressed in terms of a stress concentration factor (T/σ) or as a biaxiality parameter B following Levers and Radon (1983).

$$B = \frac{T\sqrt{\pi a}}{K} \tag{3.48}$$

Some results for some important through crack geometries have been given by Sham(1991), Levers and Radon (1983), and Kfouri (1986), while Wang and Parks (1991) have given results for surface cracked panels.

The first evidence that the constraint of the crack tip field evolves from small scale yielding could have been inferred, with hindsight, from the work of Larsson and Carlsson (1973). The shapes of the plastic zones, illustrated in Fig. 3.10, show that compressive T stresses both enlarge the maximum radius of the plastic zone and cause the plastic lobes to swing forward. In contrast tensile, or positive, T stresses cause the plastic zone to decrease in size and rotate backwards. With the benefit of hindsight it now seems clear that if non-singular stresses affect the size and shape of the plastic zone then they are likely to affect the local stresses within the plastic zone.

Crack tip plasticity has been widely discussed, for non-hardening materials in terms of plane strain slip line fields. Rice has shown that for a non-hardening material under plane strain conditions, the crack tip stress state must satisfy either

$$\frac{\partial \sigma_{rr}}{\partial \theta} = 0 \text{ or } \frac{\partial(\sigma_{rr} + \sigma_{\theta\theta})}{\partial \theta} \tag{3.49}$$

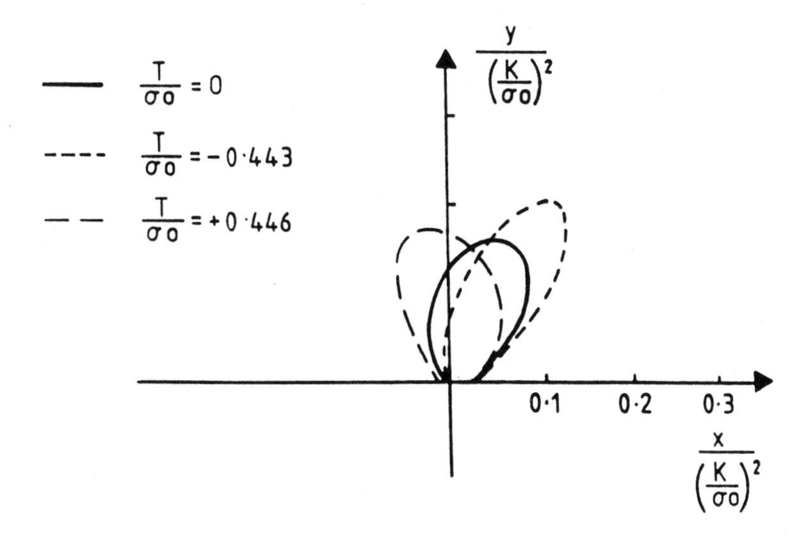

FIGURE 3.10. The effect of the T stress on the plastic zone shape in small scale yielding, after Du and Hancock (1991).

In sectors in which the first condition applies the equilibrium equations combined with the yield criterion leads to stress states of the form

$$\sigma_{r\theta} = + - k \text{ and } \sigma_{r\theta} = \sigma_{\theta\theta} = \sigma_{zz} \tag{3.50}$$

This corresponds to a centered fan, such as that shown in region II of Figure 3.11. In regions where the second condition applies the crack tip stress states correspond to a homogeneous deformation field in which the cartesian stresses are independent of angular co-ordinate θ. Within the angular span of the plastic zone at the crack tip the slip line field can only consist of assemblies of centered fans and constant stress regions, although the general case also allows the incorporation of elastic wedges. However assuming that the full angular span is at yield and the stresses are continuous, there is only one possible solution, which is the Prandtl field shown in Figure 3.11.

Rice (1968a,b) has advanced forceful arguments for the validity of this slip line field, as a representation of the limiting stress state at the crack tip in small scale yielding. The slip line field, is consistent with the dominant singularity solutions of Hutchinson (1968a,b) and Rice and Rosengren (1968) in the non-hardening limit of non-linear behavior. It's relevance to small scale yielding has been further reinforced by the numerical solutions of Levy et al. (1971) and Rice and Tracey (1973) which show the essential features of the Prandtl field.

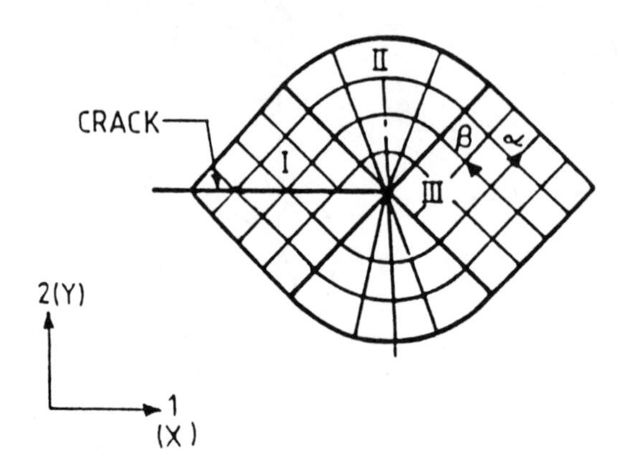

FIGURE 3.11. The Prandtl slip line field.

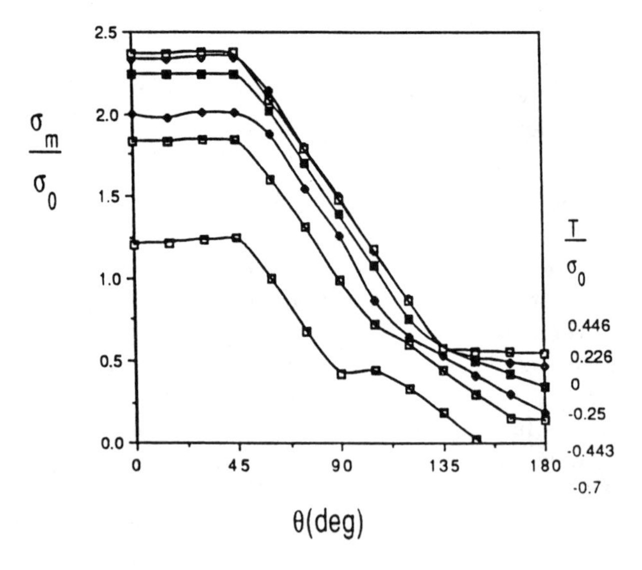

FIGURE 3.12. Angular variation of the mean stress around a crack tip as a function of T/σ_0 after Du and Hancock (1991).

Like the HRR fields, of which it is a limiting example, the Prandtl field is based on the assumption that plasticity completely surrounds the crack tip. On this basis the stresses can be solved starting from the traction free crack surface region, denoted I in Fig. 3.11. In this region the yield criterion and the free surface require that the stress field is a homogeneous tensile (or possibly compression) field parallel to the crack flanks.

$$\sigma_{11} = 2k \tag{3.51a}$$

$$\sigma_{22} = 0 \tag{3.51b}$$

$$\sigma_{33} = \sigma_m = k \tag{3.51c}$$

Here k is the yield stress in shear, and σ_{33} is the out of plane stress, which is equal to the mean stress, σ_m.

It is convenient to work in cylindrical co-ordinates (r, θ) centered at the crack tip, when the stress field in region I can be written as:

$$\sigma_{rr} = k(1 + \cos 2\theta) \tag{3.52a}$$

$$\sigma_{\theta\theta} = k(1 - \cos 2\theta) \tag{3.52b}$$

$$\sigma_{r\theta} = k \sin(2\theta) \tag{3.52c}$$

$$\sigma_m = k. \tag{3.52d}$$

The stresses in the rest of the field can be deduced from the Hencky equations (Hill 1952), which express the equilibrium requirements in terms of the rotation of the slip lines. The straight slip lines in region I thus imply a homogeneous stress state in the triangle. Following a slip line into the centered fan, denoted II in Figure 3.11, gives the stress state in this region

$$\sigma_{\theta\theta} = \sigma_{rr} = \sigma_{zz} = \sigma_m = k(1 + 3\pi/2 - 2\theta) \tag{3.53a}$$

$$\sigma_{r\theta} = k \tag{3.53b}$$

Finally in the diamond ahead of the crack, denoted III in Figure 3.11, the stress system consists of the simple stress state

$$\sigma_{\theta\theta} = k(\pi + 1 + \cos 2\theta) \tag{3.54a}$$

$$\sigma_{rr} = k(\pi + 1 - \cos 2\theta) \tag{3.54b}$$

$$\sigma_{zz} = \sigma_m = k(1 + \pi) \tag{3.54c}$$

$$\sigma_{r\theta} = k \sin 2\theta \tag{3.54d}$$

The stresses, which are illustrated in Figure 3.11, are independent of the radial distance r in accord with the limiting form of the HRR field for a non-hardening solid.

The stress field in small scale yielding may be examined by a technique used by Tracey (1976) and Rice and Tracey (1973) in which the elastic stress or displacement distributions obtained from the K field are applied as boundary conditions to a region surrounding the crack tip. Such models have become known, perhaps erroneously as boundary layer formulations. Analyses which employ the first two terms K and T of the Williams expansion are known as modified boundary layer formulations. Such modified boundary layer formulations have been used by Du and Hancock (1991) to determine the finite crack tip stresses for non-hardening plane strain deformation. The angular distribution of the stress around the tip shown in Figure 3.12 was used to construct a slip line field for small strain deformation, recalling the necessary elements of such a field.

Directly ahead of the crack, and extending to an angular span of $\pi/4$, is a constant stress region which corresponds to the diamond shaped region denoted III in Figure 3.13. This is contiguous with a centered fan in which the mean stress decreases linearly with angle and the shear stress $\sigma_{r\theta}$ has a constant value equal to k. The extent of the fan, as measured by the angular range in which in $\sigma_{r\theta}$ equals k, depends upon the T stress. For $T = 0$ the fan extends close to 130°. For greater angles, the mean stress falls, but the finite element calculations indicate that the remaining wedge is elastic. On this basis, the form of the slip line field for $T = 0$, constructed from the numerical data is given in Figure 3.13. In this figure the angular span of the fan has been estimated from the region in which $\sigma_{r\theta}$ equals k. However it was also noted that the stresses within the plastic zone were tensile. In order for all the principal stresses to be tensile at the crack tip the mean stress must be greater than k. Thus starting from the calculated mean stress directly ahead of the crack, and using the Hencky equations, it was possible to calculate orientation at which this condition was met, within an assumed plastic zone. This orientation is indicated in Figure 3.13 by a broken line, and is similar to the angular span of the centered fan until the most compressive T stress levels.

Tensile T stresses with $T \geq 0.433\sigma_0$ produced deformation corresponding to the full Prandtl field. It is worth noting that the Prandtl field features a tensile stress on the crack flanks within the plastic zone, and it is perhaps not surprising that this should be associated with a tensile T stress in the surrounding elastic field. As plasticity surrounds the crack tip there is only one possible form of solution for the problem, given that incompressible plastic deformation dominates. The Prandtl field and the non-hardening limit of the HRR field occur in the presence of tensile T stresses. Because plasticity envelops the tip, any further increase in the T stress does not change the stresses in the plastic field. This explains the observation of Hancock and Betegón (1991) that J dominance is characterized by positive T stresses, but that any further increase in the T stress does not change the local field once plasticity is established on the crack flanks. The HRR field thus emerges as a special crack tip field which evolves from fields with positive T stresses which cause yielding on the crack flanks.

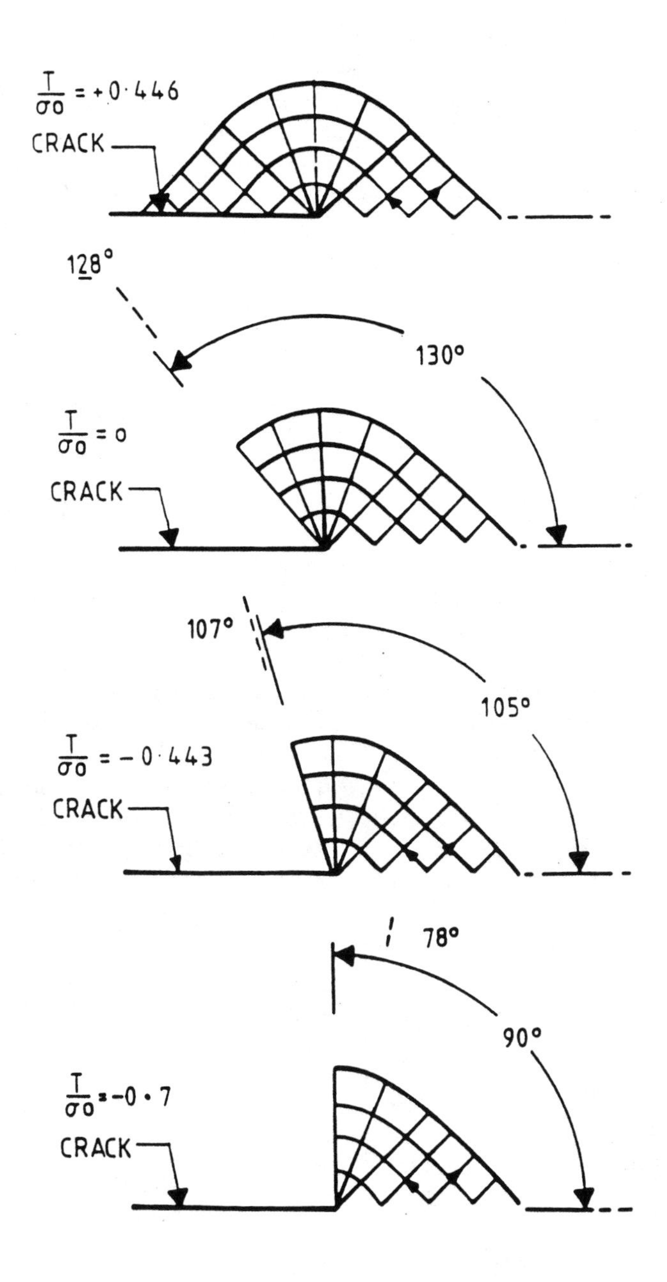

FIGURE 3.13. Effect of T/σ_0 on the crack tip slip line field in small scale yielding, after Du and Hancock (1991).

In contrast, compressive T stresses significantly change the crack tip deformation. The effect of compressive T stresses is to decrease the angular span of the centered fan, giving an incomplete Prandtl field in which plasticity does not envelop the tip. As the fan is the region of strain concentration in the small strain solution, this corresponds to the forward rotation of the lobes of the plastic zone. The diamond ahead of the tip maintains a uniform stress state in which the hydrostatic stress is less than the fully constrained Prandtl value. For moderate compressive T stresses the data is simply expressed by:

$$(\sigma_m/\sigma_0)_{(r,T)} = (\sigma_m/\sigma_0)_{(r,T=0)} + T/\sigma_0 \quad (-0.5 \leq T/\sigma_0 \leq 0). \tag{3.55}$$

Over the complete range for which data is available the results are expressed by

$$(\sigma_m/\sigma_0)_{(r,T)} = (\sigma_m/\sigma_0)_{(r,T=0)} + 0.712T/\sigma_0 - 1.1(T/\sigma_0)^2 \quad (T/\sigma_0 \leq 0). \tag{3.56}$$

Similar expressions have been given by Wang (1991a,b) covering a wider range of T stresses, but again using the $T = 0$, or small scale yielding field, as the reference.

Bilby et al. (1984) and Cardew et al. (1987), and O'Dowd and Shih (1991a,b) have examined the effect of the T stress on the large geometry change crack tip field. The results presented in Figure 3.14 are those of Hancock Reuter and Parks (1991) for a material following a Ramberg-Osgood relation with a hardening exponent, $n = 13$. Assuming that the crack blunts smoothly into a circular arc, the local deformation can be represented by the introduction of a log-spiral slip-line field detail at the crack tip. The extent of the log spiral can be seen by matching the stress state at the end of the spiral with that of the homogeneous region III. The mean stress in the log-spiral is given by:

$$\sigma_m = k(1 + \frac{2}{\sqrt{3}} \ln[1 + \frac{2x}{\delta}]) \tag{3.57}$$

For the full Prandtl field, the log-spiral thus extends to 1.96δ. On the basis that the mean stress in the diamond is simply reduced by T the extent of the log-spiral directly ahead of the surface of the blunt crack tip is given by:

$$x = \frac{\delta}{2}(\exp[\frac{\pi}{2} + \frac{T}{2k}] - 1) \quad \left(-0.5 \leq \frac{T}{\sigma_0} \leq 0\right). \tag{3.58}$$

The extent of the log-spiral region is thus expected to decrease with compressive T stresses, and this effect is confirmed, albeit for a weakly strain hardening material in Figure 3.14.

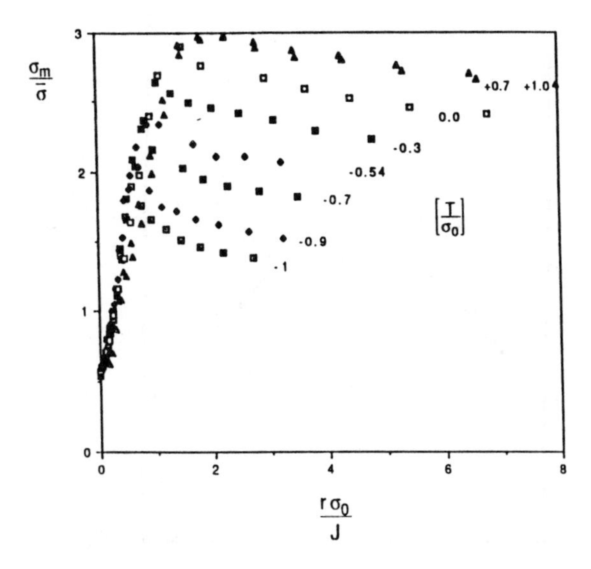

FIGURE 3.14. Crack tip constraint in a large strain modified boundary layer formulation as a function of the T stress, after Hancock, Reuter and Parks (1991).

3.5.3 Fully Plastic Flow Fields

The restrictions of linear elastic fracture mechanics are relaxed by non-linear elastic-plastic fracture mechanics. Recent developments in this area have been reviewed by Parks (1991a,b). As in the case of linear elastic deformation the crack tip field can be expressed as an asymptotic series. Within the framework of small strain deformation Li and Wang (1986) and Sharma and Aravas (1991) have sought asymptotic expansions of the form:

$$\frac{\sigma_{ij}(r,\theta,n)}{\sigma_0} = K_s r^s f_{ij}(\theta,n) + K_t r^t g_{ij}(\theta,n) + \ldots \tag{3.59}$$

Here s is the radial dependence of the first term and t the dependence of the second term in the expansion. The first term is identified with the HRR solution as the leading singular term.

$$\frac{\sigma_{ij}(r,\theta,n)}{\sigma_0} = \left(\frac{J}{e_0\sigma_0 I_n r}\right)^{1/(n+1)} \tilde{\sigma}_{ij}(\theta,n) \tag{3.60a}$$

$$s = 1/(n+1) \tag{3.60b}$$

I_n and $\tilde{\sigma}_{ij}(\theta,n)$ are tabulated functions of the strain hardening exponent n, and where appropriate the angular co-ordinate θ. The strength of the singular field is characterized by the J integral, introduced by Rice (1967), which provides the basis for the characterization of crack tip deformation.

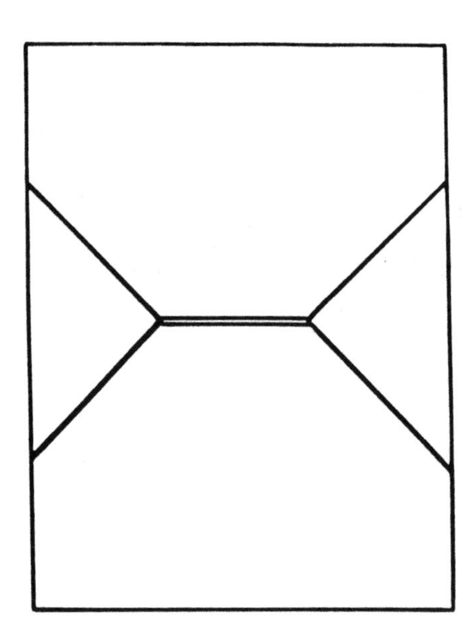

FIGURE 3.15. The slip line field for a center cracked panel.

Sharma and Aravas (1991) give numerical calculated values for, t, the exponent of the second term, although the amplitude of the second term remains undetermined from global analysis. For weakly hardening materials t is of the order of 1/20, which allows contact to be made with the observation of Betegón and Hancock (1991) that the second term could be regarded to first approximation as being independent of r.

Single parameter elastic-plastic fracture mechanics is thus based on the assumption that fracture processes occur sufficiently close to the crack tip that they can be characterized by J through the HRR field. This requires that the size of the zone described by the HRR field encompasses the fracture process zone and is independent of specimen or component geometry.

However in a key development, McClintock (1968) pointed out that in the absence of strain hardening the fully plastic deformation field is not unique, and that the appropriate slip line field depends both on the geometry of the body and the way that it is loaded. As an example the slip line field for a center cracked panel is shown in Figure 3.15. This may be compared with the field for a deeply cracked bend bar given by Green (1953) shown in Figure 3.16. For shallow cracked bars plasticity initially extends to the cracked face in a form discussed by qualitatively by Green (1953) and quantitatively by Ewing (1968). Although Ewing did not explicitly consider cracked geometries, the critical depth for plasticity to be confined to the ligament for a notch with an included angle of 3.21° was $(a/W = 0.3)$.

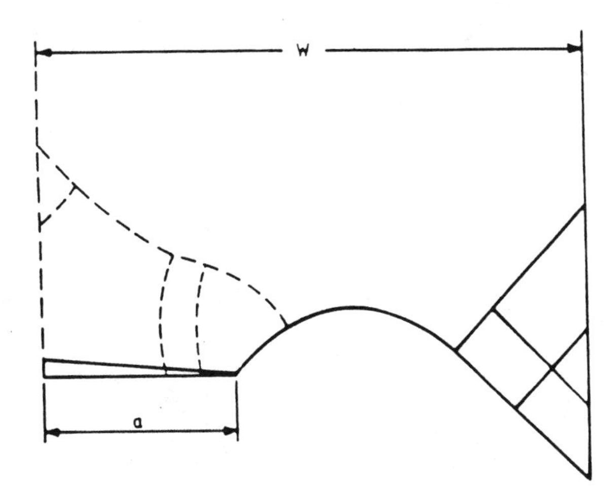

FIGURE 3.16. The slip line field for a deeply cracked bar, proposed by Green (1953). The extension suggested by Ewing(1968) for shallow cracks is indicated with broken lines.

The transition from deeply cracked to shallow cracked behavior has been examined computationally by Al-Ani and Hancock (1991) for a weakly strain hardening material, and in pure bending plasticity was limited to the ligament for ($a/w \geq 0.3$) in accord with Ewing's calculations. In force loaded tension similar behavior occurs and Al-Ani and Hancock (1991) found plasticity to be limited to the ligament of edge cracked bars for ($a/W \geq 0.55$). A similar transition in flow fields is found in double edge cracked bars. For ($a/W \geq 0.9$) plasticity is restricted to the ligament and results in the well known Prandtl field shown in Figure 3.17(a). However at shallower crack depth plasticity extends out to the shoulder of the specimen as discussed by Ewing and Hill (1967) in accord with the slip line field shown in Figure 3.17(b).

Although fully plastic flow fields are clearly not unique, Hancock and co-workers (Betegón and Hancock (1991), Du and Hancock (1991), and Al-Ani and Hancock (1991)) have argued that the lack of uniqueness is not associated with the sudden development of the fully plastic flow field but rather evolves gradually from small scale yielding where constraint is controlled by T.

Evidence in support of this suggestion arises from the relation between transitions the form of the fully plastic slip line field and transitions in the sign of the T stress. Deep crack behavior is found when plasticity is limited to the uncracked ligament. The transition from shallow to deeply cracked

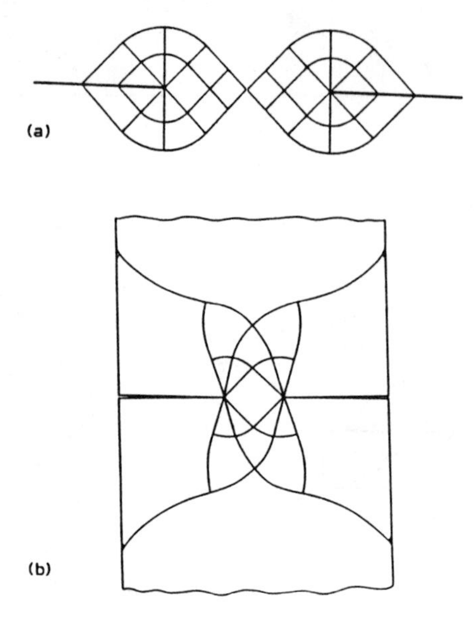

FIGURE 3.17. The Prandtl slip line field for a deeply double edge cracked bar, and the slip line field for a shallow edge cracked bar proposed by Ewing and Hill (1967).

behavior occurs for edge cracked bars at $a/W = 0.3$ in bending and 0.55 in tension, and matches the change in T stress from negative to positive in both cases. Shallow double edge cracked bars show unconstrained flow fields in which plastic deformation is not limited to the ligament, and in this range of geometries T is negative. For deeper cracks T goes to zero as plasticity is limited to the ligament in a constrained flow field. The center cracked panel, however exhibits strongly negative T values for all (a/W) ratios, and shows similar unconstrained flow fields for all values of (a/W). All the specimens thus exhibit the feature in which unconstrained plastic flow fields develop from elastic fields with negative or compressive T stresses.

Geometries which evolve from small scale yielding to an unconstrained flow field are thus all identified by negative T stresses. The question that remains in the context single parameter fracture mechanics is the specimen size requirements such that the crack tip deformation retains the character of small scale yielding. In a phrase coined by McMeeking and Parks (1979), these are the conditions for "J dominance". The choice of a J dominance criterion is somewhat subjective, particularly as there as a gradual, rather than abrupt change in the character of the fields. A dimensional argument shows that J dominance is maintained under conditions which depend on the size of a critical dimension, c, which can be identified with the ligament or the crack length.

$$c \geq \frac{\mu J}{\sigma_0} \tag{3.61}$$

For deeply cracked tensile geometries such as center cracked panels McMeeking and Parks (1979) and Shih and German (1981) suggest μ should be taken as 200, with the motive that this broadly corresponds to the first development of plasticity across the ligament. Similarly when the crack length is the controlling dimension Al-Ani and Hancock (1991) pointed out that for shallow edge cracked bars in tension or bending, J dominance was lost before $a\sigma_0/200$, which again approximately corresponds to the condition in which plasticity breaks through to the cracked face of short edge cracked bars.

However neither result is conservative. Given that constraint is lost as the field develops from small scale yielding into full plasticity, Betegón and Hancock (1991) suggested that the J dominance criteria for could be unified by a criterion of the form:

$$T \leq -0.2\sigma_0 \tag{3.62}$$

This was based on the criterion that the stress field should be within 90% of the HRR field at a distance $r = 2(J/\sigma_0)$. On this basis the size limitations for J dominated flow were given by Betegón and Hancock (1991). These requirements are exceptionally severe and lead to the conclusion that central cracked panels lose J dominance within the requirements of L.E.F.M. as codified in both the British and American standards (A.S.T.M. (1970)). When comparison is made with unmodified boundary layer formulations, as representations of the small scale yielding field, the results are less severe, but exhibit the same trends as shown in Figure 3.18. Specimens with a negative T stress lose J dominance as they develop into full plasticity. Conversely specimens with positive T stresses maintain J dominance until the crack tip opening becomes a significant fraction of the ligament when J dominance is lost at $25J/\sigma_0$, due to the compressive part of the bending field intruding into the crack tip field. In this context it must be emphasized that T stress based criteria have never been intended to describe loss of constraint arising from the compressive part bending field intruding into the crack tip field. Similarly the criteria are all based on an assumption of ductile failure occurring in the finite strain region ahead of a blunting crack tip, and are not intended to represent J dominance conditions for cleavage failure.

3.5.4 TWO PARAMETER CONSTRAINT BASED FRACTURE MECHANICS

Boundary layer formulations provide a formal framework for examining plasticity in small scale yielding. However Betegón and Hancock (1991)

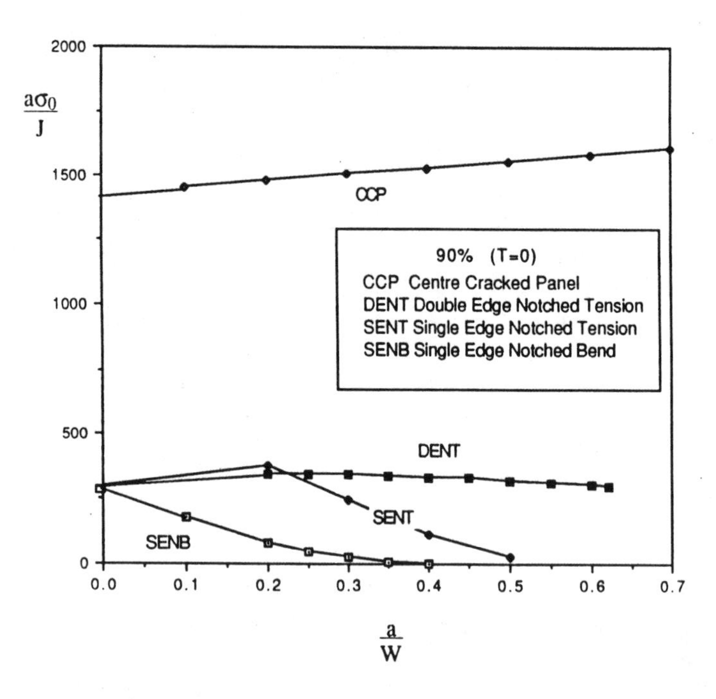

FIGURE 3.18. Limits of J dominance.

and Al-Ani and Hancock (1991) attempted to correlate modified boundary layer formulations a diverse range of geometries into full plasticity .

Although the T stress is an elastic parameter, the correlation was made by identifying T with the applied load or the elastic component of J. On this basis it was possible to correlate crack tip deformation for edge cracked bars in tension and bending from (a/W) 0.03 to 0.9 , as well as center cracked panels and double edge cracked bars.

The ability of an elastic parameter to correlate fully plastic flow fields of such a diverse range of geometries can be explained qualitatively. At infinitesimally small loads, plasticity is only a minor local perturbation of the leading term of the elastic field, allowing crack tip deformation to be represented by single parameter characterization in a boundary layer formulation with the K field displacements imposed on the boundaries. As the load increases within contained yielding, the outer elastic field can be characterized by K and T, both of which are rigorously defined. Within the plastic zone the crack tip field now evolves in a way that is correctly represented by a modified boundary layer formulation with both K and T as boundary conditions. The initial evolution of the crack tip field is thus rigorously determined by T. Geometries with negative T stresses start to lose crack tip constraint, while those which have positive T stresses maintain the character of the small scale yielding field. For simplicity it is appropriate

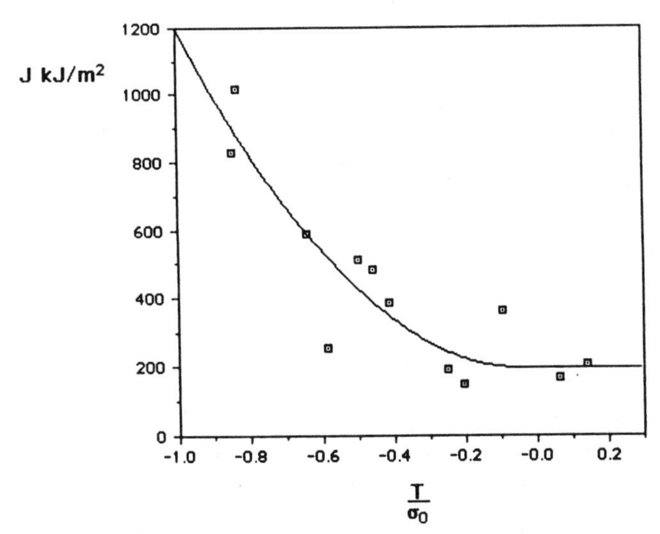

FIGURE 3.19. Toughness of edge cracked bend bars after Betegón (1990), and Betegón and Hancock (1990) as a function of T.

to restrict discussion to small geometry change perfect plasticity when the appropriately non-dimensionalized crack tip field reaches a steady state, independent of deformation. At limit load, the value of T calculated from the applied load or equivalently the elastic component of J, also remains constant. When T, as calculated from the limit load, is used to make contact with the modified boundary layer formulations, the predicted stress field also reaches a steady state.

Although the accuracy of this approach has been questioned by O'Dowd and Shih (1991a,b), the simplicity and robustness of the technique as an engineering tool seems beyond doubt. As regards the accuracy and usefulness of the methodology, the author is somewhat biased. However the accuracy to which it is necessary to characterize crack tip fields is essentially a question to be resolved by experiment rather than computation. As an example, the effect of (a/W) ratio on the toughness of a brittle experimental steel tested by Betegón and Hancock (1990) is shown in Figure 3.19. For the deeply cracked geometries $(a/W \geq 0.3)$ the toughness is independent of geometry as failure occurs at the same level of constraint in basically similar flow fields under flow fields which evolve from small scale yielding fields with a positive T stress. However, for $(a/W \leq 0.3)$ plasticity extends initially to the cracked face and constraint is lost resulting in flow fields which evolve from compressive T stress fields resulting in an increased toughness for shallow cracks.

Hancock, Reuter and Parks (1991) tested a wide range of through cracked geometries of an A710 steel. The discussion of the results was couched in terms of the initiation of crack extension and the subsequent resistance to tearing as parameterized by T. In this, and other tough materials the start of crack tip extension is capable of a number of rather arbitrary definitions. Firstly it is possible to extrapolate resistance curves back to the point of zero crack extension. Alternatively it is possible to define a critical value of J or CTOD at a small amount of extension, for example the current ASTM standard (1983) gives J at an extension of $200\mu m$. Finally it is also possible to give the toughness at some multiple of the CTOD or equivalently J/σ_0, using the intersection of a resistance curve with a blunting line.

From a physical point of view crack extension is usually regarded as a discrete process in which voids formed at second phase particles grow and coalescence with the crack tip. From this viewpoint it is problematic to think of crack extension over micro-structurally insignificant distances, and indeed such measurements are rather subjective. There is of course a sense in which the random distribution of particles along the crack front allows crack extension at sections in which the nearest void happens to be statistically closer to the tip than average, however crack extension is rarely measured in the corresponding statistical manner. In the present context it seems less than practical to pursue a statistical definition of initiation as a toughness at $\Delta a = 0$. These arguments detract from the interpretation of resistance curves at points of zero extension. Nevertheless it is clearly a practical procedural definition of initiation, and as such has been shown in Fig. 3.20. The salient point is that on this definition of initiation A710 shows little effect of crack tip constraint on toughness. Indeed, paradoxically the toughness of the unconstrained center cracked panel and surface cracked tension panel, $(a/2c)= 0.5$, is significantly lower than that of the fully constrained deeply cracked bend bars and CTS specimens. This is in no small part to the rather subjective nature of crack extension measurements over small distances in geometries with very low constraint.

A simpler and preferable procedure is to follow the current ASTM standard and define toughness at a small but significant amount of crack extension, such as $200\mu m$, as given in Fig. 3.20. In this case the results show a significant effect of constraint on toughness. Thus the J and CTOD values for center cracked panels are approximately 4 times greater than that of highly constrained deeply cracked bend bars and CTS specimens. The effect is even more marked if extension is compared at $\Delta a = 400\mu m$ as shown in Figure 3.20, however caution must be used here as the extension now starts to alter the geometry and associated constraint, particularly for short cracks.

The important point is that the toughness of all the geometries is correctly ordered by the T stress. Geometries with positive T stresses show a basically geometry independent toughness as the crack tip are capable of single parameter characterization by J or CTOD. Similarly geometries with negative T stresses show a geometry dependent toughness in which the crack tip constraint and associated toughness are correctly parameter-

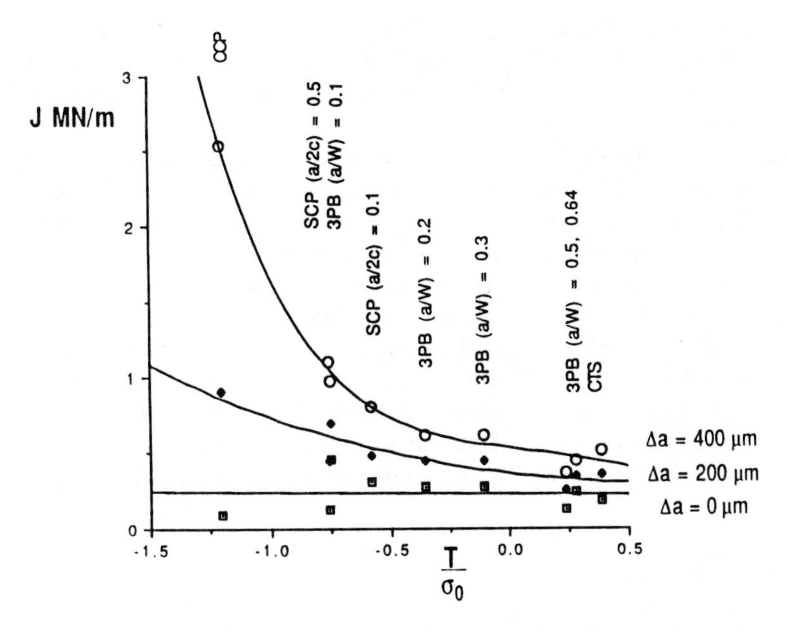

FIGURE 3.20. The toughness of a range of through and part through crack geometries parameterized by T, after Hancock Reuter and Parks (1991).

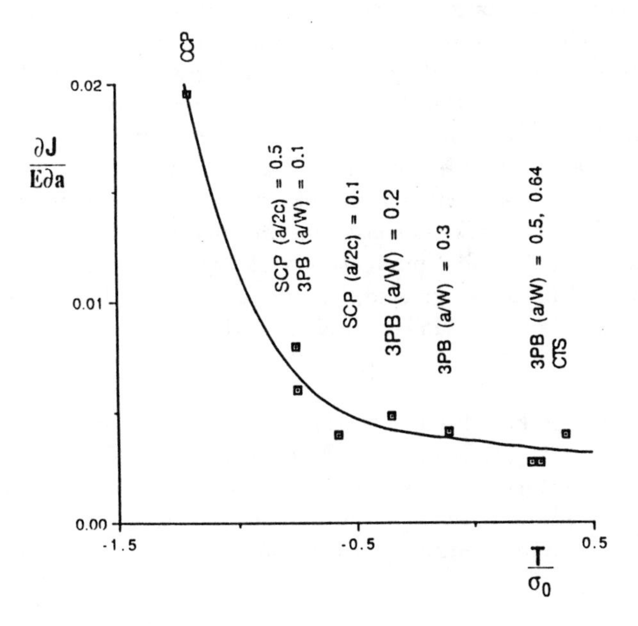

FIGURE 3.21. The resistance to rearing of a range of through and part through crack geometries parameterized by T, after Hancock Reuter and Parks (1991).

ized by T. The reason for the strong effect of the amount of crack extension on the constraint sensitivity of toughness is due to the effect of constraint on the slope of the resistance curves. For example the slope of the $\delta - \Delta a$ plot of the center cracked panels is so high ($\frac{\partial \delta}{\partial a} = 4.2$) it exceeds that of the usual blunting line construction ($\frac{\partial \delta}{\partial a} = 1$) and prevents its use as a practical definition of initiation In comparison the deeply cracked bend bars and CTS specimens exhibit slopes ($\frac{\partial \delta}{\partial a}$) ≈ 0.75. Similarly the T stress correctly orders the resistance to crack tip tearing. Geometries with negative T stresses show an enhanced resistance to tearing which is correctly parameterized by T. As a simple specific illustration it is appropriate to compare the behavior of surface cracked panel ($a/2c = 0.1$) tested in tension and the edge cracked bar ($a/W = 0.2$) tested in three point bending. Both configurations have similar T stresses and exhibit closely similar behavior in terms of initiation and toughness. It would be difficult to have made such a connection without the use of a constraint based methodology.

3.6 Conclusion

The recognition that failure criteria are dependent on the state of stress, originates largely from the pioneering work Professor F. A. McClintock at MIT. The experimental techniques that he originated demonstrated ways to measure the ductility of materials in multi-axial states of stress. These experimental developments were paralleled by an understanding the physics of failure processes such as hole growth and coalescence, which also provide mechanisms of crack extension. McClintock's observation of the lack of uniqueness of non-hardening crack tip fields, demonstrated the significance of constraint in crack extension processes. In an attempt to tackle this problem a branch of fracture mechanics has developed which tries to recognize the effects of crack tip constraint on toughness. It has not been difficult for researchers to find productive work trying to exploit concepts that Professor McClintock originated, but the real advance lies in the originality and timeliness of the fundamental ideas themselves.

Acknowledgments: Acknowledgements are due to Alexander Mackenzie, who was a student of Professor F. A. McClintock at MIT. Dr. Mackenzie, in his turn, encouraged colleagues and students at the University of Glasgow in Scotland. He brought to the study of the deformation and fracture an enthusiasm and attention to detail which his colleagues struggle to emulate.

REFERENCES

Al-Ani, A. and Hancock, J. W. (1991). J. Mech. Phys Solids 1:23–43.

Amizigo, J. C. Some Mathematical Problems of Elastic-Plastic Crack Growth (editor, Burridge R.) Fracture Mechanics 12:125.

Anand, L. and Spitzig. (1980) J. Mech. Phys. Solids, 28:113.

Argon, A.S., Im and Safoglu (1975) Metallurgical Transactions, 6A:825–837.

Asaro, R. J and Rice, J. R. (1977). J. Mech. Phys. Solids, 25:309–338.

A.S.T.M. (1983). Standard Methods for Plane Strain Toughness Testing of Metallic Materials, E399, 83:487–511.

A.S.T.M. (1987). A Standard Method for J1c, a Measure of Fracture Toughness, E813-88, 29:968–990.

Beremin, F. M. (1983). Met Trans, A, 24A:2272–2287.

Berg, C. A. Proc 4th U.S. National Congress of Applied Mechanics, Berkeley, CA editor, Rosenberg, R. M., ASME.

Berg, C. A. (1970). In Kanninen, M., Adler, W. F., Rosenfield, A. R. and Jaffee, R.I., editors, *Inelastic Behavior of Solids*, pages 171-210, McGraw-Hill, New York.

Betegón, C. PhD Thesis, Department of Construction, University of Oviedo, Spain.

Betegón, C. and Hancock, J. W. Fracture Behavior and the Design of Materials and Structures", ECF 8, 2:999–1002, Firrao, D., editor, EMAS UK.

Bilby, B. A., Cardew, G. E., Goldthorpe, M. R. and Howard, I. C. (1987). The Stability of Cracks in Tough Materials, Department of Mechanical Engineering Report, University of Sheffield.

Bishop, J.F.W. and Hill R. (1951). Phil. Mag. 42:414–427.

Bridgman, P. W. (1952). *Studies in Large Plastic Flow and Fracture*, McGraw-Hill, New York.

Brown, L. M. and Embury, J. D. (1973). Inst. Met., 1:164.

Budiansky, B., Hutchinson, J.W. and Slutsky, S. (1981). In Hopkins, H. G. and Sewell, M. J., editors, *The Mechanics of Solids*, the Rodney Hill 60th Anniversary Volume, Pergamon Press Oxford, pages 13–45.

Burke, M. A. and Nix, W. D. (1979). Int. J. Solids and Structures, 15:379–393.

Cardew, G. E., Goldthorpe, M. D., Howard, I.C. and Kfouri A. P. (1984). in *Fundamentals of Deformation and Fracture*, (Eshelby Memorial Symposium), Cambridge University Press, 456.

Clausing, D. P. (1970). Int. J. Fracture Mech., 6(1):71.

Cowper and Onat (1962). Proc 4th US National Congress on Applied Mechanics, 2:1023, ASME, New York.

Drucker, D. C. (1960). J. Mecan., 3:253–249.

Edelson, B. I. and Baldwin, W. M. Jr. (1962). Trans. QASM, 55:230.

Eshelby, J. D. (1957). Proc. Roy. Soc. A241:376–396.

Ewing D. J. F. (1968). J. Mech. Phys. Solids, 16:305.

Ewing, D. J. F. and Hill, R. (1967). J. Mech. Phys. Solids, 15:115.

Goodier, J. N. (1933). ASME Applied Mechanics Magazine, 55:39–44.

Goods, S. H. and Brown, L. M. (1979). Acta Metallurgica, 27:1–15.

Green, A. P. (1953). Quart. J. Mech. Appl. Math., 6:223.

Gurson, A. L. (1977). Trans. ASME Journal of Engineering Materials Technology, 99:2–15.

Haigh (1923). Thermodynamic Theory of Fatigue and Hysterisis in Metals, Report of the Brit. Assoc., page 358.

Hancock, J. W. and Brown, D. K. (1983). J. Mech. Phys. Solids, 31:1.

Hancock, J. W. and Cowling, M. J. (1980). Metal Science, 293.

Hancock, J. W. and Mackenzie, A. C. (1977). J. Mech. Phys. Solids, 14:147–169.

Hancock, J. W. , Reuter, W. A. and Parks, D. M. (1991). ASTM Symposium on Constraint Effects in Fracture, Indianapolis.

Hill, R. (1950). *The Mathematical Theory of Plasticity*, Oxford, (Clarendon Press).

Huang, Y., Hutchinson, J. W. and Tvergaard, V. (1991). J. Mech. Phys. Solids, 39:223–241.

Hutchinson, J. W. (1968a). J. Mech. Phys.Solids, 16:13–32.

Hutchinson, J. W. (1968b). J. Mech. Phys. Solids, 16:337–348.

Hutchinson, J. W. (1989). *Micro-Mechanics of Damage in Deformation and Fracture*, Technical University of Denmark.

Hutchinson, J. W. (1981). Int. J. Solids and Structures, 17:451–470.

Johnson, W., Sowerby, R. and Venter, R. D (1982). *Plane Strain Slip Line Fields for Metal Deformation Processes*, Pergamon Press, Oxford.

Kirkcaldy, D. (1860). *Results of a Comparative Inquiry into the Comparative Strengths and other Properties of Wrought Iron and Steel*, Proc. Scot. Shipbuilders Assoc., Appendix, page 74.

Kfouri, A. P. (1986). Int. J. Fracture, 30:301.

Levers, P. S. and Radon, J. C. (1983). Int. J. Fracture, 19:311–325.

Lode, W. (1926). Z. Physik, 36:913–939.

Ludwik, P. and Scheu, R. Stahl und Eisen, 43:999–1001.

Larsson, S. G. and Carlsson, A. J. (1973). J. Mech. Phys. Solids, 21:263–277.

Levy, N., Marcal, P. V., Ostergren, W. J. and Rice, J. R. (1971). Int. J. Fracture Mech. 7:143.

Li, Y. and Wang, Z. (1986). Scientia Sinica, A29:942.

Li, F. Z., Shih, C. F. and Needleman, A. (1985). Eng. Fract. Mech., 21:405–421.

Mackenzie, A. C., Hancock, J. W. and Brown, D. K. (1977). J. Engineering Fracture Mechanics, 9:167–188

McClintock, F. A. (1968). J. Appl. Mech., 35:362–371

McClintock, F. A. (1967). in *Ductility*, 255–278, A.S.M., Ohio, U. S. A.

McClintock, F. A. (1969). Plasticity Aspects of Fracture, in *Fracture, an Advanced Treatise*, Liebowitz, H., editor, Academic Press, New York, pages 47–225.

McClintock, F. A. and Rhee, S. S. (1962). Proc. 4th US National Congress on Applied Mechanics, 2:1002.

McClintock, F. A., Kaplan, S. A. and Berg, C. A. (1966). Int. J. Fract. Mech., 29(4):614.

McMeeking, R. M. and Parks, D. M. (1979). in *Elastic Plastic Fracture*, ASTM STP 668, Landes, J. D. et al., editors, ASTM Philadelphia, 175–194.

McMeeking, R. M. and Rice, J. R. (1975). Int. J. Solids and Structures 11, 601–616.

Nagpal, V., McClintock, F. A., Berg, C. A. and Subudhi, M. (1973). In Sawczuk, A., editor, *Foundations of Plasticity*, Noordhooo, Leyden, page 365.

Neimark, J. E. (1959). *The Initiation of Ductile Fracture in Tension*, ScD Thesis MIT, Cambridge, Mass.

Needleman, A. J. and Rice, J. R. (1978). *Limits to Ductility Set by Localization*, In Koisteinen and Wang, editors, *Mechanics of Sheet Metal Forming*, Plenum Press, pages 237–265.

Needleman, A. and Tvergaard, V. (1984). Finite Element Analysis of Localization in Plasticity, Chapter 3 of *Finite Elements: Special Problems in Solid Mechanics*, volume V, Oden, J. T. and Carey, G. F., editors, Prentice-Hall, 94.

Norris, Jr. D. M., Moran, B., Scudder, J. K. and Quinones, R. F. (1978). J. Mech. Phys. Solids, 26:1–19.

O'Dowd, N. P. and Shih, C. F. (1991a). J. Mech. Phys. Solids (in press).

O'Dowd, N. P. and Shih, C.F. (1991b). J. Mech. Phys. Solids (in press).

Ohno, N. and Hutchinson, J. W. (1984). J. Mech. and Phys. of Solids, 32:63–85.

Onat, E. T. and Prager, W. (1954). J. Appl. Phys., 25:491–493.

Orr, J. and Brown, D.K. (1974). Engineering Fracture Mechanics, 6:261–274.

Orowan, E. (1945). Trans. Inst. Eng. Shipbuilders Scot, 1063:165.

Parry, T. V. and Wronski, A. S. (1985). J. Mat. Sci., 20:2141-2147.

Parry, T. V. and Wronski, A. S. (1986). J. Mat. Sci., 21:4451–4455.

Parks, D. M. (1991). Engineering Methodologies for Assessing Crack Front Constraint in *Proc. Spring Meeting, S.E.M., Milwaukee.*

Parks, D. M. (1977). Computer Methods in Appl. Mech. and Eng., 12:353.

Parks, D. M. (1991a). Defect Assessment in Components–Fundamentals and Applications, ESIS/EGF9, Blauel, J. G. and Schwalbe, K. H., editors, Mechanical Engineering Publications, London, pages 205–231.

Prager, W. and Hodge, P. G. (1951). *Theory of Perfectly Plastic Solids,* Wiley, New York.

Prandtl, L. (1920). Nachr. Ges. Wiss., Gottingen, 74.

Rice, J. R. (1968). J. Appl. Mech., 35:379–386.

Rice, J. R. (1968a). Mathematical Analysis in the Mechanics of Fracture, in Liebowitz, H., editor, *Fracture, an Advanced Treatise*, 2, Academic Press, New York.

Rice, J. R. (1976). The Localization of Plastic Deformation, In Koiter, W., editor, *Theoretical and Applied Mechanics*, North-Holland 207.

Rice, J. R. and Johnson, M. A. (1970). In Kanninen, M., Adler, W. F., Rosenfield, A. R. and Jaffee, R. I., editors, *Inelastic Behaviour of Solids,* Battelle, McGraw-Hill, New York, 641–672.

Rice, J. R. and Rosengren, G. F. (1968). J. Mech. Phys Solids, 16:1–67.

Rice, J. R. and Tracey, D. M. (1969). J. Mechanics and Physics of Solids, 17:201–217.

Rice, J. R. and Tracey, D. M. (1973). In *Numerical and Computational Methods in Structural Mechanics.*

Rice, J. R. and Tracey, D. M. (1974). J. Mech. Phys. Solids, 22:17–26.

Rudnicki, J. W. and Rice, J. R. (1975). J. Mech. Phys. Solids, 23:371–399.

Saje, M., Pan, J. and Needleman (1982). Void Nucleation in Porous Plastic Solids, Int. J. Fract., 19:163–182.

Sham, T.-L. (1991). Int. J. Fracture, 48, 81–102.

Shih, C. F. and German, M. D. (1981). Int. J. Fract., 17:27–43.

Tipper, C. F. (1949). Metallurgia, 39:133–137.

Thomason, P. F. (1968). J. Inst. Metals, 96:360.

Thomason, P. F. (1968). Int. J. Mech. Sci., 10:501.

Thomason, P. F. in Sih, G. C., van Elst, H. C. and Broek, D., editors, *Prospects of Fracture Mechanics*, Nooordhoff International, pages 3–19.

Thomson, R. D. and Hancock, J. W. (1984a). Int. Journal of Fracture, 24:209–228.

Thomson, R. D. and Hancock, J. W. (1984b). Int. Journal of Fracture, 26:99–112.

Tracey, D. (1976). ASME J. Eng. Mats. and Tech., 98:146–151.

Tvergaard, V. and Needleman, A. (1984). Acta Metall., 32:157–169.

Tvergaard, V., Needleman, A. and Lo, K. K. (1981). J. Mech. Phys. Solids, 29:115.

Tvergaard, V. (1982). Journal of the Mechanics and Physics of Solids, 30:265–286.

Wang, Y. (1991a). PhD thesis department of Mechanical Engineering, MIT.

Wang, Y. (1991b). ASTM Symposium on Constraint Effects in Fracture, Indianapolis.

Wang, Y. and Parks, D. M. (1991). Int. J. Fract. in press.

Yamamoto, H. (1978). Int. J. Fracture, 14:347.

Zok, F., Embury, J. D., Ashby, M. F. and Richmond, O. (1988). Proc Ninth Riso International Symposium on Metallurgy and Materials Science, Andersen, S. I. et al., editors, *Mechanical and Physical Behavior of Metallic and Ceramic Composites*, Riso, pages 517–526.

4

Void Growth in Plastic Solids

A. Needleman, V. Tvergaard and J.W. Hutchinson

ABSTRACT An overview is given of the continuum mechanics of void growth pertaining to room temperature ductile fracture processes. Analyses of the growth of isolated voids and of void interaction effects are reviewed. A framework for phenomenological constitutive relations for porous plastic solids is discussed. Calculations of localization and failure in porous plastic solids are reviewed that illustrate the progressive development of ductile failure. Additional considerations, including the effect of the constraint provided by contact between the growing void and the void-nucleating particle, cavitation, and the effect of non-uniform porosity distributions are briefly noted.

4.1 Introduction

The role played by void nucleation, growth and coalescence in ductile fracture was identified in the 1940's, Tipper (1949). However, it was not until the 1960's that the phenomenology of this process was well documented, Rogers (1960), Puttick (1960), Beachem (1963), and Gurland and Plateau (1963). In structural metals deformed at room temperature, the voids generally nucleate by decohesion of second phase particles or by particle fracture, and grow by plastic deformation of the surrounding matrix. Void coalescence occurs either by necking down of the matrix material between adjacent voids or by localized shearing between well separated voids.

McClintock's (1968) analysis of the expansion of a long cylindrical hole in an ideally plastic solid marks the beginning of a now extensive literature on the micromechanics of ductile fracture. More broadly, this paper showed that a precise mechanics analysis of a carefully chosen continuum model could serve to quantify complex microstructural behavior. Micromechanics analyses of void nucleation and coalescence, as well as of void growth, have now been carried out. Such analyses have served as the basis for phenomenological constitutive relations for porous plastic solids, which in turn have been used to predict macroscopic ductile fracture behavior. An overview of these developments is given by Tvergaard (1990). Another overview of void growth studies is given in Gilormini, Licht and Suquet (1988).

Our focus in this paper is on issues in the mechanics of void growth relevant to room temperature ductile fracture processes. We begin by reviewing analyses of the growth of an isolated void. A key point in this regard concerns the hydrostatic stress dependence of the void growth rate. The phenomenological framework for constitutive relations for porous plastic solids introduced by Gurson (1975) is outlined, since there has been considerable interplay between analyses based on this constitutive relation and the development of more accurate void growth models.

4.2 Growth of an Isolated Void

Predicting ductile fracture behavior requires knowing the relation between the growth of a void and the imposed stress and strain histories. Of particular importance is the dependence of the rate of void growth on the remote stress triaxiality, which is revealed by tension tests on notched tensile bars, Hancock and Mackenzie (1976), Hancock and Brown (1983). Although the maximum strain occurs at the root of the notch, for certain combinations of notch acuity and stress state (plane strain or axisymmetric), ductile failure initiates inside the specimen where the triaxiality of the stress state is greater than near the surface.

To investigate the stress state dependence of the void growth rate, McClintock (1968) considered a long cylindrical void in an ideally plastic solid, extending in the direction of its axis, as shown in Fig. 4.1. The cylindrical void has current radius b and the aim is to determine the dependence of the void growth rate $\dot{b}/b$ on the imposed axial strain rate $\dot{\epsilon}$ and the transverse stress σ_∞. McClintock's (1968) analysis is summarized here.

Generalized plane strain and cylindrical symmetry are assumed, so that with polar coordinates (r, θ, z), all field quantities are functions solely of r. Equilibrium in the current configuration is written as

$$\frac{d\sigma_r}{dr} + \frac{\sigma_r - \sigma_\theta}{r} = 0 \tag{4.1}$$

The radial displacement is denoted by $u(r)$ and is the only non-vanishing in-plane displacement. With a superposed dot denoting the time derivative, the strain rate-displacement rate relations are

$$\dot{\epsilon}_r = \frac{d\dot{u}}{dr} \qquad \dot{\epsilon}_\theta = \frac{\dot{u}}{r} \tag{4.2}$$

and incompressibility implies

$$\dot{\epsilon}_r + \dot{\epsilon}_\theta + \dot{\epsilon} = 0 \tag{4.3}$$

The flow rule for a Mises solid requires

$$\dot{\epsilon}_r = \frac{3}{2}(\sigma_r - \sigma)\frac{\dot{\bar{\epsilon}}}{Y} \qquad \dot{\epsilon}_\theta = \frac{3}{2}(\sigma_\theta - \sigma)\frac{\dot{\bar{\epsilon}}}{Y} \tag{4.4}$$

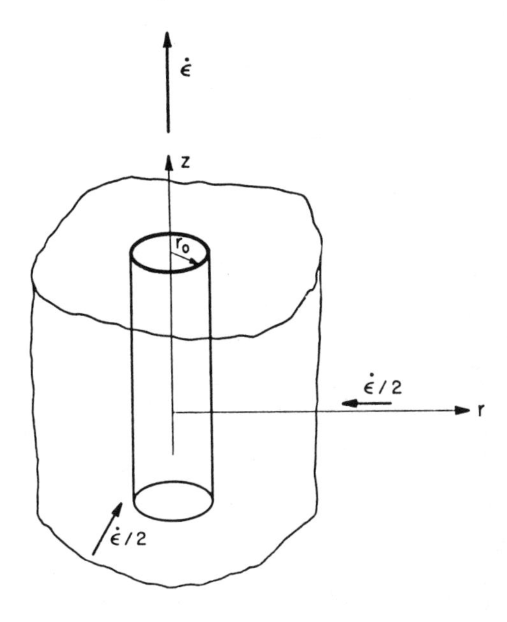

FIGURE 4.1. Long cylindrical void extended along its axis.

where Y is the tensile flow strength and

$$\sigma = \frac{1}{3}(\sigma_r + \sigma_\theta + \sigma_z) \qquad \dot{\bar{\epsilon}} = \sqrt{\frac{2}{3}(\dot{\epsilon}_r^2 + \dot{\epsilon}_\theta^2 + \dot{\epsilon}^2)} \tag{4.5}$$

Here, σ is the mean stress and $\dot{\bar{\epsilon}}$ is the Mises equivalent strain rate.

The boundary conditions are that the radial stress vanishes on the surface of the void and that it takes on its remote value as $r \to \infty$.

$$\sigma_r(b) = 0 \qquad \sigma_r(\infty) = \sigma_\infty \tag{4.6}$$

Solving the flow rule Eq. 4.4 for $(\sigma_r - \sigma_\theta)$ gives,

$$\frac{\sigma_r - \sigma_\theta}{r} = \frac{2Y}{3r}\frac{\dot{\epsilon}_\theta - \dot{\epsilon}_r}{\dot{\bar{\epsilon}}} \tag{4.7}$$

Using Eq. 4.7 in the equilibrium equation and integrating we obtain,

$$\frac{\sigma_\infty}{Y} = \frac{2}{3}\int_b^\infty \left[\frac{\dot{\epsilon}_\theta - \dot{\epsilon}_r}{\dot{\bar{\epsilon}}}\right]\frac{dr}{r} \tag{4.8}$$

The incompressibility condition Eq. 4.3 gives the velocity field $\dot{u}(r)$ in terms of the imposed strain rate $\dot{\epsilon}$ and the void expansion rate $\dot{b}/b$ as,

$$\dot{u} = b^2\left[\frac{\dot{b}}{b} + \frac{\dot{\epsilon}}{2}\right]\frac{1}{r} - \frac{\dot{\epsilon}}{2}r \tag{4.9}$$

so that

$$\dot{\epsilon}_r = -\frac{b^2}{r^2}[\frac{\dot{b}}{b} + \frac{\dot{\epsilon}}{2}] - \frac{\dot{\epsilon}}{2}$$

$$\dot{\epsilon}_\theta = \frac{b^2}{r^2}[\frac{\dot{b}}{b} + \frac{\dot{\epsilon}}{2}] - \frac{\dot{\epsilon}}{2}$$

$$\dot{\bar{\epsilon}}^2 = \frac{4}{3}[\frac{\dot{b}}{b} + \frac{\dot{\epsilon}}{2}]^2 \frac{1}{r^4} + \dot{\epsilon}^2 \tag{4.10}$$

Using these expressions in Eq. 4.8, with the substitutions

$$x = \alpha \frac{b^2}{r^2} \qquad \alpha = \frac{2}{\sqrt{3}}[\frac{\dot{b}}{\dot{\epsilon} b} + \frac{1}{2}] \tag{4.11}$$

gives

$$\frac{\sigma_\infty}{\sqrt{3}Y} = \int_0^\alpha \frac{dx}{\sqrt{1+x^2}} \tag{4.12}$$

Integrating Eq. 4.12 and rearranging terms gives the dependence of the relative void growth rate, $\dot{b}/\dot{\epsilon}b$, on the transverse stress σ_∞,

$$\frac{\dot{b}}{\dot{\epsilon} b} = \frac{\sqrt{3}}{2} \sinh[\frac{\sigma_\infty}{\sqrt{3}Y}] - \frac{1}{2} \tag{4.13}$$

McClintock's (1968) exact solution exhibits an exponential increase in the void growth rate with positive transverse stress. This can be contrasted with the linear increase that is predicted for a linear viscous material, Berg (1962). For a spherical void, subject to a general remote stress state, the problem is no longer one dimensional and exact solutions are not available. Rice and Tracey (1969) used a Rayleigh-Ritz method to obtain approximate solutions for the growth of an isolated spherical void surrounded by an ideally plastic matrix, as sketched in Fig. 4.2. At high triaxiality, i.e. for large values of the ratio of remote mean stress, σ_h^∞, to matrix flow strength, Y, their numerical results were well-approximated by

$$\frac{\dot{R}_0}{\dot{\epsilon} R_0} \approx 0.283 \exp[\frac{3\sigma_h^\infty}{2Y}] \approx 0.566 \sinh[\frac{3\sigma_h^\infty}{2Y}] \tag{4.14}$$

Furthermore, Rice and Tracey (1969), found that Eq. 4.14 was a very good approximation even at low triaxialities.

Similar explicit expressions showing the effect of material strain hardening on void growth rates have not yet been obtained for rate independent plastic solids. However, the assumption of non-linear viscous, rather than rate independent strain hardening, material behavior facilitates the analysis. Within this context, Budiansky, Hutchinson and Slutsky (1982) have obtained a high triaxiality approximation for the growth rate of an initially

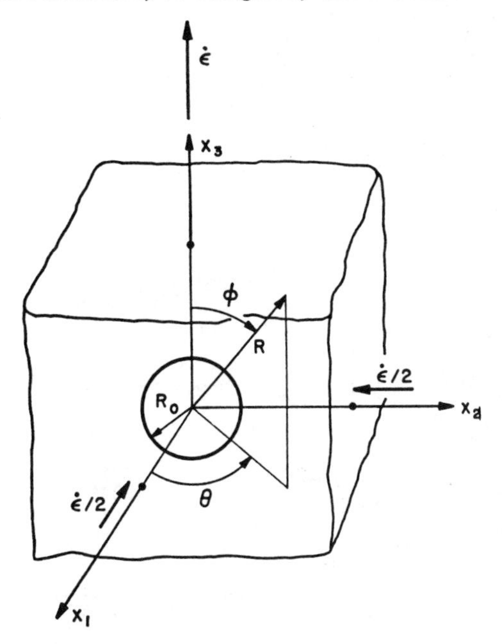

FIGURE 4.2. Spherical void in a remote simple tension strain rate field.

spherical void in a power law viscous solid, i.e. one for which the uniaxial response is $\dot{\epsilon} \propto \sigma^n$, as

$$\frac{\dot{R}_0}{\dot{\epsilon} R_0} = \frac{1}{2}\left[\frac{3\sigma_h^\infty}{2n\sigma_e^\infty} + \frac{(n-1)(n+0.4319)}{n^2}\right]^n \tag{4.15}$$

In the limit $n \to \infty$, the response of the power law viscous solid approaches that of a rigid-ideally plastic Mises solid. In this limit, Eq. 4.15 reduces to Eq. 4.14.

This consistent picture persisted until 1989. In the course of a study of cavitation instabilities in elastic-plastic solids Huang, Hutchinson and Tvergaard (1991) obtained finite element results for the growth of an isolated spherical void in ideally plastic solid that were about 50% higher than expected based on the Rice-Tracey formula Eq. 4.14. This discrepancy was investigated further by Huang (1989) who carried out a Rayleigh-Ritz analysis using a much larger number of terms than Rice and Tracey (1969) and Budiansky et al. (1982). Huang (1989) found that a surprisingly large number of terms were needed to obtain a converged value for the void growth rate and that when a sufficient number of terms were included in the analysis, the Rayleigh-Ritz results were consistent with finite element solutions in Huang et al. (1991). Huang's (1989) high triaxiality approximation is

$$\frac{\dot{R}_0}{\dot{\epsilon} R_0} \approx 0.427 \exp\left[\frac{3\sigma_h^\infty}{2Y}\right] \tag{4.16}$$

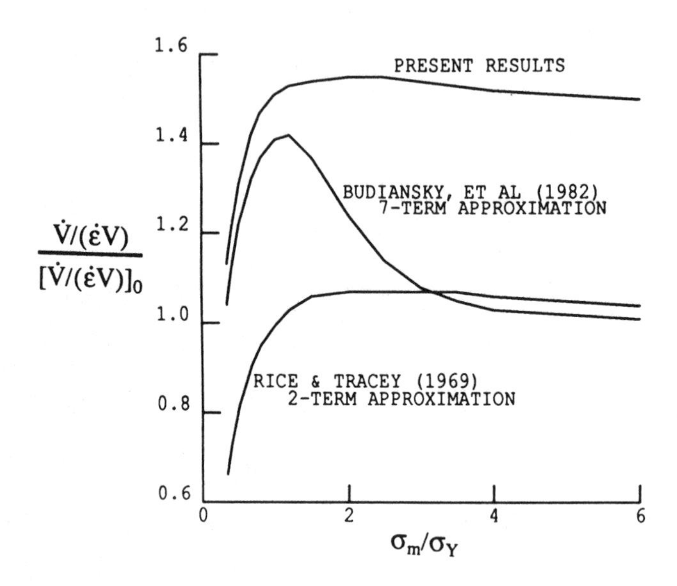

FIGURE 4.3. Dilation rate in a rigid-perfectly plastic solid as computed at various levels of approximation. The volumetric growth rate is normalized by that obtained from Eq. 4.14 using $(\dot{V}/\dot{\epsilon}_0 V)_0 = 3\dot{R}_0/\dot{\epsilon}_0 R_0$ (from Huang, 1989).

This correction is important because the expression for the growth rate of an isolated spherical void is used explicitly in direct predictions of ductility, e.g. Marini, Mudry and Pineau (1985), or implicitly in phenomenological constitutive relations for porous plastic solids, e.g. Gurson (1975). Figure 4.3 compares the void growth rates for a perfectly plastic solid predicted by the analyses of Rice and Tracey (1969), Budiansky et al. (1982) and Huang (1989).

4.3 Constitutive Relations for Porous Plastic Solids

In general, the flow potential for a porous plastic solid can be written in the form

$$\Phi(\boldsymbol{\sigma}, \mathcal{H}, \mathcal{F}) = 0 \tag{4.17}$$

where $\boldsymbol{\sigma}$ denotes the Cauchy stress tensor, $\mathcal{H}$ are a set of properties characterizing the plastic flow of the matrix and $\mathcal{F}$ are a set of properties characterizing the porosity. In practice, only special cases of the form

$$\Phi(\sigma_e, \sigma_h, \bar{\sigma}, f) = 0 \tag{4.18}$$

and satisfying

$$\Phi = \frac{\sigma_e^2}{\bar{\sigma}^2} - 1 = 0 \qquad \text{when } f = 0 \tag{4.19}$$

have been considered, e.g. by Gurson (1975), Shima and Oyane (1976), Guennouni and Francois (1987). Here, the hydrostatic stress, σ_h, and the Mises effective stress, σ_e, are given by

$$\sigma_h = \frac{1}{3}\sigma_{kk} \quad \sigma_e = \sqrt{\frac{3}{2}\sigma'_{ij}\sigma'_{ij}} \tag{4.20}$$

In Eq. 4.20 and subsequently, Cartesian tensor notation is employed.

The most widely used porous plastic constitutive relation for analyzing ductile fracture phenomena is that due to Gurson (1975), which is based on averaging techniques similar to those used by Bishop and Hill (1951). Gurson (1975) approximated a solid with a volume fraction f of voids by a homogeneous spherical body with a spherical cavity. An approximate rigid-plastic limit analysis of this situation was used to develop the yield condition

$$\Phi = \frac{\sigma_e^2}{\bar{\sigma}^2} + 2f\cosh\left(\frac{3\sigma_h}{2\bar{\sigma}}\right) - 1 - f^2 = 0 \tag{4.21}$$

where $\bar{\sigma}$ is an internal variable representing the average strength of the matrix material. Figure 4.4 sketches the shape of the yield surface at various levels of porosity. When $f = 1$, the yield surface shrinks to a point and the material's stress carrying capacity vanishes. The yield function Eq. 4.21 has the following characteristics; (i) it reduces to that for a Mises solid when $f = 0$, (ii) the dependence on void volume fraction is linear when $\sigma_h = 0$, (iii) the dependence on stress triaxiality, $\sigma_h/\bar{\sigma}$, is exponential.

The plastic part of the rate of deformation tensor, D^p_{ij}, is given by the flow rule

$$D^p_{ij} = \dot{\Lambda}\frac{\partial\Phi}{\partial\sigma_{ij}} \qquad D_{ij} = \frac{1}{2}\left(\frac{\partial \dot{u}_i}{\partial x_j} + \frac{\partial \dot{u}_j}{\partial x_i}\right) \tag{4.22}$$

where u_i are the Cartesian components of the displacement vector and a superposed dot denotes partial differentiation with respect to time.

Evolution equations need to be specified for the internal variables f and $\bar{\sigma}$. In general, the evolution of the void volume fraction results from both nucleation of new voids and from growth of existing voids, but void nucleation is not considered here. The growth rate of existing voids is determined by requiring the matrix material to be plastically incompressible. With V_v denoting the void volume and V_m the matrix volume, $f = V_v/(V_v + V_m)$. Differentiating with respect to time and using $\dot{V}_m = 0$, so that $D^p_{kk} = \dot{V}_v/(V_v + V_m)$,

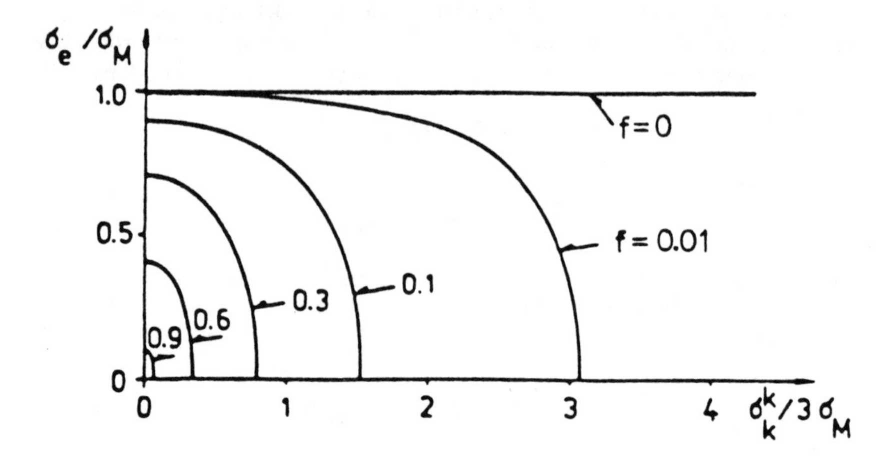

FIGURE 4.4. Yield surface dependence on the hydrostatic tension for various values of porosity, f, in Eq. 4.21.

gives,

$$\dot{f} = (1 - f) D^p_{kk} \tag{4.23}$$

The plastic work rate for the porous solid is set equal to the matrix plastic work rate. Accordingly,

$$\sigma_{ij} D^p_{ij} = (1 - f)\, \bar{\sigma}\, \dot{\bar{\epsilon}} \tag{4.24}$$

From Eq. 4.24, the plastic flow proportionality factor, $\dot{\Lambda}$, in the flow rule Eq. 4.22 is determined to be,

$$\dot{\Lambda} = \frac{(1 - f)\, \bar{\sigma} \dot{\bar{\epsilon}}}{\sigma_{kl} \dfrac{\partial \Phi}{\partial \sigma_{kl}}} \tag{4.25}$$

Here, $\dot{\bar{\epsilon}}$ is the matrix Mises equivalent strain rate, which is determined from the matrix strain hardening relation

$$\dot{\bar{\epsilon}} = (\frac{1}{E_t} - \frac{1}{E})\dot{\bar{\sigma}} = \frac{1}{\bar{h}} \dot{\bar{\sigma}} \tag{4.26}$$

where E and E_t are, respectively, the Young's modulus and the tangent modulus, the slope of the uniaxial true stress-logarithmic strain curve, of the matrix material.

Substituting Eq. 4.26 in Eq. 4.24 and rearranging gives the evolution equation for $\dot{\bar{\sigma}}$ as

$$\dot{\bar{\sigma}} = \frac{\bar{h} \sigma_{ij} D^p_{ij}}{(1 - f)\, \bar{\sigma}} \tag{4.27}$$

The flow rule for the porous aggregate is obtained from the consistency condition that $\Phi(\sigma_{ij}, \bar{\sigma}, f) = 0$ for continued plastic loading. Hence,

$$\dot{\Phi} = \frac{\partial \Phi}{\partial \sigma_{ij}} \dot{\sigma}_{ij} + \frac{\partial \Phi}{\partial \bar{\sigma}} \dot{\bar{\sigma}} + \frac{\partial \Phi}{\partial f} \dot{f} = 0 \tag{4.28}$$

Substituting (4.22), (4.23), (4.25) and (4.27) in (4.28) gives

$$D^p_{ij} = \frac{1}{h} \left(\frac{\partial \Phi}{\partial \sigma_{kl}} \dot{\sigma}_{kl} \right) \frac{\partial \Phi}{\partial \sigma_{ij}} = \frac{1}{h} \left(\frac{\partial \Phi}{\partial \sigma_{kl}} \hat{\sigma}_{kl} \right) \frac{\partial \Phi}{\partial \sigma_{ij}} \tag{4.29}$$

where $\hat{\sigma}_{kl}$ is the Jaumann derivative of Cauchy stress and

$$h = - \left[(1 - f) \frac{\partial \Phi}{\partial f} \frac{\partial \Phi}{\partial \sigma_{kk}} + \frac{\bar{h}}{(1 - f)\bar{\sigma}} \frac{\partial \Phi}{\partial \bar{\sigma}} \left(\sigma_{kl} \frac{\partial \Phi}{\partial \sigma_{kl}} \right) \right] \tag{4.30}$$

The expression (4.29) pertains to plastic loading, which is when $(\sigma_{kl} \partial \Phi / \partial \sigma_{kl})/h > 0$. Otherwise, $D^p_{ij} = 0$.

For an elastic-plastic solid, we write

$$D_{ij} = D^e_{ij} + D^p_{ij} \tag{4.31}$$

In circumstances where the elastic strains remain small, although the plastic strains may be large, it is convenient to use a hypoelastic approximation for D^e_{ij} that has the form

$$\hat{\sigma}_{ij} = \frac{E}{1 + \nu} \left[D^e_{ij} + \frac{\nu}{(1 - 2\nu)} D^e_{kk} \delta_{ij} \right] = L^e_{ijkl} D^e_{kl} \tag{4.32}$$

where E is Young's modulus, ν is Poisson's ratio and δ_{ij} is Kronecker's delta.

Combining (4.29), (4.31) and (4.32) gives, for plastic loading,

$$\hat{\sigma}_{ij} = \left[L^e_{ijkl} - \frac{P_{ij} P_{kl}}{h + P_{mn} L^e_{mnpq} P_{pq}} \right] D_{kl} \tag{4.33}$$

where

$$P_{ij} = L^e_{ijkl} \frac{\partial \Phi}{\partial \sigma_{kl}} \tag{4.34}$$

In ductile fracture processes, the porosity is typically very small until just prior to fracture, when the material's stress carrying capacity is reducing rapidly. In such circumstances, the effects of porosity on plastic response dominate the effects on elastic response, so that the dependence of E and ν on porosity can be neglected. However, when appropriate, the dependence of the elastic properties on porosity can be accounted for, see Fleck et al. (1991), as in continuum damage mechanics, Lemaitre (1985).

Figure 4.5 shows curves of σ_e and of f versus overall effective strain for a material element subject to uniaxial tension with a superposed hydrostatic stress. The stress ratio $T = \sigma_h/\sigma_e$ is constant throughout the deformation history. The initial porosity is $f_0 = 0.002$ and the matrix is assumed ideally plastic. The elastic properties of the matrix are given by $\sigma_0/E = 0.004$ and $\nu = 0.3$. The large effect of stress triaxiality on void growth and aggregate stress-strain response is evident in this figure.

It is worth separating the general features of the framework introduced by Gurson (1975) from the specifics of the particular flow rule. The general features are (i) the one parameter characterization of porosity; (ii) the use of matrix incompressibility to obtain the evolution equation for the void volume fraction and (iii) the use of the equivalence of aggregate and matrix rate of plastic work to relate aggregate hardening to matrix plastic response. The particulars are the specific form of the yield function (Eq. 4.21) and the characterization of the matrix material as a rate independent isotropically hardening solid in Eq. 4.26.

4.4 Localization in Porous Solids

During the deformation of ductile solids it is frequently observed that a rather smoothly varying deformation pattern grows into a pattern involving highly localized deformations in the form of shear bands. Then, final shear fracture often occurs at an overall strain, which is only slightly larger than that at the onset of localization, and failure tends to occur by a void-sheet mechanism, where small voids coalesce inside the shear band as the localized strains grow large. When voids are represented in terms of a ductile porous material model, such as the Gurson model, the onset of flow localization can be studied by a simple model analysis, and such analyses predict that the critical strain for localization is significantly reduced by the presence of voids in the elastic-plastic material (Rice, 1977; Yamamoto, 1978).

For rate independent solids, the onset of localization can be formulated as a bifurcation problem within the framework of a material instability analysis, Hadamard (1903), Hill (1962), Mandel (1966), Rice (1977), with the mode of localization corresponding to a shear band. All-around displacement conditions are imposed so as to rule out geometric instabilities. An element of a solid is considered subject to displacement boundary conditions that in a homogeneous (and homogeneously deformed) solid would give rise to a uniform deformation gradient field. Conditions are sought under which bifurcation into a localized band mode can occur. Within this framework the onset of localization coincides with the loss of ellipticity of the equations governing rate equilibrium.

Current values of field quantities and material properties inside and outside the band are presumed identical so that one possible solution for the

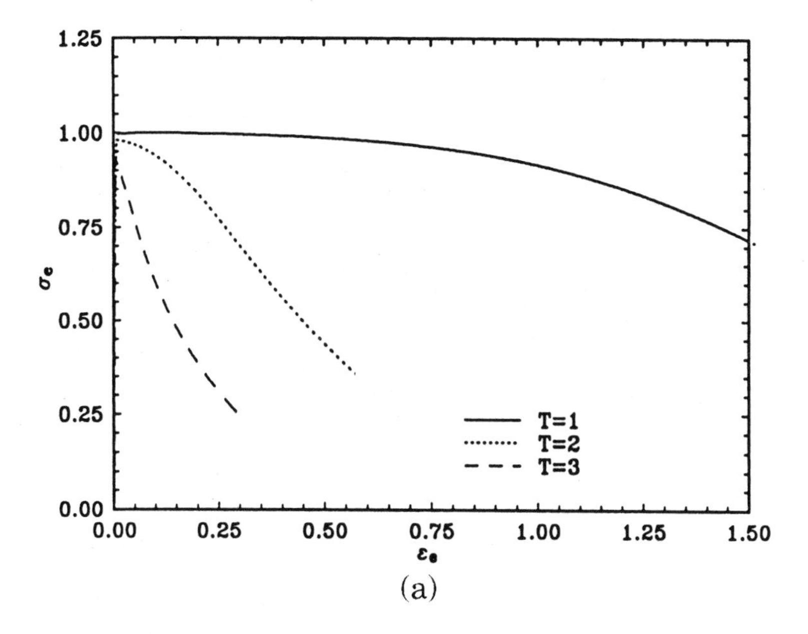

(a)

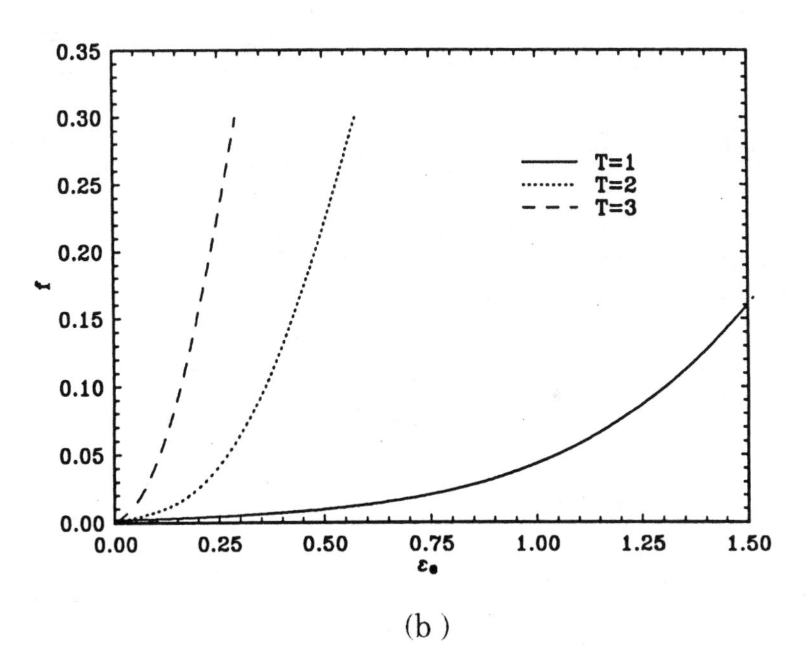

(b)

FIGURE 4.5. (a) Aggregate effective stress, σ_e, versus effective strain, ϵ_e, and (b) void volume fraction, f versus effective strain, ϵ_e, for a material element subject to uniaxial tension with a superposed hydrostatic tension, σ_h, with $T = \sigma_h/\sigma_e$. The original Gurson yield function relation, Eq. 4.21, is used, the matrix is ideally plastic and $f_0 = 0.002$.

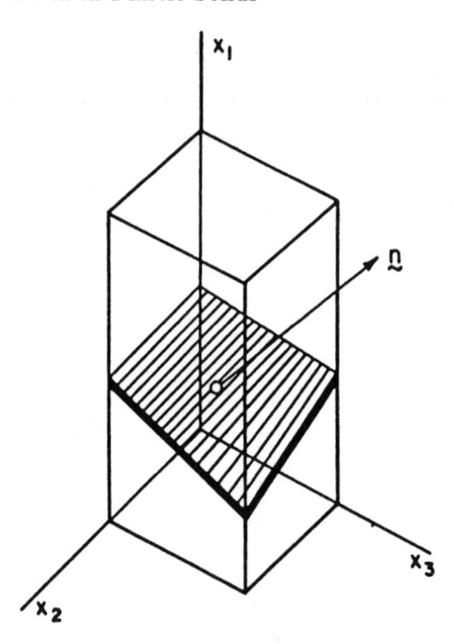

FIGURE 4.6. Body with an incipient band or plane of imperfection.

incremental quantities corresponds to the homogeneous one. At the considered stage of the deformation history, suppose that, as sketched in Fig. 4.6, within a thin planar band of orientation $\mathbf{n}$ in the reference configuration incremental field quantities are permitted to take on values differing from the uniform values outside the band. The band is presumed sufficiently narrow to be regarded as homogeneously deformed.

Two requirements must be satisfied across the band interface. First, compatibility requires continuity of the displacement rate. With the reference and current configurations taken to coincide, this implies,

$$\frac{\partial \dot{u}_i^b}{\partial x_j} = \frac{\partial \dot{u}_i^o}{\partial x_j} + \dot{c}_i n_j \tag{4.35}$$

Next, incremental equilibrium requires

$$n_j \dot{\sigma}_{ji}^b = n_j \dot{\sigma}_{ji}^o \tag{4.36}$$

The material stress rate in Eq. 4.36 is related to the Jaumann stress rate by

$$\hat{\sigma}_{ij} = \dot{\sigma}_{ij} + \sigma_{ik}\Omega_{kj} - \Omega_{ik}\sigma_{kj} \qquad \Omega_{ij} = \frac{1}{2}\left(\frac{\partial \dot{u}_i}{\partial x_j} - \frac{\partial \dot{u}_j}{\partial x_i}\right) \tag{4.37}$$

Substituting (4.35), (4.33) and (4.37) into (4.36) results in three homogeneous algebraic equations for the three unknowns $\dot{c}_i$. Setting the determinant of coefficients equal to zero gives the condition for the onset of a localization bifurcation. The onset of localization depends on stress state as well as on material properties.

With attention restricted to circumstances where $|\boldsymbol{\sigma}|/E$, f and $f\exp(3\sigma_h/2\bar{\sigma})$ are small, Needleman and Rice (1978) solved for the value of the matrix hardening rate at bifurcation. For an initially porous solid, with no nucleation, under plane strain conditions they obtain

$$\left(\frac{\bar{h}}{\bar{\sigma}}\right)_c = \frac{3}{2} f \cosh\left(\frac{3\sigma_h}{2\bar{\sigma}}\right) \sinh\left(\frac{3\sigma_h}{2\bar{\sigma}}\right) \tag{4.38}$$

while for axisymmetric straining

$$\left(\frac{\bar{h}}{\bar{\sigma}}\right)_c = -\frac{E}{4\bar{\sigma}} + \frac{3}{2} f \cosh\left(\frac{3\sigma_h}{2\bar{\sigma}}\right) \sinh\left(\frac{3\sigma_h}{2\bar{\sigma}}\right) \tag{4.39}$$

The matrix hardening rate, $\bar{h}$, varies during the deformation history. Typically, $\bar{h}$, has some positive value at initial yield and then decreases, so that larger values of $\bar{h}$ correspond to earlier stages of the deformation history. For a fully dense solid, $f \equiv 0$, the localization results for an isotropically hardening Mises solid are recovered; an ideally plastic state, $\bar{h}_c = 0$, is needed for localization under plane strain conditions and a strongly negative hardening rate is required under axisymmetric conditions. With an initial void volume fraction, a localization bifurcation is possible under plane strain conditions while the matrix material is hardening. The value of the critical matrix hardening rate depends sensitively on the stress triaxiality. Even with an initial porosity, a localization bifurcation under axisymmetric conditions is effectively excluded for a workhardening matrix, $\bar{h} > 0$.

Yamamoto (1978) showed that an initial imperfection, in the form of a band with slightly increased porosity, could trigger localization in a porous solid under axisymmetric conditions. Figure 4.7, taken from Yamamoto (1978), shows critical strains to localization in a porous solid under both plane strain and axisymmetric conditions. As expected, based on the bifurcation analysis, localization occurs at much lower strains under plane strain conditions than under axisymmetric straining conditions. Also shown in Fig. 4.7 (b) is the band orientation in plane strain, where $\theta_0 = 0$ corresponds to a band orthogonal to the tensile axis. As the critical strain increases, i.e. for smaller imperfections, θ_0 decreases. The tendency for θ_0 to decrease with increasing localization strains also holds under axisymmetric conditions. A similar analysis has been used by Needleman and Triantafyllidis (1978) to study porosity induced local necking in thin sheets.

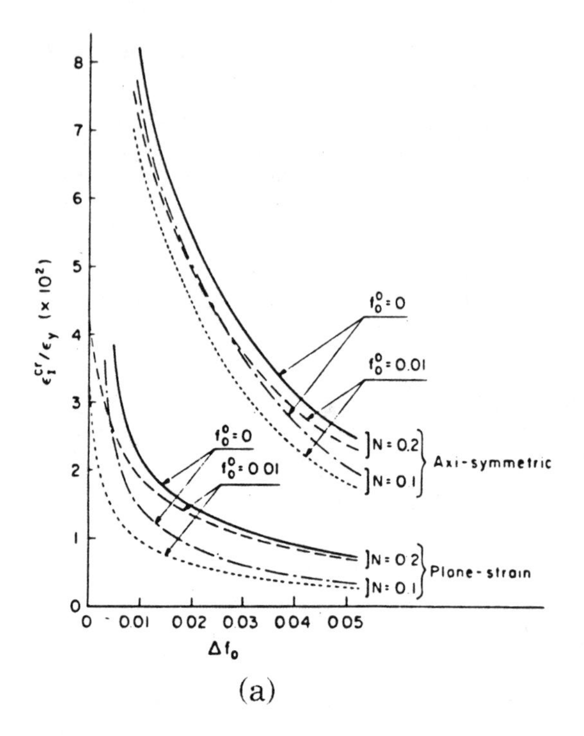

(a)

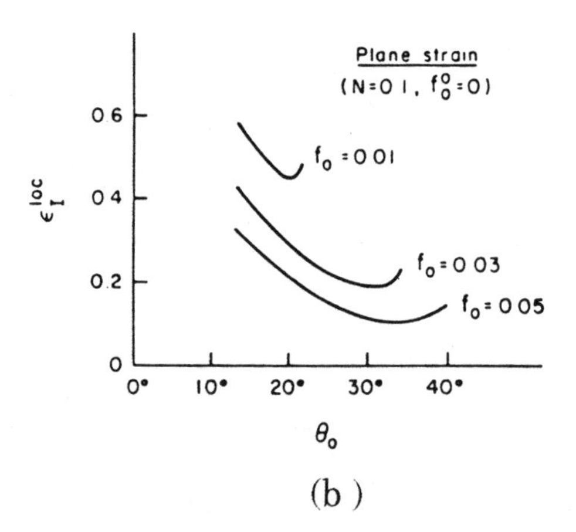

(b)

FIGURE 4.7. (a) Curves of critical strain for localization as a function of the initial void concentration in the band, f_0. (b) Curves of critical strain for localization versus initial band angle in plane strain. N is the hardening exponent of the matrix material (from Yamamoto, 1978).

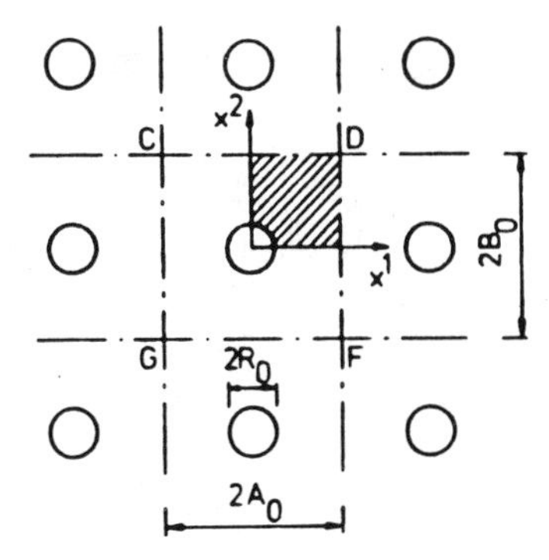

FIGURE 4.8. Doubly periodic array of circular cylindrical voids.

4.5 Void Interaction Effects

An early investigation in which interaction effects are rigorously accounted for is the numerical analysis of Needleman (1972) for an elastic-plastic solid containing a doubly periodic array of circular cylindrical voids, as shown in Fig. 4.8. Due to symmetries only the region hatched in Fig. 4.8 had to be analyzed, and finite strain effects were fully accounted for in the computations. Thus, final failure by necking of the ligaments to zero thickness could be approached in the numerical solutions, and linear extrapolations of the overall stress-strain curves were used to estimate the average strain at coalescence. The assumption of an array of parallel cylindrical voids is a strong idealization, but is very convenient computationally, since this allows for a planar analysis. It is expected that the results based on this idealization give quite a useful indication of actual void coalescence behavior.

A numerical finite element study by Andersson (1977) for voids in a rigid-perfectly plastic material has been continued to a stage very near final coalescence with neighboring voids. Here the focus was on voids just ahead of a moving crack tip, and therefore a state of uniaxial straining was assumed, so that the voids grew under high hydrostatic straining. These voids were taken to be initially spherical, and an axisymmetric model problem was solved, representing a periodic distribution of voids in the layer ahead of the crack tip. The analyses were used to estimate the total work of separation per unit area of new crack surface when crack growth occurs by a void coalescence mechanism. It was also found that in the uniaxial strain state analyzed the initially spherical voids grow into oblate spheroids rather than elongating in the main tensile direction.

An approximate representation of a material with a certain volume fraction of spherical voids can be obtained in terms of a spherical unit cell containing a concentric spherical void, where the loading or displacements on the outer surface is chosen in agreement with the overall stress or strain state of the material. This type of analysis accounts for the interaction between neighboring voids in a very approximate manner. Upper bound rigid-plastic analyses for such model problems have been used by Gurson (1975), to obtain the approximate overall yield condition for a void containing ductile material, as described in Section 4.3.

A detailed understanding of the effect of voids on the occurrence of material instabilities requires a micromechanical analysis that accounts for the nonuniform stress and strain fields around voids and the interaction between neighboring voids. Early studies of this type were carried out by McClintock, Kaplan and Berg (1966) and by Nagpal, McClintock, Berg and Subudhi (1972). Tvergaard (1981) carried out a full bifurcation analysis for the doubly periodic array of circular cylindrical voids previously analyzed by Needleman (1972). Here, the solution obtained by assuming symmetry conditions on all four edges of the region hatched in Fig. 4.8 represents the fundamental solution with a homogeneous macroscopic strain state, but taking full account of the nonuniform strain fields on the micro level around voids. The possibility of bifurcation into another periodic pattern was analyzed on the basis of Hill's (1958) general theory of uniqueness and bifurcation for elastic-plastic solids. In this bifurcation analysis several symmetry properties of the repetitive pattern were employed so that again only the hatched region in Fig. 4.8 had to be analyzed, and the analysis also incorporated the necessary conditions of equilibrium and compatibility on the shear band interface, which could only be satisfied on the average between the two different periodic deformation patterns. In Fig. 4.9 the first critical bifurcation points and the corresponding angles of inclination of the shear bands are marked on curves of nominal traction versus overall strain, for the model material subject to uniaxial plane strain tension. In predictions based on a continuum model bifurcation into a shear band mode is associated with loss of ellipticity of the governing differential equations, but in this full bifurcation analysis accounting for discrete voids the equations remain elliptic.

Bifurcation into a localized mode has also been studied for a periodic array of spherical voids (Tvergaard, 1982), based on an axisymmetric analysis for a circular cylindrical unit cell containing a spherical void. The outer surface of the unit cell had to remain cylindrical throughout the deformations, to approximately represent compatibility with neighboring cells, and a fixed ratio of the average true principal stresses was prescribed. This axisymmetric model cannot represent shear bands with arbitrary angle of inclination to the direction of tension, as was possible in the planar analysis (Tvergaard, 1981), but the special case of bands perpendicular to this direction was analyzed, as is mainly relevant under high triaxial tension.

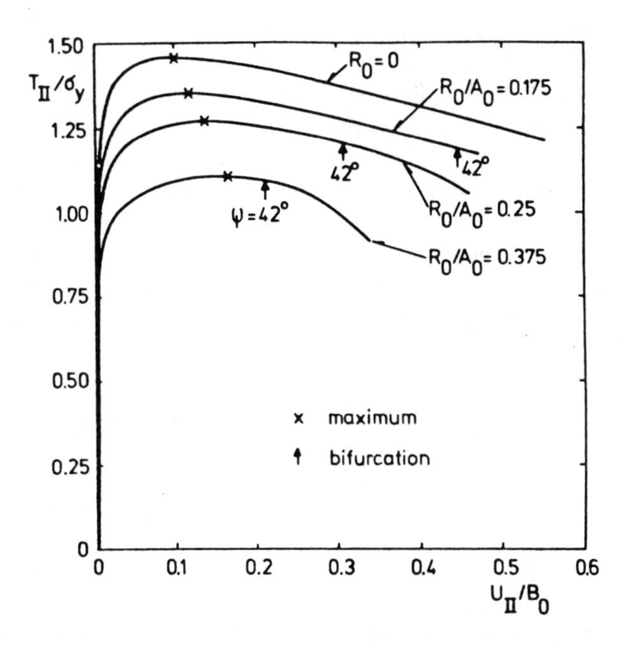

FIGURE 4.9. Nominal traction versus average strain for material containing a doubly periodic array of circular voids (from Tvergaard, 1981).

The analyses made use of the fact that during localization a slice of material undergoes deformation, while equilibrium and compatibility with the undeforming material outside this slice of material is retained. Based on a comparison of these bifurcation results with corresponding results using the material instability analysis described in Section 4.4, Tvergaard (1981, 1982) proposed modifying the yield condition (4.21) to

$$\Phi = \frac{\sigma_e^2}{\bar{\sigma}^2} + 2q_1 f \cosh\left(\frac{3q_2\sigma_h}{2\bar{\sigma}}\right) - 1 - q_3 f^2 = 0 \tag{4.40}$$

with

$$q_1 = 1.5 \qquad q_2 = 1 \qquad q_3 = q_1^2 = 2.25 \tag{4.41}$$

Analyses based on the same axisymmetric cell model problem have been carried out by Koplik and Needleman (1988) with the purpose of determining critical conditions for the onset of failure by coalescence of neighboring voids. These calculations show a shift from the state of axisymmetric elastic-plastic deformations to a mode of uniaxial straining where the plastic deformations localize to the ligament between neighboring voids. This localization is closely related to the bifurcations found by Tvergaard (1982). The two events would occur simultaneously if a longer cylindrical cell with two voids was analyzed, and earlier localization was also found by Koplik and Needleman (1988) in cases where a longer unit cell was considered.

Figure 4.10 shows results of these analyses for the values 1, 2 and 3 of the macroscopic stress triaxiality, $T = \Sigma_m/\Sigma_e$. Based on their cell model calculations, Koplik and Needleman (1988) suggested $q_1 = 1.25$ and $q_2 = 1$ in Eq. 4.40. It is seen in Fig. 4.10 that after the onset of localization the macroscopic effective strain E_e grows very little, while the macroscopic Mises stress Σ_e decays rapidly, and the rapid increase of the void volume fraction f marks the development of coalescence. It is also seen that the failure strain is significantly reduced for increasing stress triaxiality. In order to account for this rapid drop in stress carrying capacity, Tvergaard and Needleman (1984) suggested replacing f in Eq. 4.40 with the function $f^*(f)$, so that

$$\Phi = \frac{\sigma_e^2}{\bar{\sigma}^2} + 2q_1 f^* \cosh\left(\frac{3q_2\sigma_h}{2\bar{\sigma}}\right) - 1 - (q_1 f^*)^2 = 0 \qquad (4.42)$$

where

$$f^* = \begin{cases} f & f \le f_c \; ; \\ f_c + \dfrac{f_u^* - f_c}{f_f - f_c}(f - f_c) & f \ge f_c \end{cases} \qquad (4.43)$$

Here, f_f is the void volume fraction at which the stress carrying capacity vanishes, so that $f^*(f_f) = f_u^* = 1/q_1$.

The cell model studies in Koplik and Needleman (1988) and similar studies in Becker et al. (1988) suggested that f_c and f_f vary slowly with stress triaxiality and with matrix strain hardening, but depend strongly on the initial porosity. Figure 4.11, from Tvergaard (1990), shows the dependence of f_c on initial porosity according to these cell model studies. Figure 4.12 compares the overall stress strain behavior and void growth based on the modified Gurson constitutive relation with corresponding results based on Gurson's original yield function (Eq. 4.21) for the case with $T = 3$ in Fig. 4.5. The modified constitutive relation gives failure at realistic values of the void volume fraction.

The studies of void interaction effects discussed here are all based on simplifying assumptions that allow for planar analyses with cylindrical voids or axisymmetric analyses with spherical voids. More realistic models would generally require full three-dimensional numerical solutions, but such computations put very large requirements on computer time and storage. Full 3D computations have been carried out by Hom and McMeeking (1989) and Worswick and Pick (1990) for the growth of initially spherical voids in periodic cubic arrays. Very good agreement was found between the predictions of these full three dimensional analyses and corresponding predictions of axisymmetric cell models. The results of the 3D studies have been mainly used to evaluate the validity of approximations involved in various dilatant plasticity continuum models for porous materials. Tvergaard (1988) has analyzed 3D cubic arrays of larger voids in a matrix containing smaller voids. In this analysis the formation of the larger voids with approximately

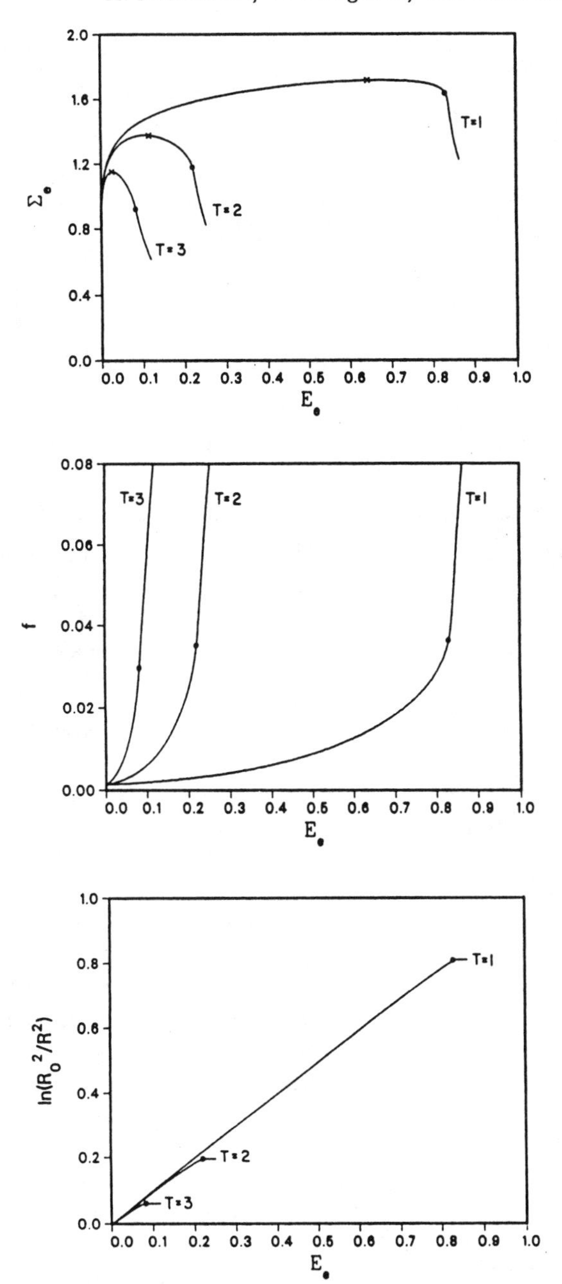

FIGURE 4.10. Finite element results for a cylindrical cell model with a spherical void and an initial void volume fraction $f_0 = 0.0013$, for different values of the stress triaxiality T (from Koplik and Needleman, 1988).

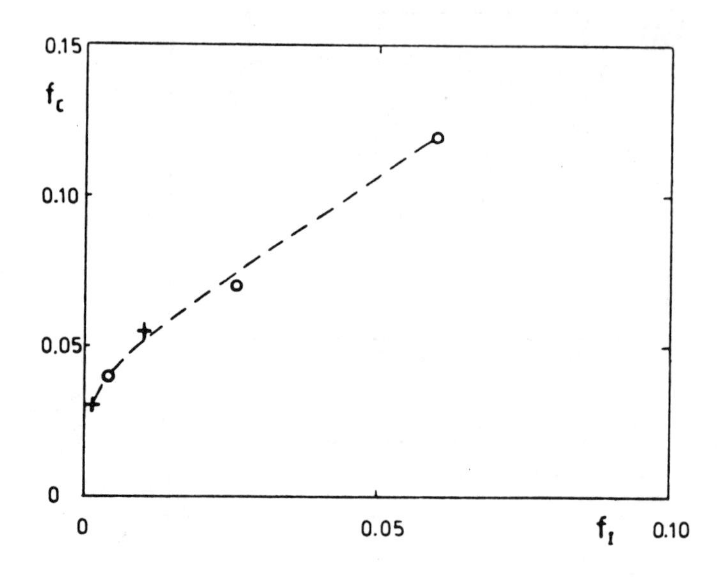

FIGURE 4.11. Dependence of the critical value f_c in Eq. 4.43 on the initial void volume fraction f_I according to various cell model studies (from Tvergaard, 1990).

spherical shape is formulated in terms of a porous ductile material model, and the focus is on the prediction of final fracture.

4.6 Boundary Value Problem Solutions

Calculations of localization and failure in a homogeneously deformed material element, provide much insight into the behavior of the constitutive relation. However, such calculations cannot capture a key feature that determines the observed ductile failure mode in test specimens and structural components; the redistribution of stress and deformation accompanying progressive micro-rupture. Here, some calculations that illustrate the development of failure modes in full specimens will be reviewed.

A characteristic feature of ductile fracture in structural metals is the contrast between the shear fracture mode observed in plane strain tensile specimens and the cup-cone mode observed in axisymmetric tensile specimens. In both cases, the deformations remain essentially homogeneous up to the maximum load point after which a diffuse neck develops. In the round bar tension test, diffuse necking is followed by a cup-cone type fracture, while typically in a plane strain tensile test of the same material, the deformation mode shifts to one involving localized shearing while the diffuse neck is rather shallow.

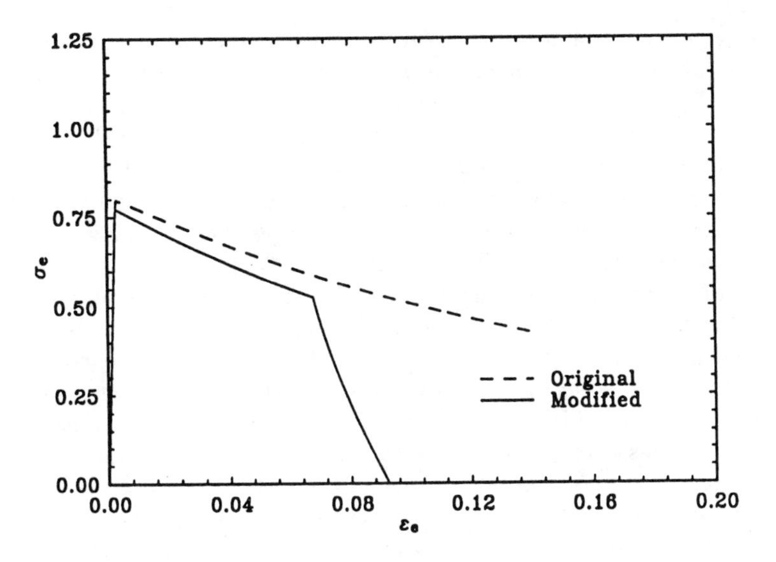

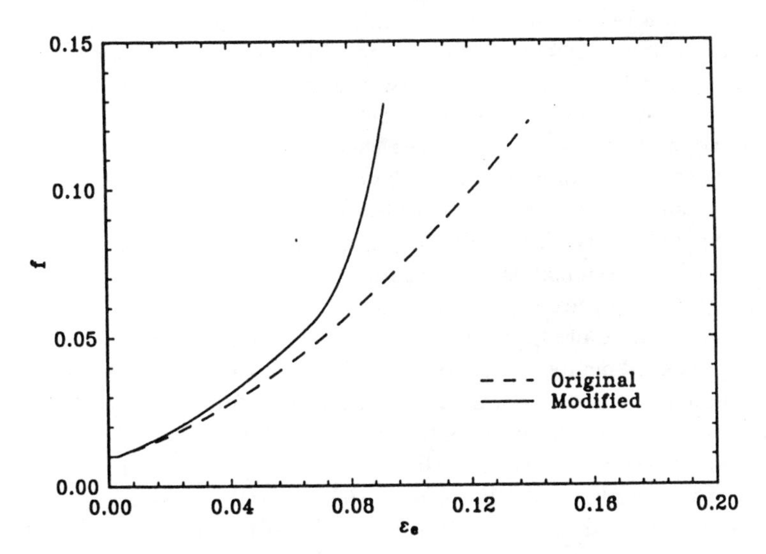

FIGURE 4.12. Comparison of overall stress strain behavior and void growth predictions based on the original, Eq. 4.21, and modified, Eq. 4.42, Gurson constitutive relations for a material element subject to uniaxial tension with a superposed hydrostatic tension, σ_h, with $T = \sigma_h/\sigma_e$. The matrix is ideally plastic and $f_0 = 0.002$.

Figures 4.13 and 4.14 show the fracture behaviors obtained from finite element calculations based on the modified Gurson constitutive relation, Eq. 4.42. In the calculations symmetry conditions are imposed so that only one quadrant of the specimen needs to be analyzed numerically and deformed finite element meshes are shown for this quadrant. Figure 4.13 is from Tvergaard and Needleman's (1984) finite element analysis of necking and failure in the tensile test, which incorporates a model for void nucleation into the formulation. The predicted fracture mode reproduces the essential features of a cup-cone fracture. Among the features typically observed in axisymmetric tensile tests of ductile metals is that voids initiate and grow in the center of the neck finally coalescing to form a central crack. There is a tendency for the crack to zig-zag, which is a consequence of shear localization being inhibited by the additional plastic work associated with the hoop strains that accompany shearing in the axisymmetric geometry. As the free surface is approached this axisymmetric constraint is relaxed, permitting the cone of the cup-cone fracture to form. This analysis shows how the interaction of the tendency to localization in a material weakened by void nucleation and growth together with a constraining geometrical effect lead to the cup-cone fracture. The development of failure has a significant effect on the overall load-displacement response of the specimen as shown in Fig. 4.13 (a). There is no corresponding geometrical constraint in plane strain tension so that, once initiated, a shear band can propagate across the entire specimen as illustrated in Fig. 4.14, which were obtained by Becker and Needleman (1986) using a rate dependent material model where Eq. 4.42 serves as a flow potential. In this case, the load drop is associated with the formation of the shear band and failure subsequently ensues as strain accumulates in the band.

These results illustrate the ability to qualitatively predict observed ductile failure behaviors. A meaningful quantitative comparison between predictions and experiment is more complex because of the path dependent and progressive nature of ductile fracture. Becker et al. (1988) have compared quantitative predictions of void growth, strength and ductility with detailed measurements in round notched bars. Various notch geometries were studied in order to obtain different stress histories. The tensile specimens were machined from partially consolidated and sintered iron powder compacts. The only experimentally determined quantities input into calculations in Becker et al. (1988) were the uniaxial stress-strain curve for the matrix material and the initial void volume fraction. The parameters q_1, q_2, f_c and f_f entering the porous plastic constitutive description were chosen to provide a reasonable fit of both strength and void growth predictions with results of micro-mechanical models of periodic arrays of voids as discussed in Section 4.5.

Figure 4.15 (a) shows a finite element mesh illustrating the axisymmetric specimen geometry and predicted evolution of the minimum section area,

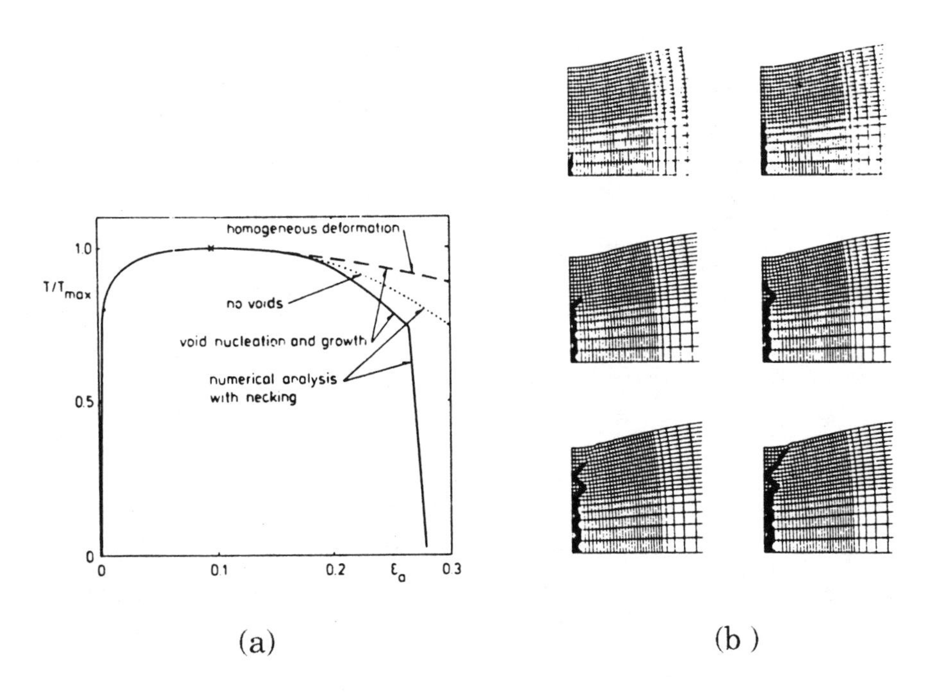

FIGURE 4.13. (a) Curves of load versus imposed strain and (b) crack growth in the neck for axisymmetric tension. In (b), triangular elements that have undergone a complete loss of stress carrying capacity are painted black (from Tvergaard and Needleman, 1984).

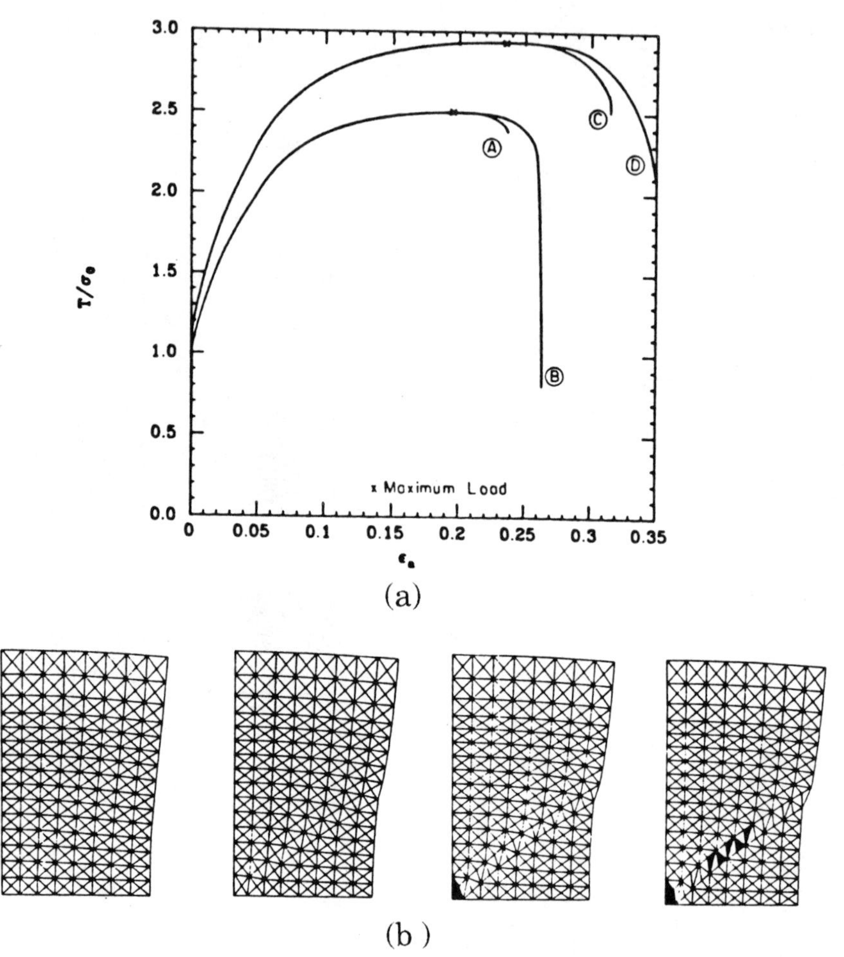

FIGURE 4.14. (a) Curves of load versus imposed strain and (b) crack growth in the neck for plane strain tension. The curve marked B in (a) pertain to the deformed meshes shown in (b), where triangular elements that have undergone a complete loss of stress carrying capacity are painted black (from Becker and Needleman, 1986).

$\ln(R_0^2/R^2)$, and the void volume fractions, f, measured at the center of the specimens. Also shown in Figs. 4.15 (b) and 4.15 (c) are measurements for eight specimens with 7% initial void volume fraction and one specimen with a 1% initial void fraction. The predictions are quite good at lower strains but the model somewhat under predicts void growth at large strains. Similar good agreement was found in comparisons made between prediction and experiment for other notch geometries.

4.7 Additional Considerations

When voids deform in a material subject to low triaxial tension the void volume grows little or the voids may even collapse into shapes like cracks or needles, as has been found by Budiansky, et al. (1982) for isolated voids. Often voids nucleate from a particle by debonding of the particle-matrix interface, and in such circumstances contact between the particle and the void surface may have a significant effect on void growth (Fleck, Hutchinson and Tvergaard, 1989). Figure 4.16 shows curves of normalized void dilatation rates versus stress triaxiality for an isolated void that has nucleated from a rigid spherical inclusion, in an elastic-perfectly plastic material subject to an axisymmetric stress state. At values of the triaxiality parameter around zero it is seen that the interaction with the particle enforces a void volume increase, because the contact with the particle gives rise to an internal pressure on the void surface in the transverse direction. For triaxialities higher than about 0.6 the presence of a particle inside the void has no influence, as the void expands in all directions.

Contact between particle and void surface is an important effect in simple shear, where an unfettered void closes as it deforms. This has been studied by Fleck et al. (1989) in terms of a planar elastic-perfectly plastic analysis for a row of voids in a shear field, with the geometry (Fig. 4.17) chosen to model experiments of Cowie, Azrin and Olson (1987). The development of the average shear stress, Σ_{12}, and the cross-sectional void area, A_v versus shear strain, γ, are shown in Fig. 4.18 for a superposed tensile or compressive stress, Σ_{22}. It is seen that for negative Σ_{22} the void volume still increases in the presence of a particle, but decays when there is no particle. The same type of behavior has been found for pure shear, $\Sigma_{22} = 0$.

When the stress triaxiality is high, material elasticity can have a major effect on the course of void evolution, even though the elastic strains remain small. This is because of the existence of a cavitation instability driven by the elastic energy in the surrounding material. For an isolated void in an elastic-plastic solid subject to remote hydrostatic tension there is a critical stress at which the void growth rate becomes unbounded, Bishop, Hill and Mott (1945), Hill (1950). This instability does not occur for a rigid-plastic solid. There has been much recent interest in cavitation instability

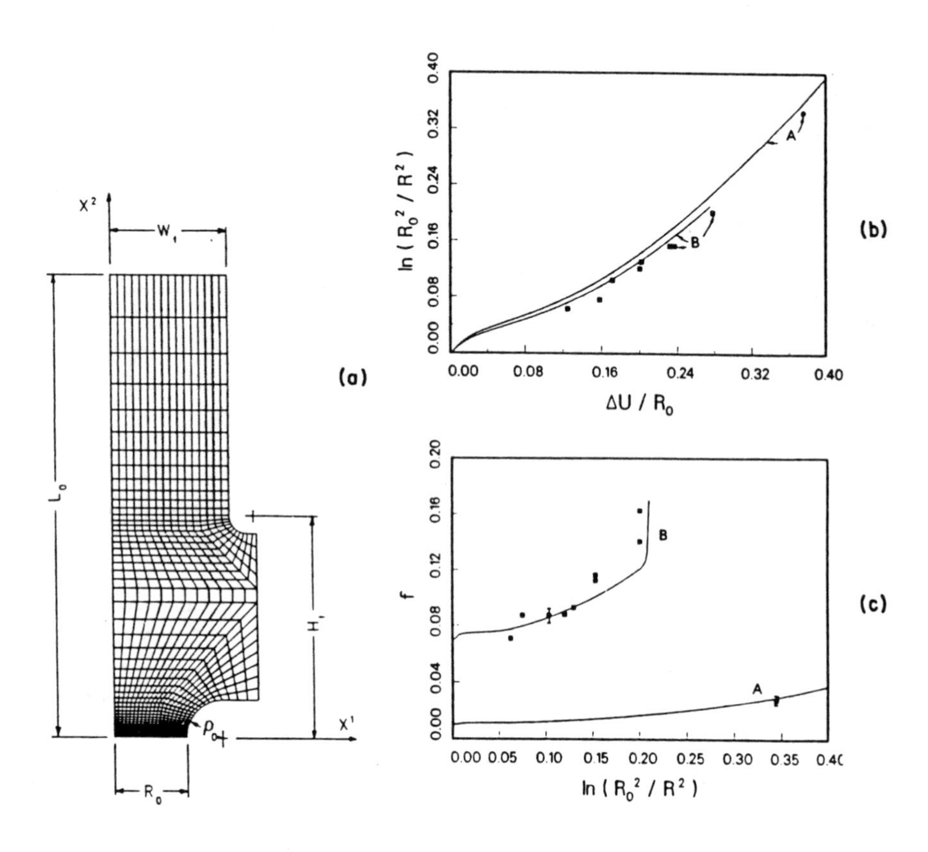

FIGURE 4.15. (a) Specimen geometry and finite element mesh. (b) and (c) respectively show the experimental and calculated development of the minimum section and porosity evolution as a function of strain for 1% and 7% initially porous notched bars from (Becker et al., 1988).

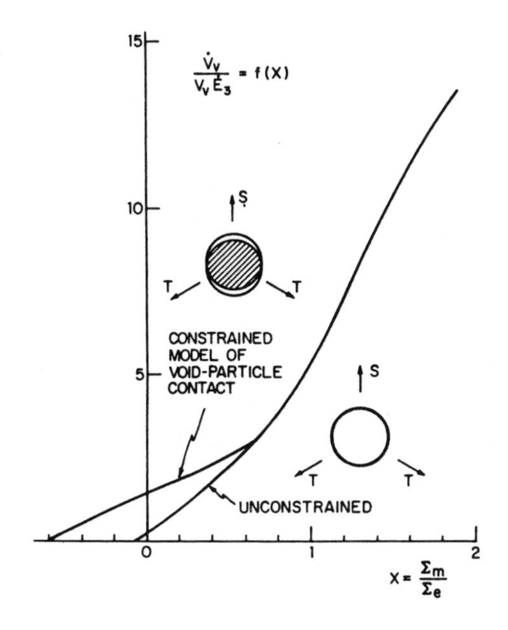

FIGURE 4.16. Normalized dilation rate following nucleation as a function of stress triaxiality (from Fleck et al., 1989).

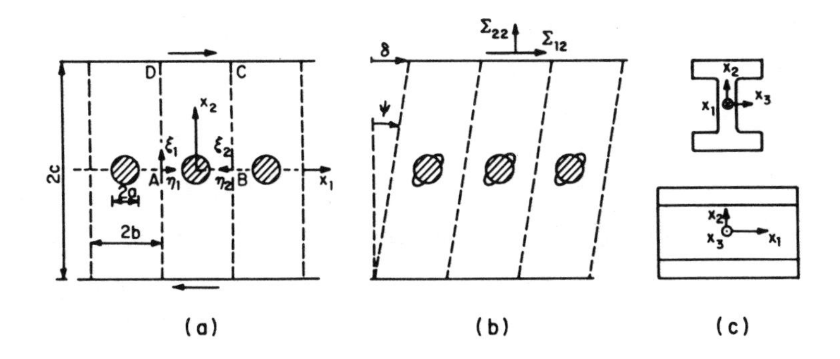

FIGURE 4.17. Model for the nucleation and growth of 2D voids in simple shear. (a) Starting geometry. (b) Subsequent to nucleation. (c) Test specimen modelled (from Fleck et al., 1989).

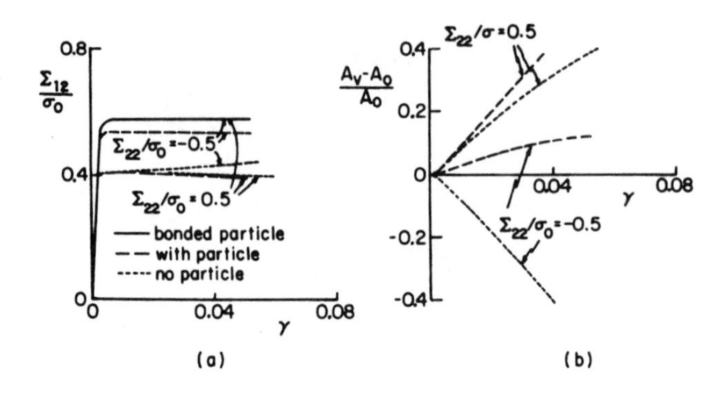

FIGURE 4.18. Effect of a superposed normal stress, Σ_{22}, for $a/b = 0.25$ in Fig. 4.17 (from Fleck et al., 1989).

phenomena from a variety of perspectives, e.g. Ball (1982), Horgan and Abeyaratne (1986), Abeyaratne and Hou (1989). Here, we illustrate the phenomenon by outlining the classic analysis for an isolated void in an isotropic, incompressible elastic-plastic solid subject to remote hydrostatic tension. The development and notation follows that in Huang et al. (1991).

We define R as the distance, in the current state, of a material point from the void center, R_i is the initial void radius and R_0 is the current void radius.

Equilibrium is expressed by

$$\frac{d\sigma_r}{dR} + \frac{2(\sigma_r - \sigma_\theta)}{R} = 0 \tag{4.44}$$

and incompressibility implies

$$\epsilon_r = -2\epsilon_\theta = \frac{2}{3}\ln[1 - (R_0^3 - R_i^3)/R^3] \tag{4.45}$$

The stress-strain relation for a Mises solid is of the form

$$\frac{\bar{\sigma}}{\sigma_y} = f(\bar{\epsilon}) \tag{4.46}$$

where

$$\bar{\sigma} = |\sigma_r - \sigma_\theta| \qquad \bar{\epsilon} = |\epsilon_r| \tag{4.47}$$

The boundary conditions are

$$\sigma_r(R_0) = 0 \qquad \sigma_r(\infty) = \sigma_\infty \tag{4.48}$$

From equilibrium and incompressibility,

$$\frac{\sigma_\infty}{\sigma_y} = -2\int_1^\infty f\left[\frac{2}{3}\ln\left[1 - \frac{(1-(R_i/R_0)^3}{x^3}\right]\right]\frac{dx}{x} \tag{4.49}$$

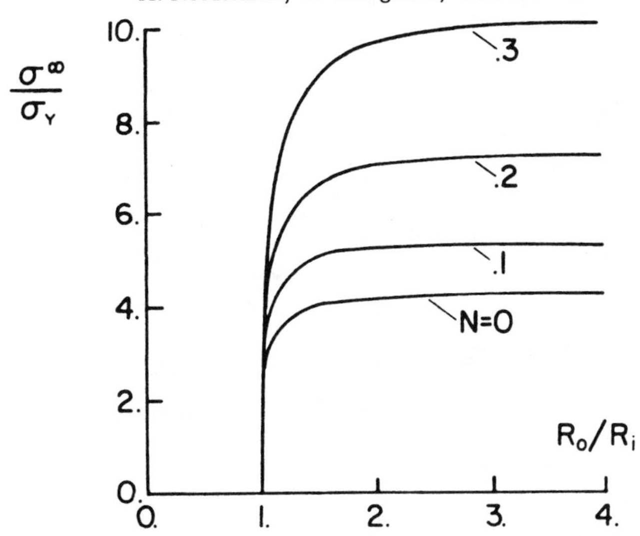

FIGURE 4.19. Remote stress versus radius of the cavity for hardening and non-hardening solids under spherically symmetric loading (from Huang et al., 1991).

The cavitation limit, $\sigma_\infty \rightarrow S$, is reached when $R_0/R_i \rightarrow \infty$. From Eq. 4.49,

$$\frac{S}{\sigma_y} = -2\int_1^\infty f\left[\frac{2}{3}\ln[1-x^{-3}]\right]\frac{dx}{x} \tag{4.50}$$

For a perfectly plastic solid, Eq 4.50 gives the result, Hill (1950),

$$\frac{S}{\sigma_y} \approx \frac{2}{3}\left[1 + \ln(\frac{2}{3\epsilon_y})\right] \tag{4.51}$$

Figure 4.19, from Huang et al. (1991), shows curves of remote stress versus cavity radius for a power law solid, $\bar{\sigma} \propto \bar{\epsilon}^N$. Because of the cavitation limit, the void growth rate at high stress triaxiality can be significantly greater than predicted based on a rigid-plastic analysis, as shown in Fig. 4.20.

Micromechanical analyses of porous solids have generally presumed that the voids are distributed in some regular pattern. However, void distribution effects have been shown to play a significant role in limiting ductility, both experimentally, Dubensky and Koss (1987), and theoretically, Becker (1987). Becker (1987) analyzed void growth and coalescence in a small material element, with the material characterized by the modified Gurson (1975) constitutive relation, but with a nonuniform initial distribution of the void volume fraction. Becker (1987) found a significantly smaller fracture strain for a solid with a non-uniform porosity distribution than for a solid with a uniform porosity at the average value. There was a negligible

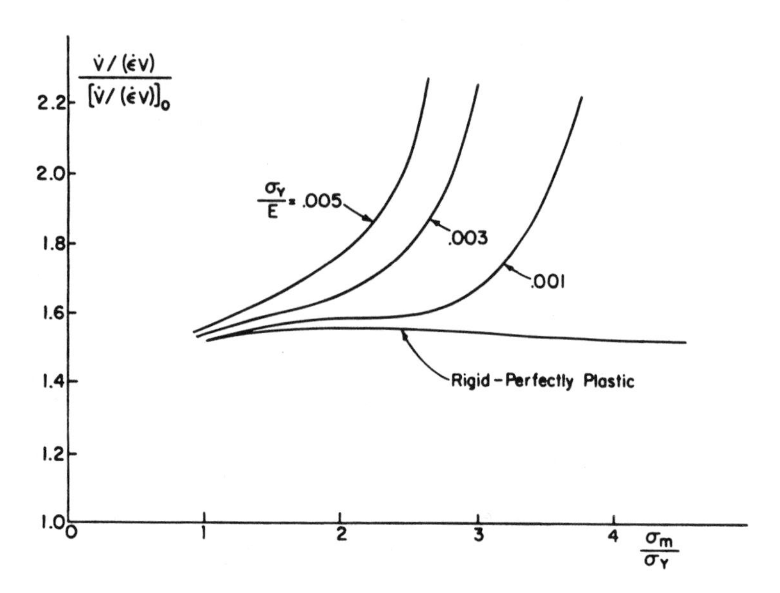

FIGURE 4.20. Effect of elasticity on the dilation rate of nominally spherical voids. The volumetric growth rate is normalized by that obtained from Eq. 4.14 using $(\dot{V}/\dot{\epsilon}_0 V)_0 = 3\dot{R}_0/\dot{\epsilon}_0 R_0$ (from Huang et al., 1991).

effect of the nonuniformity prior to the localization that precedes fracture. The quantification of distribution effects on ductile fracture processes remains a challenge for future research.

The phenomenological yield surfaces that have been proposed for porous plastic solids have all been smooth. Because of the non-homogeneous stress state around a void, a change in loading direction can result in unloading of part of the current plastic zone. This leads to a vertex on the yield surface as illustrated in Mear (1990). The vertices found by Mear (1990) were "blunt" so that the response did not differ greatly from what would be predicted based on a smooth yield surface.

Acknowledgments: A.N. is grateful for the support provided by the Brown University Materials Research Group on the Micromechanics of Failure-Resistant Materials, funded by the National Science Foundation. The work of JWH was supported in part by the Materials Research Laboratory (Grant NSF-DMR-89-20490) and the Division of Applied Sciences.

References

Abeyaratne, R. and Hou, H.-S. (1989). Growth of an infinitesimal cavity in a rate dependent solid. *J. Appl. Mech.*, 56:40.

Andersson, H. (1977). Analysis of a model for void growth and coalescence ahead of a moving crack tip. *J. Mech. Phys. Solids,* 25:217.

Ball, J. M. (1982). Discontinuous equilibrium solutions and cavitation in nonlinear elasticity. *Phil Trans. R. Soc. London,* A306:557.

Bishop, R. F. and Hill, R. (1951). A theory of the plastic distortion of a polycrystalline aggregate under combined stresses. *Phil. Mag. Ser. 7,* 42:414.

Bishop, R. F., Hill, R. and Mott, N. F. (1945). The theory of indentation and hardness tests. *Proc. Phys. Soc.,* 57:147.

Beachem, C. D. (1963). An electron fractographic study of the influence of plastic strain conditions upon ductile rupture processes in metals. *Trans. ASM* 56:318.

Becker, R. (1987). The effect of porosity distribution on failure. *J. Mech. Phys. Solids,* 35:577.

Becker, R. and Needleman, A. (1986). Effect of yield surface curvature on necking and failure in porous plastic solids. *J. Appl. Mech.,* 53:491.

Becker, R., Needleman A., Richmond, O., and Tvergaard, V. (1988). Void growth and failure in notched bars. *J. Mech. Phys. Solids,* 36:317.

Berg, C. A. (1962). The motion of cracks in plane viscous deformation. In Rosenberg, R. M., editor, *Proc. Fourth U.S. National Congress of Applied Mechanics,* ASME, NY, 2:885.

Budiansky, B., Hutchinson, J. W. and Slutsky, S. (1982). Void growth and collapse in viscous solids. In H. G. Hokins and M. J. Sewell, editors, *Mechanics of Solids,* Pergamon Press, Oxford, 13.

Cowie, J. G., Azrin, M. A. and Olson, G. B. (1990). Micro-void formation during shear deformation of ultrahigh strength steels. In Olson, G. B. et al., editors, *Innovations in Ultrahigh-Strength Steel Technology, Proceedings of the 34th Sagamore Army Materials Research Conference,* U.S. Government Publication, 357.

Dubensky, E. M. and Koss, D. A. (1987). Void/pore distributions and ductile fracture. *Met. Trans.,* 18A:1887.

Fleck, N. A., Hutchinson, J. W. and Tvergaard, V. (1989). Softening by void nucleation and growth in tension and shear. *J. Mech. Phys. Solids,* 37:515.

Fleck, N. A., Otoyo, H. and Needleman, A. (1991). Indentation of porous solids. *Int. J. Solids Struct*, in press.

Gilormini, P., Licht, C. and Suquet, P. (1988). Growth of voids in a ductile matrix: a review. *Arch. Mech.* 40:43.

Guennouni, T. and Francois, D. (1987). Constitutive equations for rigid plastic or viscoplastic materials containing voids. *Fatigue Fract. Engng. Mater. Struct.*, 10:399.

Gurland, J. and Plateau, J. (1963). The mechanism of ductile rupture of metals containing inclusions. *Trans. ASM*, 56:442.

Gurson, A. L. (1975). *Plastic Flow and Fracture Behavior of Ductile Materials Incorporating Void Nucleation, Growth and Interaction.* Ph.D. Thesis, Brown University.

Hadamard, J. (1903). *Leçons sur la Propagation des Ondes et les Équations de L'Hydrodynamique*, Paris, Chp. 6.

Hancock, J. W. and Brown, D. K., (1983). On the role of strain and stress state in ductile failure. *J. Mech. Phys. Solids*, 31:1.

Hancock, J. W. and MacKenzie, A. C. (1976). On the mechanisms of ductile failure in high-strength steels subjected to multi-axial stress-states. *J. Mech. Phys. Solids*, 24:147.

Hill, R. (1950). *The Mathematical Theory of Plasticity*, Clarendon Press, Oxford.

Hill, R. (1958). A general theory of uniqueness and stability in elastic-plastic solids. *J. Mech. Phys. Solids*, 6:236.

Hill, R. (1962). Acceleration waves in solids. *J. Mech. Phys. Solids*, 10:1.

Hom, C. L. and McMeeking, R. M. (1989) Void growth in elastic-plastic materials. *J. Appl. Mech.*, 56:309.

Horgan, C. O. and Abeyaratne, R. (1986). A bifurcation problem for a compressible nonlinearly elastic medium: growth of a microvoid. *J. Elast.*, 16:189.

Huang, Y. (1989) Accurate dilation rates for spherical voids in triaxial stress fields. Report Mech-155. Division of Applied Sciences. Harvard University.

Huang, Y., Hutchinson, J. W. and Tvergaard, V. (1991). Cavitation instabilities in elastic-plastic solids. *J. Mech. Phys. Solids*, 39:223.

Koplik, J., and Needleman, A. (1988). Void growth and coalescence in porous plastic solids. *Int. J. Solids Struct.*, 24:835.

Lemaitre, J. (1985). A continuous damage mechanics model for ductile fracture. *J. Engng. Mat. Tech.*, 107:83.

Mandel, J. (1966). Conditions de stabilité et postulat de Drucker. In Kravtchenko, J. and Sirieys, P. M., editors, *Rheology and Soil Mechanics*, Springer-Verlag, 58.

Marini, B., Mudry, F. and Pineau, A. (1985). Ductile rupture of A508 steel under nonradial loading. *Engng. Fract. Mech.*, 22:375.

Mear, M. E. (1990). On the plastic yielding of porous metals. *Mech. Matl.* 9:33.

McClintock, F. A. (1968). A criterion for ductile fracture by the growth of holes. *J. Appl. Mech.*, 35:363.

McClintock, F. A., Kaplan, S. M. and Berg, C. A. (1966). *Int. J. Frac. Mech.*, 2:614.

Nagpal, V., McClintock, F. A., Berg, C. A. and Subudhi, M. (1972). Traction-displacement boundary conditions for plastic fracture by hole growth. In Sawczuk, A., editor, *Foundations of Plasticity*, Noordhoff, 365.

Needleman, A. (1972). Void growth in an elastic-plastic medium. *J. Appl. Mech.*, 39:964.

Needleman, A. and Rice J. R., (1978). Limits to ductility set by plastic flow localization. In Koistinen, D. P. and Wang, N. M., editors, *Mechanics of Sheet Metal Forming*, Plenum Press, New York, 237.

Needleman, A. and Triantafyllidis, N. (1978). Void growth and local necking in biaxially stretched sheets. *J. Engng. Mat. Tech.*, 10:164.

Puttick, K. E. (1959). Ductile fracture in metals. *Phil. Mag.*, 4:964.

Rice, J. R. (1977). The localization of plastic deformation. In Koiter, W. T., editor, *Theoretical and Applied Mechanics*, North-Holland, 207.

Rice, J. R. and Tracey, D. M. (1969). On the ductile enlargement of voids in triaxial stress fields. *J. Mech. Phys. Solids*, 17:201.

Rogers, H. C. (1960). The tensile fracture of ductile metals. *Trans. Metall. Society AIME*, 218:498.

Shima, S. and Oyane, M. (1976). Plasticity theory for porous metals. *Int. J. Mech. Sci.*, 18:285.

Tipper, C. F. (1949). The fracture of metals. *Metallurgia*, 33:133.

Tvergaard, V. (1981). Influence of voids on shear band instabilities under plane strain conditions. *Int. J. Fract.*, 17:389.

Tvergaard, V. (1982). On localization in ductile materials containing spherical voids. *Int. J. Fract.*, 18:237.

Tvergaard, V. (1988). 3D-analysis of localization failure in a ductile material containing two size-scales of spherical particles. *Engng. Fract. Mech.*, 31:421.

Tvergaard, V. (1990). Material failure by void growth to coalescence. Adv. Appl. Mech., 27:83.

Tvergaard, V. and Needleman, A. (1984). Analysis of cup-cone fracture in a round tensile bar. *Acta Metall.*, 32:157.

Worswick, M. J. and Pick, P. J. (1990). Void growth and constitutive softening in a periodically voided solid. *J. Mech. Phys. Solids* 38:601.

Yamamoto, H. (1978). Conditions for shear localization in the ductile fracture of void-containing materials. *Int. J. Fract.*, 14:347.

5

Crack Blunting and Void Growth Models for Ductile Fracture

R. M. McMeeking

ABSTRACT In this chapter a review is presented of the work over a number of years in the area of crack blunting and void growth in ductile materials. These phenomena contribute to the process of crack growth by ductile fracture. The models of void growth involve both continuum averaging methods and idealized configurations involving only a few voids.

5.1 Introduction

Frank McClintock's contributions to the area of ductile fracture have been seminal. Furthermore, his influence on many researchers who have worked in this area has been profound. It is not an exaggeration to say that most of the work in the last twenty five years concerned with modeling crack blunting and void growth has been inspired by his research and influenced by his thinking. Certainly the work presented here has been. Thus it is very appropriate to devote a chapter on this important topic assessing the contributions of McClintock in retrospect.

McClintock (1968) provided the first comprehensive model of ductile fracture by the growth of holes, thereby founding this whole theoretical area. Furthermore, he recognized early the importance of crack blunting (McClintock, 1969) and his student Neimark (1968), following Lee and Wang (1954), provided analysis which can be used to understand the stresses and deformations around smoothly blunting crack tips. McClintock (1971) has also been interested in blunting by flow at vertices on the tips of cracks and on the effects of asymmetry on blunting. His model for hole growth and coalescence with a blunting crack tip (McClintock, 1969) was the first effective treatment of this process and even to this day has not been supplanted by a better analysis.

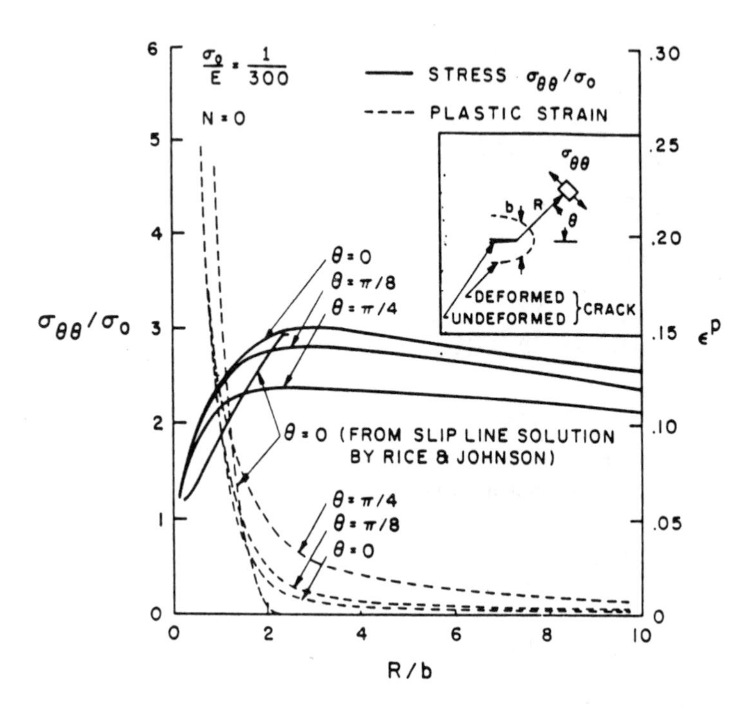

FIGURE 5.1. Stresses and tensile equivalent plastic strains near a blunting crack tip in plane strain small scale yielding in an elastic perfectly plastic material from McMeeking (1977a).

5.2 Crack Tip Blunting

Rice and Johnson (1970) studied crack tip blunting in mode I by using slip line solutions. The near tip behavior of elastic-perfectly plastic materials in plane strain approaches that of rigid-perfectly plastic materials and justifies the use of slip line theory. Their solution enforced smooth blunting of the crack tip and as a result a region of high strain develops ahead of the crack in contrast to the sharp crack case. The stress adjacent to the free surface is limited by yielding but builds up away from the tip due to triaxiality. As in the sharp crack case (Rice, 1968) the largest stress is $3\sigma_0$ where σ_0 is the yield stress in tension and is located a distance approximately $2\delta_t$ ahead of the tip where δ_t is the crack tip opening displacement. The solution indicates clearly that there is a high strain and stress region near the tip in which plastic instabilities can occur.

McMeeking (1977a) used large deformation finite element analysis to model the blunting of a sharp crack under small scale yielding in plane strain conditions. He considered both nonhardening and hardening elastic-plastic materials. The constitutive law was that of von Mises with isotropic hardening (J_2 flow theory) and accounted for rotation of the principal deformation directions. The small scale yielding solution was achieved by applying displacement boundary conditions remote from the crack tip to impose an asymptotic dependence on the mode I elastic crack tip singular stress field. Figure 5.1 shows the true stress $\sigma_{\theta\theta}$ (see inset of Fig. 5.1) normalized by the true tensile yield stress σ_0, vs the distance from the notch tip in the undeformed configuration for an elastic perfectly-plastic material. The distance is normalized by the current notch-opening, which allows results from the later increments of the finite element calculations to be plotted together. As the notch tip is approached at a given angle to the crack line in the undeformed configuration, the stress rises due to increasing strain. However, the hydrostatic stress cannot be maintained on the blunted notch surface and as a result there is a maximum for $\sigma_{\theta\theta}$, coinciding with a maximum for hydrostatic stress, some distance from the notch tip. Figure 5.1 also shows equivalent plastic strain plots from the later increments of the finite element solutions. The strains are clearly small except very close to the blunted tip. Outside the near-tip region, the large plastic strains are on lines at an angle to the crack plane and this is in agreement with the HRR solution (Rice and Rosengren, 1968, Hutchinson, 1968). For comparison, the stress and plastic strains ahead of the crack ($\theta = 0$) from the slip line solution of Rice and Johnson (1970) have been plotted in Fig. 5.1. The agreement between the finite element results and the slip line results are quite close as far as the position and magnitude of the stress maximum are concerned.

Similar finite element calculations for power-law hardening materials show that the magnitude of the stress at the maximum is higher and that the position of the stress maxima move closer to the notch tip as shown in Figure 5.2. Another hardening effect is an upturn in stress close to the notch surface, which arises from the elevation of flow stress by the large plastic strains in this area. In fact, when a sharp crack in a power-law hardening material is blunted to a finite width, infinitely large stress on the notch surface will arise, but only over a distance small compared to the blunted-crack width. This stress singularity arises because in the power-law hardening model infinite plastic strains produce infinite flow stresses, whereas a saturation to constant flow stress after large plastic strains is perhaps more physically realistic. In addition, few cracks are likely to be atomistically sharp and the finite initial radius at the tip of most real cracks will lead to only large but finite plastic strains at the tip after blunting.

Models of crack tip blunting in small scale yielding in perfect plasticity which involve corners on the crack tip and localized shearing there have been developed. McMeeking (1977b) used the methods of Rice and Johnson (1970) to develop suggestions of McClintock (1971) for near-tip slip line fields for cracks with 2 and 3 corners. They are valid near tip approximations without any untenable incompatibilities because the discontinuities on the tip are accommodated by diffuse straining in the far field. The stress

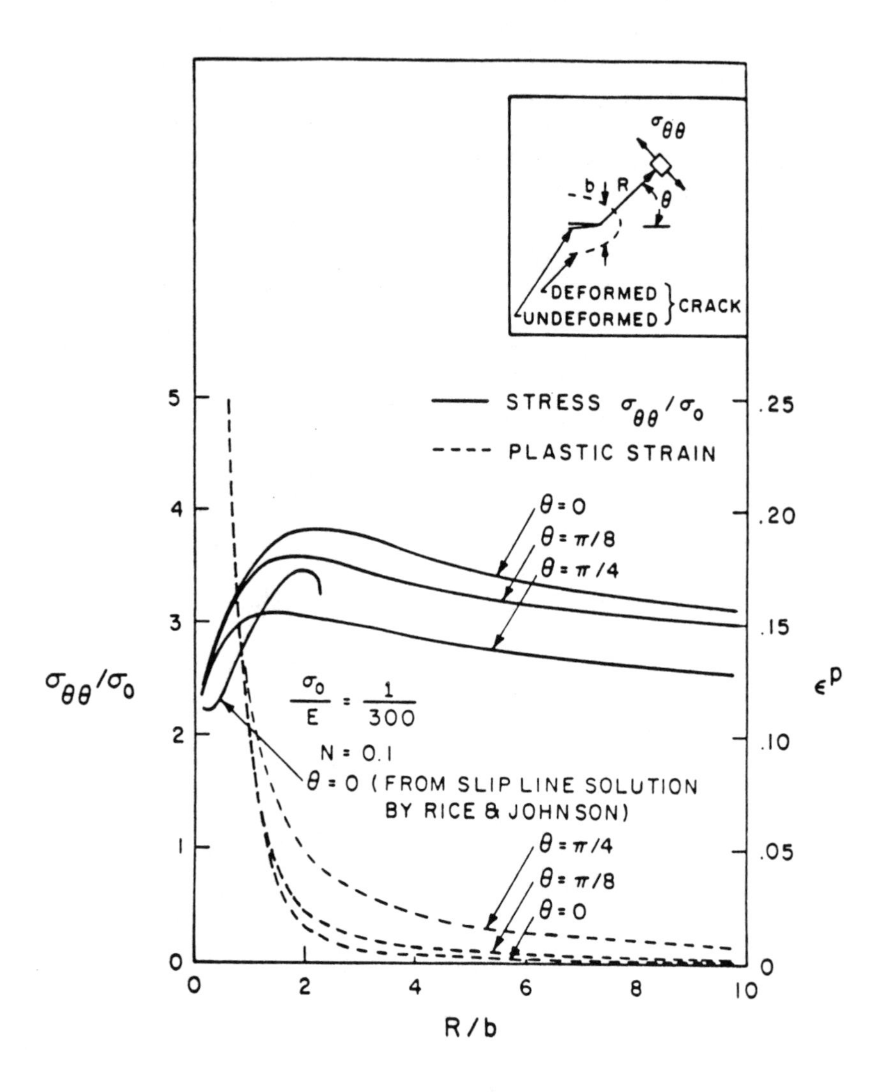

FIGURE 5.2. Stresses and tensile equivalent plastic strains near a blunting crack tip in plane strain small scale yielding in an elastic plastic material with strain hardening from McMeeking (1977a).

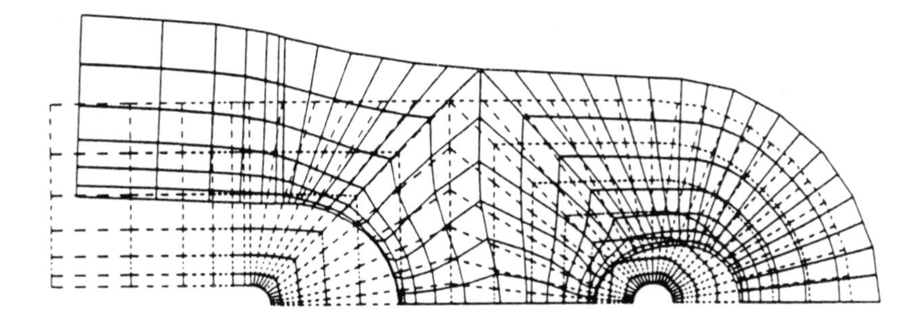

FIGURE 5.3. Near tip finite element mesh for a cylindrical hole growing near a blunting crack tip. The original mesh is in dashed lines, the deformed mesh is in full lines from Aravas and McMeeking (1985a).

fields are similar to that given by Rice and Johnson for smoothly blunting tips although they differ in detail. There are straight segments on the crack tip between each corner and the flow field in the adjacent material involves straining by small amounts. The blunting process takes place by material being brought from the interior at the corners and smeared onto the tip to create new surface. This is in contrast to the smooth blunting case where the opening takes place by a general stretching of the surface materials and the strains are intensely high in all the adjacent material.

There is a question as to which mode of blunting, the smooth or the cornered, should prevail for perfectly plastic materials. It cannot be deduced from the solution itself if it is assumed to start from a mathematically sharp crack tip. The first instant of blunting would determine the shape which would form and it seems likely that that shape would continue thereafter. However, if the starting configuration is blunt already and has cornered features then that shape might prevail thereafter. Thus the shape of a prefatigue crack, pit, or flaw is of some importance. Similarly, a crack tip with a brittle surface layer that cracks in a few locations might set off the cornered blunting. In addition, one case could be destabilized into another perhaps by features of the microstructure that eventually favor one mode over the other. Observations do not settle the issue although the smooth case seems to be seen more often than one with corners.

5.3 Ductile Rupture at Crack Tips

A mechanism of local fracture at a crack tip can be the generation of voids from inclusions and their subsequent growth and coalescence in a ductile mode. Aravas and McMeeking (1985a,b) used large deformation finite element analysis to study the near crack tip growth of long cylindrical voids aligned parallel to a mode I plane strain blunt crack under small scale yielding conditions. The results of the calculation provide a reasonable

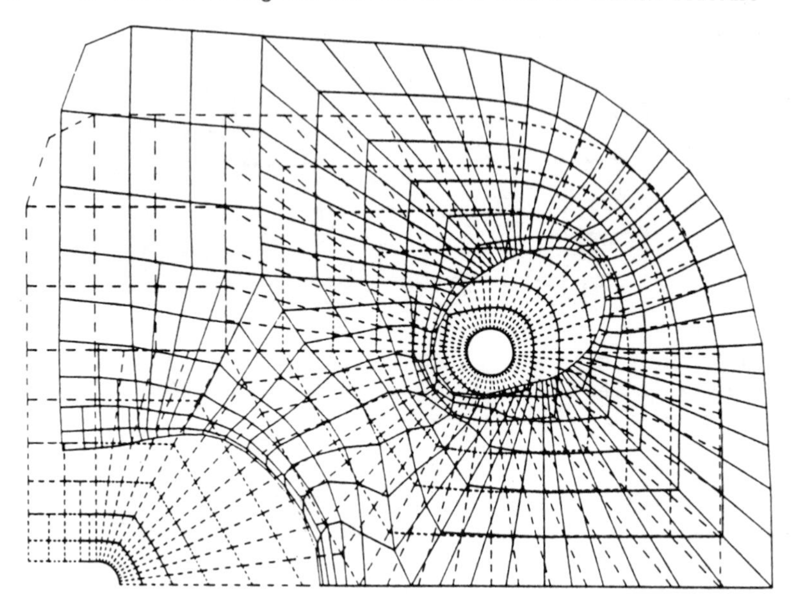

FIGURE 5.4. Near tip finite element mesh for a pair of cylindrical holes symmetrically placed near a blunting crack tip. The original mesh is dashed, the deformed mesh is full from Aravas and McMeeking (1985a).

model for the behavior of holes generated by long stringers parallel to the crack, such as those in specimens cut in the long transverse direction of a rolled plate. Two different configurations were analyzed: one with a single hole ahead of the crack and one with two holes at 30° to the crack line. Several values of the spacing to size ratio of the inclusions were considered and the effects of this ratio on the conditions for fracture initiation were examined. In one set of calculations the elastic-plastic material was assumed to be fully dense and the possibility of smaller-scale voids in the ligament between the large void and the crack tip was not taken into account. The J_2 flow theory, suitably modified to account for rotation of the principal deformation axes, was used to describe the constitutive behavior of the material. Figure 5.3 shows the deformed finite element mesh in the near tip region superposed on the undeformed one (dashed lines) for the case of the void ahead of the crack. The deformed configuration for the case of the two voids at 30° to the crack plane is shown in Fig. 5.4. In both cases the holes are pulled towards the crack tip and change their shape approximately to that of an ellipse with the major axis being radial to the crack. This shows that the effect on the hole growth of the interaction of the neighboring free surfaces is stronger than the effect of the mainly tensile stress field ahead of the crack tip. The results of Aravas and McMeeking (1985a) show that the cylindrical holes ahead of the crack grow faster than those at 30° and this is rather different from what has so far been inferred for the growth of spherical voids, for example by McMeeking (1977a).

Aravas and McMeeking (1985b) took the possible nucleation of a set of smaller-scale voids into account by using Gurson's (1975, 1977a, 1977b) equations as modified by Tvergaard and Needleman (1984), to model the constitutive behavior of the matrix material in a scale free manner. Using the modified Gurson equations and a method proposed by Tvergaard (1982) to model material failure, they studied the formation and growth of the microcrack in the ligament between the larger hole and the crack tip. In this way, the final stage of coalescence of the larger hole with the crack tip was analyzed in detail. The difference between the predicted COD at fracture initiation between the Gurson values and those based on results for the fully dense material is small. This shows that the results obtained using the fully dense elastic-plastic material together with some geometric criterion to predict localization are, numerically, quite satisfactory.

Calculations similar to those of Aravas and McMeeking (1985a,b) have been carried out by Aoki et al. (1984). They too used the Gurson laws to analyze the behavior of a failing ligament between the blunt crack and a large cylindrical hole. In their case the hole grows into a shape different from that predicted by Aravas and McMeeking and this discrepancy is unresolved. Some of the calculations of Aoki et al. concern strains due to microvoid growth around a blunt tip in the absence of a larger void. These results are probably relevant to very clean alloys or powder compacts with residual porosity. In more typical cases, the role of microvoids is in the interaction between a larger hole and the blunt crack, as analyzed by Aravas and McMeeking (1985a,b). Aoki et al. (1985) have extended their calculations to mode II (in plane shearing) and mixed mode conditions. These calculations are of interest as they give finite deformation results for mixed mode blunt cracks, including the sharpening of one side of the blunt tip due to near tip shearing.

Needleman and Tvergaard (1987) have used a blunt crack tip surrounded by the Gurson material in calculations of ductile rupture. In these calculations, however, there are islands of weak material distributed throughout the body. These islands nucleate damage at fairly low strains and should be thought of as colonies of inclusions which give rise to large voids in certain alloys such as those studied by Hancock and Mackenzie (1976). The remainder of the material nucleates voids at high strains and represents portions of the alloy containing small tenacious inclusions. The analysis is plane strain, and so in the calculation the weak colonies form long cylinders through the material. As the crack is blunted, voids effectively form in the weak regions near the tip. Eventually, the ligaments between a nearby void and the blunt crack loses strength almost entirely and the crack has grown to absorb the first void. The calculations were continued so that coalescence with the crack of voids further afield was achieved. In this way, continuing crack growth was induced. In some of the calculations, patterns of crack propagation were observed that were very similar to experimental shapes seen by Green and Knott (1976). Shear localization occurred also in the calculations. However, the predictions of toughness phrased in terms of critical crack tip opening displacement for crack growth initiation are similar to those of Aravas and McMeeking (1985a,b) and all are in reasonable agreement with the available experiments. Thus the prediction of toughness is not very sensitive to the presence of localization in these particular calculations compared to the differences among the experimental observations. However, the morphology of crack growth is reproduced more effectively in the localized shearing of the calculations of Needleman and Tvergaard (1987).

Hom and McMeeking (1989) carried out large deformation finite element calculations for the growth of initially spherical holes directly ahead of a mode I plane strain blunting crack tip in an elastic-plastic material in small-scale yielding. Elastic-perfectly plastic and strain hardening calculations were carried out, but only the non-hardening results will be reviewed here. The yield strain in uniaxial tension was 1/300 and Poisson's ratio was 0.3. A periodic array of holes parallel to the crack front was used so that the analysis could focus on a slice of material representing half the material between neighboring voids as shown in Fig. 5.5. Boundary conditions imposing the symmetry and periodicity were applied to this cell to produce an overall plane strain response with respect to the crack tip. The crack and void surfaces were traction free. Around the perimeter of the outer semi-circular boundary far from the crack, displacement boundary conditions were used to impose an asymptotic dependence on the elastic crack tip singular field of Irwin (1960). The diameter of the pre-existing voids was a_0 as was the initial COD of the crack. The distance from the center of the void to the center of the crack tip was initially D_0 and this was also the center-to-center spacing of the voids. Ratios of 10 and 4.5 were used for D_0/a_0 in the calculations.

The problems were solved using the ABAQUS (1984) finite element program. The typical mesh in Fig. 5.5 has 5,535 nodes with 16,605 degrees of freedom and 740 twenty-noded isoparametric brick elements. The mesh was made larger effectively by using an embedding technique developed by McMeeking (1977a) which makes use of a plane strain crack tip blunting, elastic-plastic solution without voids to provide results which were used for the actual boundary condition in the three-dimensional calculation. Further details on the calculations can be found in Hom and McMeeking (1989).

Figures 5.6 and 5.7 show two views of the near tip deformed finite element mesh superimposed on the undeformed mesh at a load level of $J/\sigma_0 a_0 = 3.3$ for a non-hardening material with $D_0/a_0 = 4.5$. Figure 5.6 is a section cut perpendicular to the crack front, through the center of the void, while Fig. 5.7 is a section cut in the plane of the crack. Line A–A identifies the crack front in Fig. 5.7. These figures show that the ligament between the void and the crack is contracting. As it is pulled towards the crack, the hole changes shape and becomes approximately an oblate spheroid whose

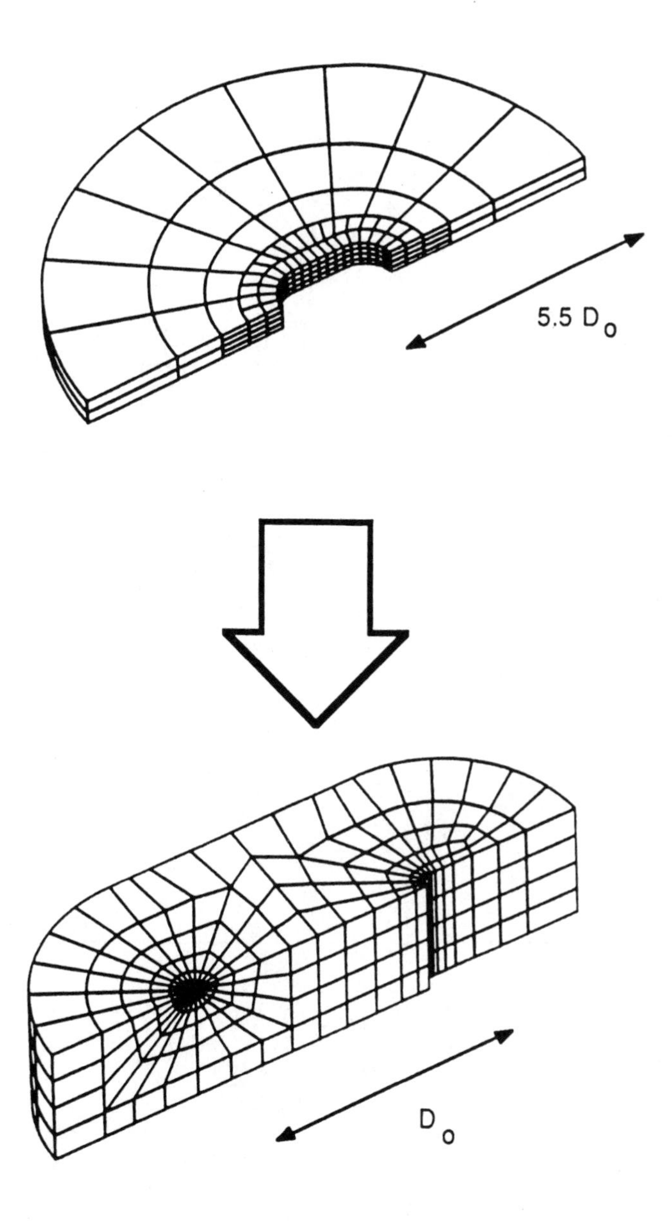

FIGURE 5.5. Typical finite element mesh used to solve the problem of three dimensional void growth before a blunting crack tip from Hom and McMeeking (1989). The mesh in Figure (a) surrounds the mesh in Figure (b).

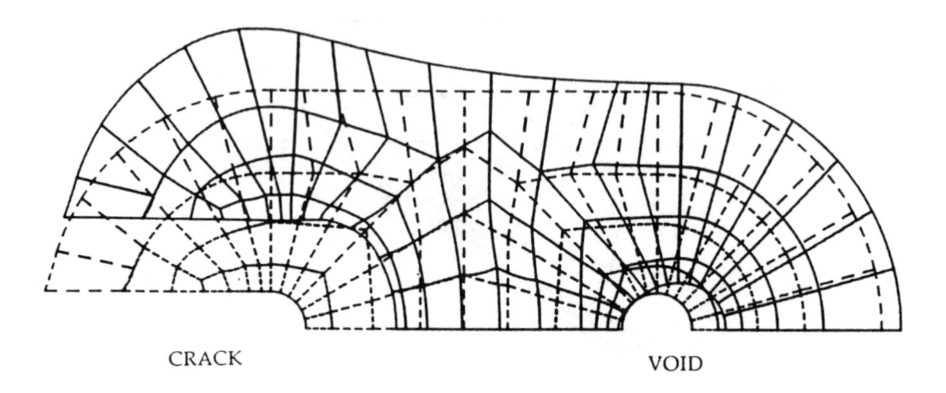

FIGURE 5.6. The deformed mesh in the near tip region for a section cut which is perpendicular to the crack and which passes through the center of the void from Hom and McMeeking (1989). The material is perfectly plastic and the load level is $J/\sigma_0 a_0 = 3.3$.

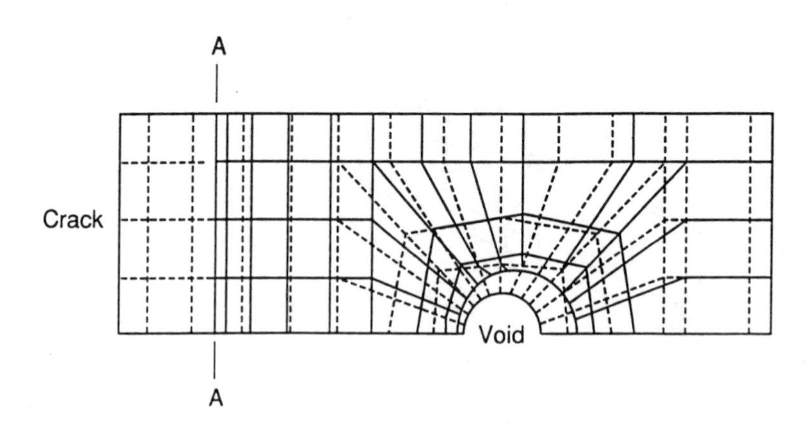

FIGURE 5.7. The deformed mesh as for Fig. 5.6 but for a section in the crack plane from Hom and McMeeking (1989).

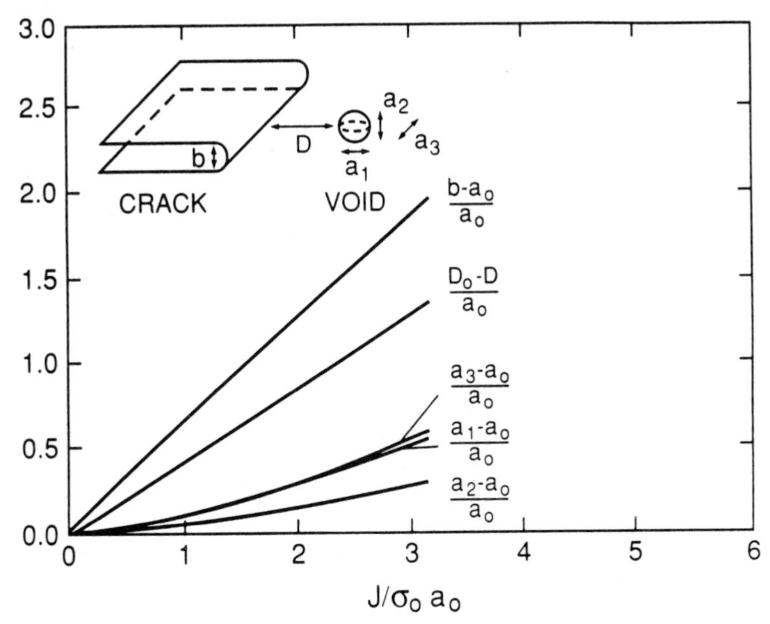

FIGURE 5.8. Plot of the COD, the ligament between the crack tip and the hole, and the void's dimensions versus applied load for a material with $N = 0$ from Hom and McMeeking (1989).

major axes are in the plane of the crack.

Figure 5.8 shows the hole growth magnitudes, the crack tip opening displacement (COD) and the ligament size between the void and the crack tip as functions of the applied load for the initial void spacing $D_0/a_0 = 4.5$ in the perfect plasticity case. Here b is the current COD, and D is the current ligament size. The quantities a_1, a_2 and a_3 are the dimensions of the hole in the deformed configuration as illustrated in the inset of the figure. The holes expand in every direction, but the voids grow fastest towards the crack tip and towards neighboring voids and thus become oblate as indicated above. The a_2 axis of the void grows at approximately half the rate of the other two axes and the effect is distinct. This shows that the interaction between the void and the crack is strong and overcomes the effect of the mainly tensile stress field ahead of the crack tip which would be expected to elongate the void in the tensile direction. Throughout the load history, a_3 is approximately equal to a_1, indicating that there is also interaction between neighboring voids ahead of the crack tip. However, the ligament between the crack and the voids is still smaller and contracting at a faster rate than the ligament between neighboring voids because of relative motion of the crack tip towards the voids induced by the void-crack interaction. Therefore the voids interact more strongly with the crack than they do with one another. Aravas and McMeeking (1985a) also found that

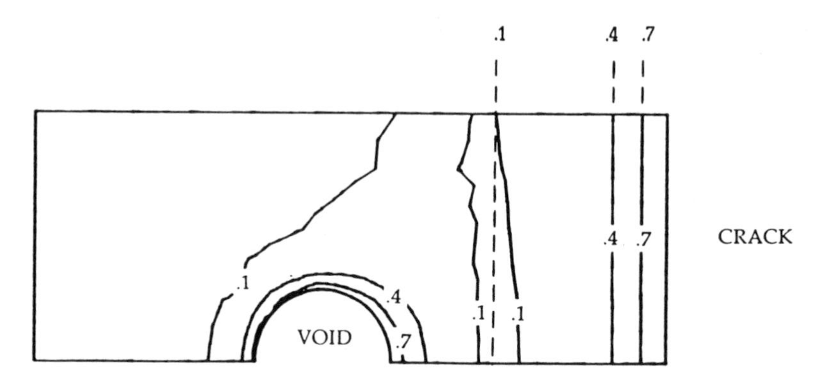

FIGURE 5.9. Contour plot of $\bar{\varepsilon}^p$ in the near crack tip region a section in the crack plane. The dashed lines indicate the contour levels when the voids are not present from Hom and McMeeking (1989).

cylindrical voids grow fastest in the direction towards the crack tip. However, their cylindrical voids grew approximately five times the rate of the initially spherical voids. This indicates that void growth strongly depends on the shape of the void, with holes which individually extend far along the front of the crack being free to grow rapidly. This effect must be a strong void-crack interaction since it is not suggested by what is known about the growth of isolated cylindrical and spherical voids McClintock (1968) and Rice and Tracey (1969). The results can be compared with predictions using the model of Rice and Tracey (1969) for an isolated spherical void under a remote uniform loading which is taken as the crack tip field. For the perfect plasticity case, the voids in the finite element calculation grow at twice the rate predicted from the Rice and Tracey model. Also in the Rice and Tracey analysis the void diameter a_2 grows faster than the other void dimensions, contrary to the more accurate finite element calculation. It is interesting to note that the predicted growth of the vertical dimension a_2 of the void is roughly the same for the analysis of Rice and Tracey and the finite element calculation. Therefore, vertical growth of the hole seems unaffected by interaction with the crack and neighboring voids.

Figure 5.9 is a contour plot of the effective plastic strain $\bar{\varepsilon}^p$ in the near tip region for the perfect plasticity case with $D_0/a_0 = 4.5$ and $J/\sigma_0 a_0 = 3.3$. There are large plastic strains near the void and in the ligament between the crack and the void. In Fig. 5.9 giving a view of the crack plane, the contour levels when the voids are not present are also plotted with dashed lines. It can be seen that around the voids the level of effective plastic strain has been greatly elevated and this effect spreads sideways between neighboring voids. Figure 5.10 is a contour plot of the hydrostatic stress $\sigma_{kk}/3$ in the near tip region for $D_0/a_0 = 4.5$ when $J/\sigma_0 a_0 = 3.3$ and when N the power-law exponent on the plastic strain is zero, i.e. the material

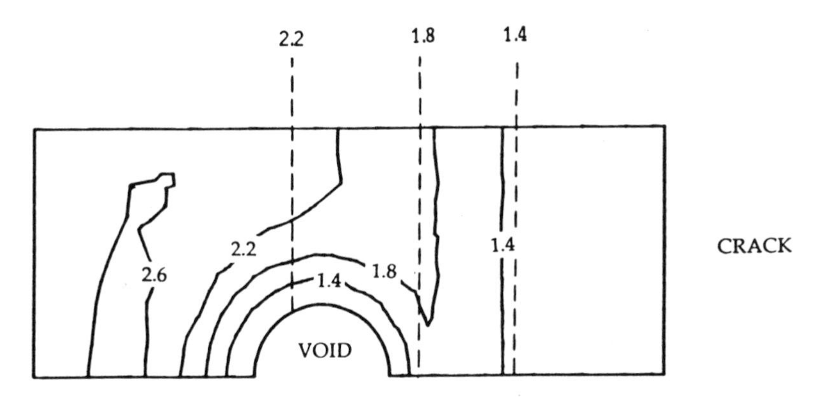

FIGURE 5.10. Contour plot of hydrostatic stress ($\sigma_{kk}/3$) in the near crack tip as for Fig. 5.9 (from Hom and McMeeking (1989)). The dashed lines indicate the contour levels when the voids are present.

is non-hardening. In the plan view of Fig. 5.10, the dashed lines indicate the contour levels when the voids are not present. The void is closer to the crack tip at this stage than the position of the maximum in hydrostatic stress when there is no void. As a consequence the hydrostatic stress in the ligament is relatively low. Comparison of the dashed lines with the full contours indicates a slight drop in the hydrostatic stress in the ligament due to the presence of the voids. Of course, around the void the hydrostatic stress is low because of the free surface of the void. Parallel to the crack front, the triaxiality has built up to nominal levels within about one current void radius. From Fig. 5.10, we can conclude that the presence of the void does not elevate the hydrostatic stress in the ligament significantly but has a more marked effect on the material between neighboring voids.

5.4 Hole Coalescence and Fracture Initiation

In metals, ductile crack advance occurs when individual voids or concentrations of voids coalesce with the crack tip. The process is often initiated when shear localization due to microvoid growth develops in the ligament between the crack tip and neighboring voids. Several criteria for determining when coalescence involving major voids occurs have been proposed based on the length of the ligament relative to the dimensions of the void. Hom and McMeeking (1989) use three of these criteria to interpret their void growth results and to predict COD values for fracture initiation. Rice and Johnson (1970) proposed that coalescence occurs when the size of the ligament between the crack tip and the void is equal to the vertical dimension of the hole. This criterion is $D = a$. Le Roy et al. (1981) formulated a similar criterion based on experimental observation of spheroidized carbon-steels under tensile strain. They proposed that void linkage occurs when the longest axis of the void is of the order of magnitude of the mean nearest neighbor spacing in the plane of the crack, or $a^{\max} = \phi D$. The parameter ϕ is an experimentally determined constant, which is 0.83 for voids nucleated from spherical particles and 1.23 for voids nucleated from elongated particles. From the computations of Hom and McMeeking (1989), a_1 is the largest dimension of the initially spherical void, so that the criterion for this model becomes $a_1 = 0.83D$. Finally, a conservative upper bound to coalescence is the criterion that the ligament must neck down to a point. This means $D = 0$.

TABLE 5.1. The ratio b_f/D_0 for various void coalescence criteria computed from the finite element results of Hom and McMeeking (1989).

	$N = 0$		$N = 0.2$	
	$D_0/a_0 = 45.$	$D_0/a_0 = 10.$	$D_0/a_0 = 4.5$	$D_0/a_0 = 10.$
$D = 0$	1.49	1.73	2.31	2.35
$a_2 = D$	1.01	1.40	1.43	1.95
$a_1 = .83D$	0.83	1.21	1.59	1.84

Estimates for the notch width b_f at fracture initiation have been obtained by Hom and McMeeking (1989) using the finite element calculations and the three coalescence criteria just discussed. Fracture initiation was identified with the coalescence of the voids separately with the crack tip. The finite element calculations were terminated well before these processes of coalescence would have commenced. However using the rates of void growth, crack tip blunting and ligament contraction prevailing when the numerical analysis was stopped, Hom and McMeeking (1989) were able to extrapolate the results to obtain values for the near tip dimensions at the time of coalescence. Table 5.1 shows the predictions for b_f in the four cases examined without ($N = 0$), and with hardening ($N = 0.2$). The results are very dependent on inclusion spacing and void size. In contrast, Aravas and McMeeking (1985a) found that for cylindrical voids b_f/D is nearly independent of D/a. Therefore, b_f is more sensitive to the size of the spherical inclusions than it is to the in plane dimensions of long cylindrical inclusions when they are controlling ductile fracture initiation.

The COD at fracture initiation has been measured by several researchers experimentally, and Fig. 5.11 is a summary of those results plotted in the form of b_f versus inclusion spacing. Most of the experimental results are for approximately spherical inclusions loosely bonded to the matrix. The predictions of the finite element calculations of Hom and McMeeking (1989) have been plotted with the criterion $D = a$ and they agree quite well with the experimental data. The results of the models of Rice and Johnson (1970) and McMeeking (1977a) have been plotted also for comparison with the finite element results. Their results tend to overestimate the COD

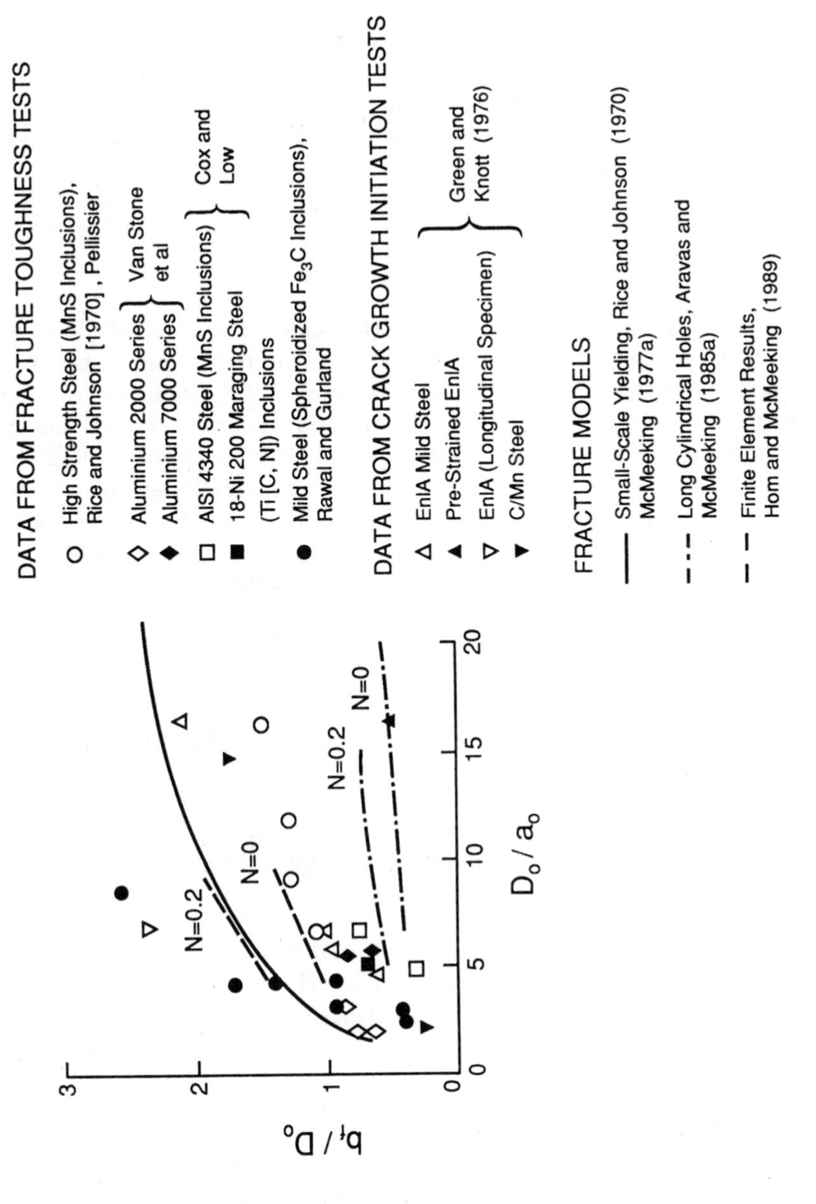

FIGURE 5.11. Experimental data and finite element predictions for the COD at initiation of crack growth or fracture, related to particle spacing D_0 and particle size a_0 (from Hom and McMeeking (1989)). Also plotted are results for fracture due to elongated voids from a model by Aravas and McMeeking (1985a), and due to spherical voids from models by Rice and Johnson (1970), and McMeeking (1977a).

at fracture initiation. The finite element results of Hom and McMeeking (1989) predict an earlier coalescence due to a strong void-crack interaction influencing the void growth. Also plotted in Fig. 5.11 are the results of Aravas and McMeeking (1985a). As expected their cylindrical voids coalesce before the initially spherical voids and lead to a lower fracture toughness for crack growth initiation.

5.5 Closure

Substantial progress has been made in the numerical analysis of crack blunting and near tip void growth. In addition, the results are in reasonable agreement with experiments measuring surface deflections of cracks and the initiation of ductile crack propagation. Substantial work remains to be done on void nucleation and coalescence including the effects of shear localization. Such work should clarify the distinctions and discrepancies between experimental data and numerical results which remain in the ductile crack propagation area. However, this area of research, inspired by McClintock's pioneering work and based on his approach, has contributed greatly to our understanding of processes of ductile fracture.

References

ABAQUS, (1984). User's Manual, Version 4.5. Hibbitt, Karlsson and Sorensen, Inc., Providence, RI 02906.

Aoki, S., Kishimoto, K., Takeya, A. and Sakata, M. (1984). Effects of Microvoids on Crack Tip Blunting and Initiation in Ductile Materials. Int. J. Fract., 24:267.

Aoki, S., Kishimoto, K., Yoshida, T. and Sakata, M. (1985). A Finite Element Study of the Near Crack Tip Deformation of a Ductile Material Under Mixed Mode Loading. Department of Physical Engineering Report, Tokyo Institute of Technology.

Aravas, N. and McMeeking, R. M. (1985a). Finite Element Analysis of Void Growth Near a Blunting Crack Tip. J. Mech. Phys. Solids, 33:25.

Aravas, N. and McMeeking, R. M. (1985b). Microvoid Growth and Failure in the Ligament Between a Hole and a Blunt Crack Tip. Int. J. Frac., 29:21.

Cox, T. B. and Lowe, J. R. (1974). Metall. Trans., 5A, 1457.

Green, G. and Knott, J. F. (1976). The Initiation and Propagation of Ductile Fracture in Low Strength Steels. Trans. ASME (J. Engg. Mater. Tech.), 98:37.

Gurson, A. L. (1975). Plastic Flow and Fracture Behavior of Ductile Materials Incorporating Void Nucleation, Growth and Interaction. Ph.D. Dissertation, Division of Engineering, Brown University, Providence, RI.

Gurson, A. L. (1977a). Porous Rigid-Plastic Materials Containing Rigid Inclusions - Yield Function, Plastic Potential and Void Nucleation. Fracture 1977 (ed. D.M.R. Talplin) Pergamon, Oxford, 2:357.

Gurson, A. L. (1977b). Continuum Theory of Ductile Rupture by Void Nucleation and Growth: Part I — Yield Criteria and Flow Rules for Porous Ductile Media. Trans. ASME (J. Engg. Mater. Tech.), 99:2.

Hancock, J. W. and Mackenzie, A. C. (1976). On the Mechanisms of Ductile Failure in High-Strength Steels Subjected to Multi-axial Stress-States. J. Mech. Phys. Solids, 24:147.

Hom, C. L. and McMeeking, R. M. (1989). Three-Dimensional Void Growth Before a Blunting Crack Tip. J. Mech. Phys. Solids, 37:395–415.

Hutchinson, J. W. (1968). Singular Behavior at the End of a Tensile Crack in a Hardening Material. J. Mech. Phys. Solids, 16:13.

Irwin, G. R. (1960). Structural Mechanics (Proceedings of First Symposium on Naval Structural Mechanics, Stanford, 1959) (ed. by J. N. Goodier and N. J. Hoff), Pergamon Press, Oxford. page 557.

Lee, E. H. and Wang, A. J. (1954). Plastic Flow in Deeply Notched Bars with Sharp Internal Angles. Proceedings of the 2nd U.S. National Congress of Applied Mechanics, ASME, 489.

Le Roy, G. J., Embury, J. D., Edward, G. and Ashby, M. F. (1981). A Model of Ductile Fracture Based on the Nucleation and Growth of Voids. Acta Metall., 29:1509.

McClintock, F. A. (1968). A Criterion for Ductile Fracture by the Growth of Holes. J. Appl. Mech., 35:363.

McClintock, F. A. (1969). Crack Growth in Fully Plastic Grooved Specimens. Physics of Strength and Plasticity (ed. A. S. Argon) MIT Press, Cambridge, Massachusetts, 307.

McClintock, F. A. (1971). Plasticity Aspects of Fracture. Fracture, An Advanced Treatise, (ed. H. Liebowitz) Academic Press, New York, 47.

McMeeking, R. M. (1977a). Finite Deformation of Crack-Tip Opening in Elastic-Plastic Materials and Implications for Fracture. J. Mech. Phys. Solids, 25:357.

McMeeking, R. M. (1977b). Blunting of a Plane Strain Crack Tip into a Shape with Vertices. Trans. ASME (J. Engg. Mater. Tech.), 99:290.

Needleman, A. and Tvergaard, V. (1987). An Analysis of Ductile Rupture Modes at a Crack Tip. J. Mech. Phys. Solids, 35(2):151–183.

Neimark, J. E. (1968). The Fully Plastic, Plane-Strain Tension of a Notched Bar. J. Appl. Mech., 35:111.

Pellissier, G. E. (1968). Engng. Fracture Mech. 1:55.

Rawal, S. P. and Gurland, J. (1976). Proc. 2nd Conf. Mechanical Behavior of Materials (Boston, 16-20 August), Federation of Materials Societies, Dearborn, Michigan.

Rice, J. R. (1968). A Path Independent Integral and the Approximate Analysis of Strain Concentration by Notches and Cracks. J. Appl. Mech., 35:379.

Rice, J. R. and Johnson, M. A. (1970). The Role of Large Crack Tip Geometry Changes in Plane Strain Fracture. Inelastic Behavior of Solids, (eds. M. F. Kanninen et al.) McGraw-Hill, New York, 641.

Rice, J. R. and Rosengren, G. F. (1968). Plane Strain Deformation Near a Crack Tip in a Power Law Hardening Material. J. Mech. Phys. Solids, 16:1.

Rice, J. R. and Tracey, D. M. (1969). On the Ductile Enlargement of Voids in Triaxial Stress States. J. Mech. Phys. Solids, 17:201.

Tvergaard, V. (1982). Material Failure by Void Coalescence in Localized Shear Bands. Int. J. Solids Struct., 18:659.

Tvergaard, V. and Needleman, A. (1984). Analysis of the Cup-Cone Fracture in a Round Tensile Bar. Acta Metall., 32:157.

Van Stone, R. M., Merchant, R. H. and Low, J. R. (1974). Fatigue and Fracture Toughness—Cryogenic Behavior, ASTM STP-556, American Society for Testing and Materials, Philadelphia, page 93.

6

Global and Local Approaches of Fracture — Transferability of Laboratory Test Results to Components

A. Pineau

ABSTRACT This paper is devoted to the application of local micromechanisms of failure to the prediction of the macroscopic fracture toughness properties of metallic materials — a field pioneered by F. McClintock. The global approach of fracture assumes that failure can be described in terms of a single parameter, such as K_{IC} or J_{IC}. This approach may yet prove to be questionable in complex situations. This is the reason why local approaches of fracture have also to be developed. An attempt is made here to review the application of these local approaches, especially those which are micro-mechanistically based and to indicate a number of research fields which necessitate further studies. The paper is divided into two main parts. In the first part, the three basic fracture modes encountered in metallic materials, i.e. cleavage, intergranular fracture, and ductile rupture are reviewed. For cleavage fracture, a statistical model based on the Weibull weakest link concept is introduced and applied to a number of low-alloy ferritic steels. Some questions relating to intergranular fracture are also discussed. For ductile rupture, the emphasis is laid on the discontinuous or continuous character of the nucleation of cavities from second phase particles, on cavity growth and coalescence. The mechanics of plastic porous materials is briefly introduced to model this mode of failure and the statistical aspects of ductile rupture. It is shown that, in spite of the large research effort devoted to ductile rupture over the past few decades, it is still necessary to use an empirical fracture criterion based on the concept of critical void growth, that was originally introduced by McClintock.
The second part of the paper is devoted to the application of these local fracture criteria to predict the fracture toughness of specimens or the fracture load of components. The concept of characteristic distances related to the microstructure of materials and that of the "process zone" are briefly discussed. Then theoretical expressions between the fracture toughness (K_{IC} or J_{IC}) for 2D cracks, tested under small-scale yielding conditions, and local criteria are introduced. The local criterion for brittle cleavage fracture is based on Weibull statistics, which gives rise to a size effect, while the criterion for ductile rupture is based on critical void growth. Finite element method (FEM) numerical simulations of compact tension

(CT) and center-cracked panel (CCP) specimens under large-scale yielding conditions were used in conjunction with these local fracture criteria to show that the ligament size requirements for "valid" J_{IC} measurements are not only dependent on the specimen geometry (crack length, tension versus bending), but also on material work-hardening exponent and, more importantly, on the microscopic modes of failure. Further applications of the local approach of fracture are also presented, including fracture toughness testing of 3D cracks, the ductile-to-brittle transition behavior for ferritic steels, and fracture tests of specimens and components under non-isothermal conditions. Finally, further developments are briefly discussed.

6.1 Introduction

The assessment of the mechanical integrity of any flawed mechanical structure requires the development of approaches which can deal not only with simple situations, such as small-scale yielding under pure Mode I loading, but also with much more complex situations involving large-scale plasticity, mixed mode cracking or non-isothermal loading. Several approaches to deal with this problem are possible.

The "global" approach which is based on the extensive developments over the past few decades of linear elastic fracture mechanics (LEFM) or elastic-plastic fracture mechanics (EPFM) assumes that the fracture resistance can be described in terms of a single parameter, such as K_{IC}, J_{IC} or crack tip opening displacement (CTOD). Rules which are uniquely based on the mechanical conditions of test specimens have been established for "valid" fracture toughness measurements, without paying attention to the failure micro-mechanisms. This raises the problem related to the transferability of laboratory test results to components. The standardization of fracture toughness testing is presently undergoing rapid development. This approach is extremely useful and absolutely necessary, but it has also some strong limitations, which can be illustrated with a simple example, that of the determination of fracture toughness, K_{IC}. The standards for K_{IC} measurements are now largely accepted in spite of the fact that, even under plane-strain small-scale yielding, which is assumed to be reached when the crack length, the ligament size and the specimen thickness are 25 times larger than the plastic zone size, there remains a size effect associated with the probability of finding a local "brittle" zone along the crack front through the specimen thickness. Therefore the fracture toughness is dependent on specimen thickness even under LEFM conditions. The size requirements, in particular the crack length and the ligament size which must be fulfilled in an EPFM test in order to measure a "valid" fracture toughness, J_{IC} are still more problematic.

Another approach which is sometimes called the "local" approach has been developed more recently. Actually, in his pioneering work on fatigue fracture, McClintock (1963, 1967) showed the lines of this approach in which it is assumed that it is possible to model macroscopic fracture behavior in terms of local fracture criteria. This was the basic idea in the first McClintock (1963) fatigue model in which he assumed that the propagation of a fatigue crack was due to the accumulation of low cycle fatigue damage over a "process" zone. These local criteria can be established from the analysis and the modelling of tests carried out on volume elements, such as the low cycle fatigue specimens in the McClintock model. Then, they are applied to the crack tip situation. Therefore, the development of this methodology requires that, at least, two conditions are fulfilled:

(i) Micro-mechanistically based models for a given physical failure process must be developed;

(ii) A perfect knowledge of the stress-strain field in front of a stationary and a propagating crack is required. This has been made possible, thanks to the advent of analytical and numerical solutions. The rapid development of numerical simulation, in particular FEM calculations, has already largely, and will still more largely, contribute to the development of this methodology.

This paper is divided into two parts. The first part is devoted to the description of local criteria for brittle and ductile fracture. A large research effort has been devoted to this topic over the last past decade. In particular a large attention has been paid to the modelling of fracture of cavitating ductile materials. In the second part, an attempt is made to show how these criteria can be incorporated into a crack tip stress-strain field analysis to model fracture toughness. It is beyond the scope of the present paper to review all the work carried out on this specific topic. Therefore, we will highlight a number of specific items which remain to be investigated more thoroughly. It should also be added that most of the examples given in this paper relate to metallic materials, such as austenitic and ferritic steels, or aluminum alloys.

6.2 Micro-mechanisms of Fracture

The three basic mechanisms of fracture, i.e. cleavage, brittle intergranular fracture, and ductile rupture are examined successively.

6.2.1 Cleavage Fracture

Many recent studies have concluded from the temperature independence of the cleavage fracture stress, σ_c, that the mechanism of cleavage fracture in "homogeneous" ferritic steels is growth-controlled (see e.g. Curry and Knott, 1976). Moreover, in many cases, a good agreement has been shown with a modified Griffith criterion:

$$\sigma_c = \sqrt{\frac{E\gamma_c}{\alpha a}} \tag{6.1}$$

where E is Young's modulus, γ_c the "effective" surface energy, α a numerical constant, and a the size of the coarsest observed second phase particles equated to the size of the microcrack nuclei.

Rather surprisingly, although the scatter in cleavage stress measurements is well established, it is only rather recently that models have been proposed to account for this scatter (For a recent review, see Wallin, 1991). For a homogeneous material in which there is no statistical spatial correlation between two adjacent areas, the Weibull weakest link concept is largely accepted to account for brittle fracture. The Weibull statistical distribution is only a particular case, as was shown recently by Jeulin (1992), who introduced other "random field" models to deal with brittle fracture. The Weibull statistics was the basis of the model proposed by Beremin (1983) and Wallin et al. (1984). The probability for failure, P_R of a specimen of volume V submitted to a homogeneous stress state, σ is given by:

$$P_R = 1 - \exp\left(-\frac{\sigma^m V}{\sigma_u^m V_u}\right) \tag{6.2}$$

where V_u is an arbitrary unit volume, σ_u is the average cleavage strength of that unit volume, while m is the Weibull exponent. Equation 6.2 is a simplified expression since no threshold parameter is introduced. In three dimensions and in the presence of smooth stress gradients, this equation can be expressed as:

$$P_R = 1 - \exp\left(-\frac{\int_{PZ} \Sigma_1^m \, dV}{\sigma_u^m V_u}\right) \tag{6.3}$$

where Σ_1 is the maximum principal stress and the volume integral is extended over the plastic zone PZ. The latter expression can be rewritten as:

$$\sigma_w = \sqrt[m]{\int_{PZ} \frac{\Sigma_1^m \, dV}{V_u}} = \sigma_u \sqrt[m]{\ln \frac{1}{(1-P_R)}} \tag{6.4}$$

where σ_w is referred to as the "Weibull stress".

The procedures used to determine the parameters σ_u, m and V_u are given in detail by Mudry (1983). Tests on volume elements are carried out on notched tensile bars containing various notch radii. Results obtained on a welded C-Mn steel studied by Fontaine et al. (1987) are shown in Fig. 6.1, where it is observed that $m = 22$. The volume V_u is introduced as the characteristic distance (a few grain sizes) used in the Ritchie, Knott and Rice (1973) model.

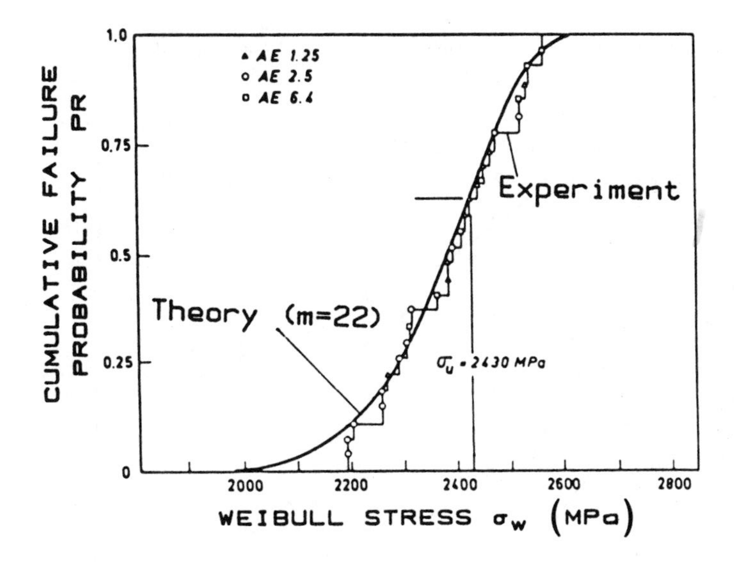

FIGURE 6.1. Welded C-Mn steel. Probability of cleavage fracture as a function of Weibull stress. Comparison between experiments and calculations (Equations 6.3 and 6.4).

Much research remains to be done to improve our understanding and the modelling of brittle cleavage fracture, besides the statistical aspect of this failure mode. In particular the influence of microstructure, of strain-induced anisotropy and dislocation mobility deserves much more research.

Influence of microstructure: While, in mild steel, this is now well established that the critical stage in cleavage fracture is growth-controlled, it is not necessarily the same situation in other metals, such as for instance zinc, or even in iron-based alloys such as fully pearlitic steels. Recent studies of Kavishe and Baker (1986), and Lewandowski and Thompson (1986), have shown that pearlite can exhibit two different cleavage mechanisms which are dependent on the strength of the material and therefore on the interlamellar spacing, Sp. For large Sp ($\geq$ 0.2 μm), the fracture is nucleation controlled and involves shear linking of microcracks nucleated from cementite particles before unstable cleavage can occur. Under these conditions, the cleavage fracture stress is dependent on temperature and increases with the yield strength of the material, as shown in Fig. 6.2. On the other hand, for smaller Sp, cracked carbides act directly as cleavage nuclei and, under these conditions where fracture is propagation controlled the cleavage stress is independent of temperature (Fig. 6.2). For a given temperature, the cleavage stress is then proportional to the yield strength of the material which is itself proportional to the inverse square root of the mean free

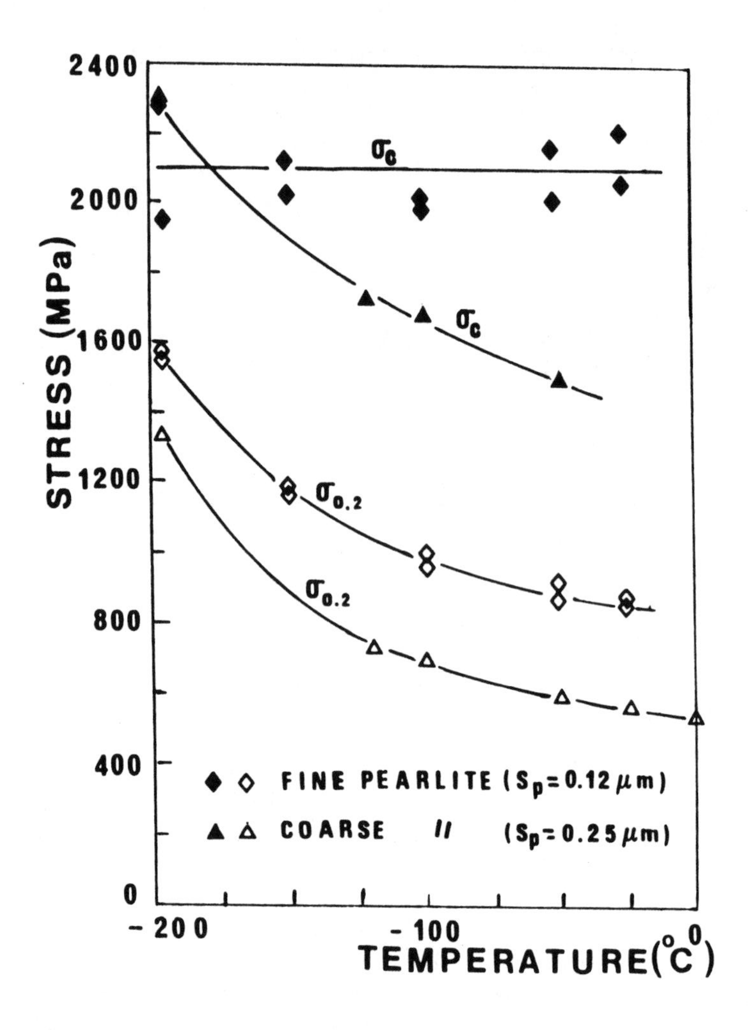

FIGURE 6.2. Variation of the yield strength ($\sigma_{0.2}$) and the cleavage stress (σ_c) as a function of temperature in an eutectoïd steel with two interlamellar spacings (Sp).

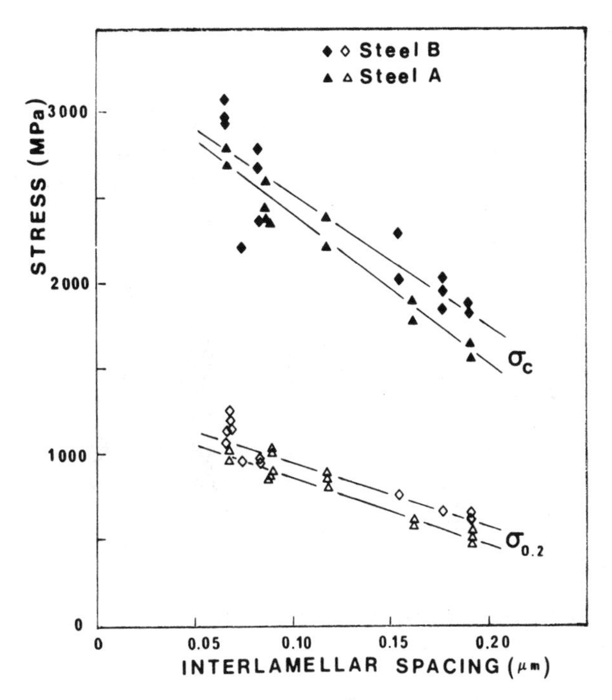

FIGURE 6.3. Variation of the yield strength ($\sigma_{0.2}$) and the cleavage stress (σ_c) in two heats of a fully pearlitic steel. Tests at room temperature.

distance in the ferrite. Recent results obtained on two types of steels of eutectoid composition shown in Fig. 6.3 indicate that, at room temperature, the proportionality factor between the cleavage stress σ_c and the yield stress σ_y is close to 2.5.

The criterion for nucleation-controlled cleavage fracture obviously must involve other terms of the stress or strain tensor than the maximum principal stress. The modelling of this failure mechanism bears, to some extent, a strong analogy with cavity nucleation in ductile rupture which is briefly examined below. This is a research field where the additional application of the theory of plasticity for inhomogeneous materials could be extremely useful.

As far as microstructural effects are concerned, some large new research effort is needed, primarily because of its practical importance. This is related to the effect of local brittle zones on the fracture toughness of welded structures. Recent results presented at the European Fracture Group (EGF9) Conference illustrate the huge scatter in CTOD measurements of the heat-affected zone (Machida et al., 1991). The results of CTOD measurements spread over one order of magnitude. This is related to the variability of fracture toughness of local brittle zones and to the probability of finding "weak" material ahead of the crack tip. The many recent devel-

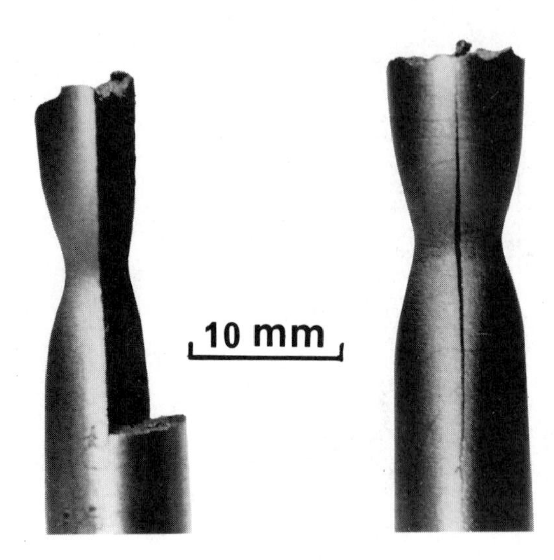

FIGURE 6.4. Optical micrographs of a low-alloy steel showing the fracture aspect of two tensile bars tested at -196° C.

opments in numerical simulations and in experimental techniques should contribute to the development of models in which the local fracture toughness could be assessed in terms of microstructural features and then distributed ahead of the crack tip to predict the global fracture toughness of weldments by using an appropriate statistical analysis.

Strain-induced anisotropy: The concept of critical stress σ_c introduced by Orowan (1948) applies essentially when growth-controlled cleavage fracture occurs after relatively small strains. In mild steel or in quenched and tempered low alloy steel, it was shown that a pre-deformation could substantially modify σ_c. (see e.g. Pineau, 1981). Large tensile strains increase σ_c, while compressive strains produce a detrimental effect on σ_c when the material is stressed in the same direction as that during pre-deformation. A similar behavior was shown when the cleavage stress was measured at low temperature in a low alloyed steel (Beremin, 1983; Mudry, 1987). In a number of circumstances, this strain-induced elevation of $\sigma_c$can produce intriguing effects, as illustrated in Fig. 6.4, where it is observed that a necked tensile bar can fracture not in the minimum section under the influence of the first principal stress but along a longitudinal plane under the influence of the radial stresses associated with neck formation. This failure mode might be also partly related to some initial anisotropy since the fracture plane is perpendicular to the short-transverse direction.

Strain-induced anisotropy of the cleavage stress may arise from several effects, such as rotation and blunting of microcrack nuclei. The modelling

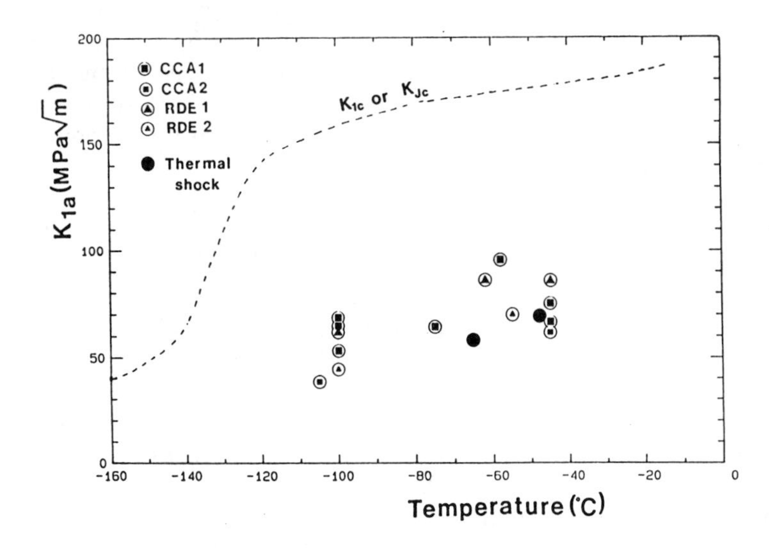

FIGURE 6.5. Static fracture toughness (dotted line) and fracture toughness at crack arrest as a function of temperature in a low alloy structural steel.

of this effect which can play an important role in very tough materials, especially in the ductile to brittle transition regime deserves also more attention.

Dislocation mobility: In materials in which the cleavage stress is growth-controlled, it is now largely accepted that the variations of yield strength with strain-rate and temperature control the temperature dependence of the fracture toughness. The elevation of the yield strength has also been used to model the irradiation embrittlement of a low alloy steel (Parks, 1976; Mundheri et al., 1989). Thus, for static fracture toughness measurements, dislocation mobility plays a key role in the variation of fracture toughness with test parameters.

The dislocation mobility can be affected by a number of factors, including effects of impurity pinning, and strain-rate effect. The influence of dislocation pinning by impurity atmospheres on the variation of cleavage stress and fracture toughness remains to be investigated more thoroughly (see e.g. Bowen et al., 1990). In particular, the dislocation mobility ahead of a crack tip can be affected by the cyclic plastic strain, introduced during fatigue pre-cracking of the specimens.

Dynamic fracture is a research field which has been extensively investigated over the last past decade and where the strain-rate effects on dislocation mobility play an important role. It is now well established that the dynamic fracture toughness, in particular the fracture toughness at crack arrest, K_{IA}, is usually much lower than the fracture toughness at crack initiation, K_{IC}, as shown in Fig. 6.5 which refers to a structural steel (DiFant et al., 1991). That the dissipated energy for cleavage fracture is a decreasing function of strain-rate, and thus of crack velocity is not surprising, but models including the variations of dislocation mobility with strain-rate and the value of the cleavage stress to account for crack arrest behaviors similar to those shown in Fig. 6.5 are lacking.

6.2.2 Brittle Intergranular Fracture

Brittle fracture in ferritic steels normally occurs by cleavage at low temperatures or high strain-rates. However, segregation of a number of solute elements, in particular impurities, such as P, Sn, Sb or As, to grain boundaries can change the brittle fracture mode to intergranular. Since this mode of failure operates at lower stresses, brittle fracture can then be observed at higher temperatures where failure by ductile rupture would otherwise take place. The classical example of this failure mode is temper-embrittlement of alloy steels. Such embrittlement is evidently caused by a lowering of intergranular cohesion, the magnitude of which depends on the quantity and type of segregated solute. The physical basis of this embrittlement effect has not yet been studied in a very quantitative manner. However, recent work by Cottrell (1989, 1990a,b) should be mentioned.

Until the present time, very few well documented studies had been performed to determine quantitatively the variation of the critical fracture stress with test parameters using procedures similar to those used to investigate the cleavage stress, especially tests on notched bars. Most often, the effect of temper-embrittlement is investigated by determining the shift in the transition temperature measured with Charpy V notched specimens, but there is no reason to believe that this shift is simply related to a decrease in the critical fracture stress. However the interesting work by Kameda and Mc Mahon (1980) should be mentioned. These authors showed that the critical local stress for fracture of a grain boundary was directly related to the Sb concentration on that boundary.

Intergranular fracture raises a number of interesting questions which should be answered before modelling the variation of fracture toughness. It is usually implicitly assumed that this failure mode is growth-controlled and can thus be described in terms of a critical stress. Detailed experiments and observations, in particular in compressive tests, might be useful to test this assumption. Grain boundaries in engineering materials routinely contain second phase particles which can act as stress raisers. However, the presence of these particles is not a necessary prerequisite to initiate brittle intergranular fracture. This is a failure mode where the application of the theory of plastic deformation of inhomogeneous materials, in particular polycrystals, should be useful. Simply stated, this theory tells us that the

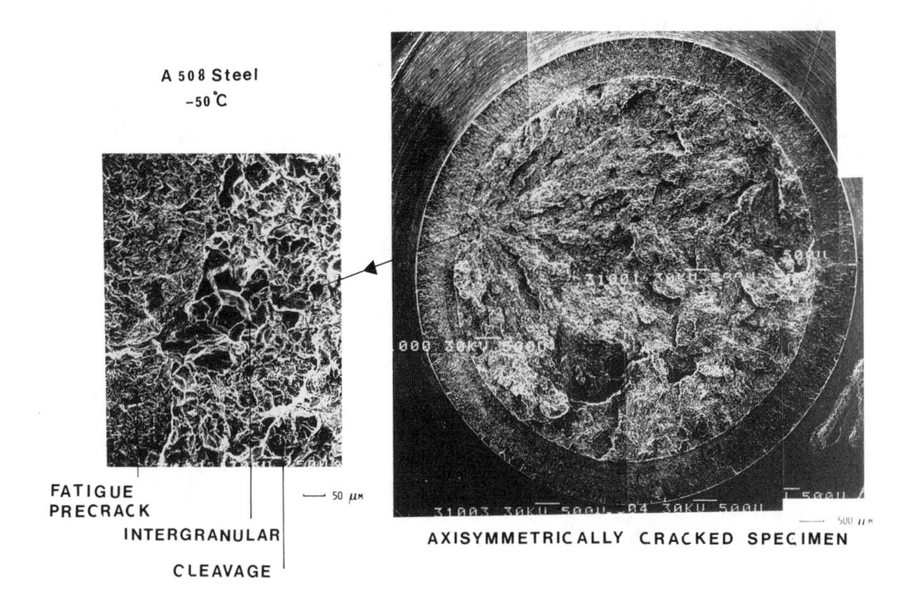

FIGURE 6.6. SEM micrographs illustrating the effect of a small area of intergranular fracture on the cleavage fracture of an axisymmetrically cracked specimen.

local stress can be written as, (see e.g. Berveiller and Zaoui, 1979):

$$\sigma = \Sigma + A\mu\,[E - \varepsilon] \tag{6.5}$$

where σ is the "local" stress, ε the "local" strain, μ the shear modulus, E the average macro strain and A a coefficient calculated by means of a self-consistent theory. Equation 6.5 in conjunction with a spatial description of the local strains, or at least their distributions, could be useful to assess the effect of the "intergranular stress".

In many situations, brittle cleavage and intergranular fracture are competing modes. This is illustrated in Fig. 6.6 where the fracture surface of a low alloy steel tested at 50°C is shown. This specimen broke at an anomalously low value of the stress-intensity factor. Figure 6.6 shows that the fracture surface is essentially covered by cleavage facets, except in a small initiation area which is intergranular. More research work should concentrate on the respective scatter of these competing failure modes. Recent results shown in Fig. 6.7 indicate that the scatter in the results of

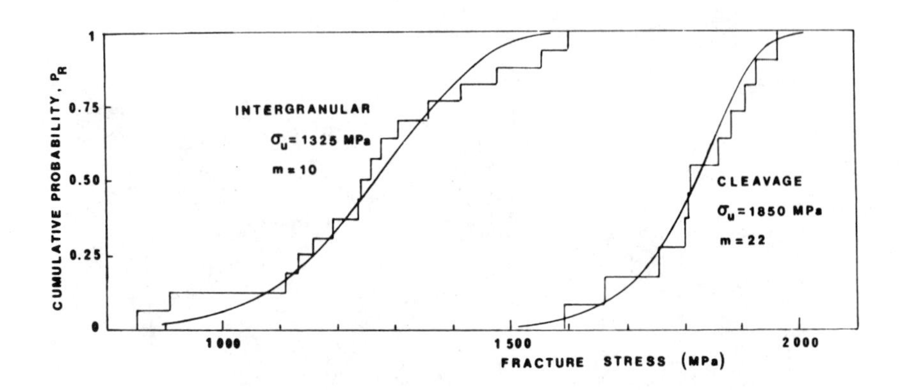

FIGURE 6.7. C-Mn-Ni-Mo steel tested after conventional heat-treatment (cleavage) or after temper-embrittlement heat treatment (intergranular). Variations of probability to fracture with fracture stress.

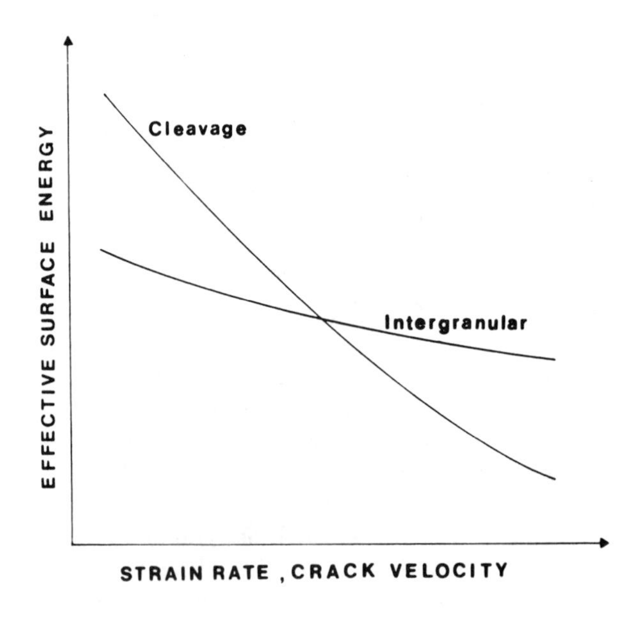

FIGURE 6.8. Schematic diagram showing the effect of strain-rate on the "effective" surface energy for both cleavage and intergranular fracture.

intergranular fracture stress measurements is much larger than those corresponding to cleavage fracture. If these results obtained on a low alloy C-Mn-Ni-Mo steel submitted to a step- cooling heat treatment are interpreted in terms of a Weibull statistical distribution (Eq. 6.2), it is found that $m \simeq 10$, which is a value much lower than that measured for cleavage fracture in the same material ($m \simeq 22$). Similar results on a 2.25 Cr-1Mo steel were reported recently by Holzmann et al. (1991). Another feature which deserves also more attention is the respective strain-rate sensitivity of both brittle failure modes. Intergranular fracture often involves plastic deformation, but there is no reason to believe that the variations of the dissipated energy, γ, as a function of strain-rate and thus crack velocity is the same for both mechanisms. This might produce interesting results, as shown schematically in Fig. 6.8 where it is suggested that a brittle cleavage crack could be initiated by intergranular fracture. Experiments based on well designed specimen geometries leading to different stress-intensity histories would be useful to test this assumption.

6.2.3 Ductile Rupture

This failure mode occurs by the formation and subsequent growth and coalescence of voids and cavities. If cavity nucleation could be delayed, large improvements in ductility and fracture toughness could obviously be achieved. It is therefore important to investigate the metallurgical and mechanical factors controlling this first stage of ductile rupture. Cavity growth and coalescence are also briefly discussed.

Cavity nucleation: Cavity nucleation sites are usually associated with second phase particles or non-metallic inclusions. For large ($\geq 1\mu$m) and widely spaced particles, discontinuous nucleation can be described in terms of continuum mechanics. A nucleation model based on a critical stress, σ_d, was thus proposed by Argon et al. (1975) as:

$$\sigma_d = \sigma_{eq} + \sigma_m \qquad (6.6)$$

where σ_{eq} is the local equivalent von Mises stress and σ_m the hydrostatic stress. In Eq. 6.6, the inhomogeneity in plastic deformation between the matrix and the inclusions does not appear explicitly. This effect is directly related to the difference between σ_{eq} and the yield strength σ_Y. Another expression derived from the application of the theory of plastic deformation for inhomogeneous materials was proposed to account for cavity formation from Mn S inclusions in a low alloy steel (Beremin, 1981):

$$\sigma_d = \Sigma_1 + k(\sigma_{eq} - \sigma_Y) \qquad (6.7)$$

where k is a function of particle shape.

These expressions can be used to model discontinuous nucleation, but usually void nucleation occurs continuously over a wide range of strains,

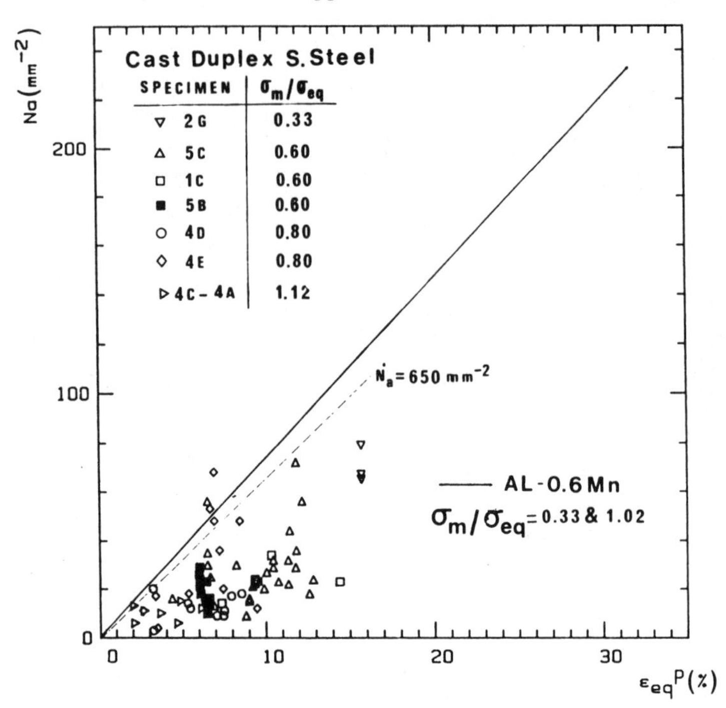

FIGURE 6.9. Cast duplex stainless steel and an aluminum alloy. Crack density as a function of applied strain. Measurements on smooth and notched specimens.

sometimes after only a certain nucleation strain has been reached. This may arise from several reasons, as discussed recently by Kwon and Asaro (1990): (i) large particles have a greater probability of containing volume or surface flaws; (ii) large particles have more difficulty to relax large stress concentrations; (iii) interactions effects between inclusions of larger than average size and spacing. This results in an increase in the number of cavity sites as a function of plastic strain.

Whether continuous nucleation is a strain or stress-controlled phenomenon is still widely discussed in the literature. Kwon and Asaro (1990) concluded that an interfacial stress-controlled nucleation criterion was more realistic than a strain-controlled criterion to characterize the nucleation behavior of spheroidized steel. On the often hand, results obtained on Al alloys by Walsh et al. (1989) and on a cast duplex stainless steel by Pineau and Joly (1991); and Joly et al. (1990) indicate that cavity nucleation is essentially strain-controlled, as shown in Fig. 6.9. These results were obtained on notched specimens with different stress-triaxiality ratio (σ_m/σ_{eq}). In both cases, no clear effect of this parameter is observed. In cast duplex stainless steel, the cavities are initiated from cleavage fracture of the embrittled ferrite phase. In this material it was shown that void nucleation

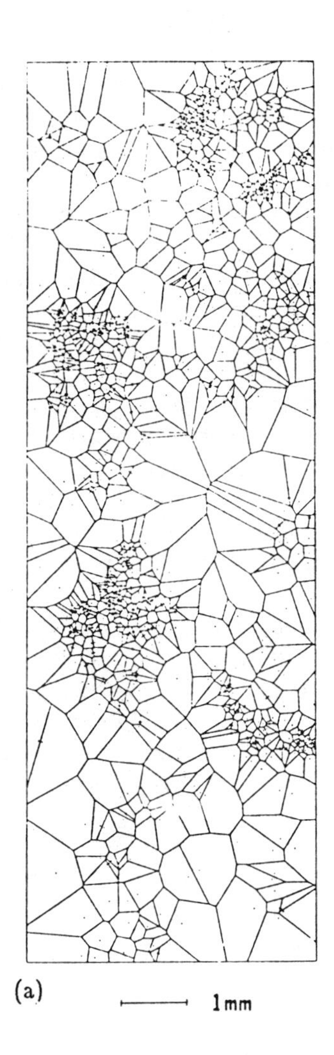

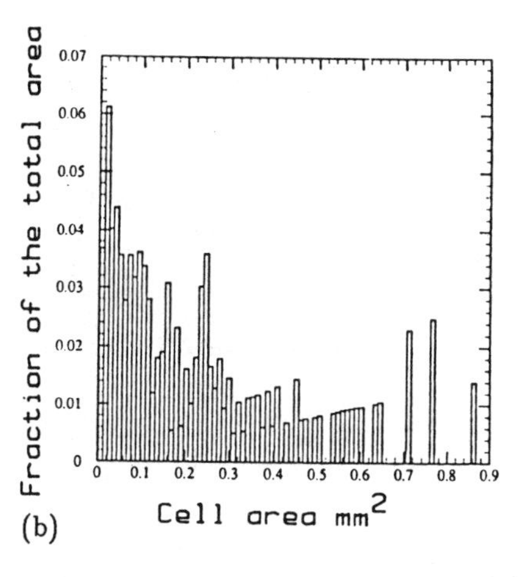

Cast Duplex Stainless Steel

FIGURE 6.10. Cast duplex stainless steel. Damage measurements in a smooth bar ($\varepsilon = 0.16$). (a) Voronoi cells showing the distribution of cleavage microcracks initiated in the ferrite phase; (b) Distribution of cell sizes.

was related to the orientation of the austenite matrix and occurred preferentially in austenite grains oriented for single slip, which can be considered to illustrate the "intergranular stress" effect described by Eq. 6.5. In both cases, a wide scatter in the results of cavity density measurements is observed (Fig. 6.9). This effect also needs modelling. In spheroidized steel, it was assumed that the nucleation rate could be described with a normal distribution about a mean interfacial stress, but there are very few experimental results to validate this assumption (Kwon and Asaro, 1990). In duplex stainless steel, local cavity densities were measured using interrupted tests, as shown in Fig. 6.10, where clusters of cavity sites represented by Voronoï cells are clearly observed. The histogram of the cell sizes shows that a very small fraction of the surface area ($\simeq$ a few percent) in the micrograph is occupied by clusters leading to extremely large local nucleation rates ($\simeq$ 600 voids/mm^2/100% plastic strain). This clustering effect which is particularly pronounced in this material is another feature of cavity nucleation that should be investigated more thoroughly. Examination of the same areas at different strains would be extremely useful to described more rigorously cavity nucleation and, in particular, to provide the basis of an autocatalytic model for void formation.

Cavity growth: Substantial progress in the understanding of hole growth has been made through the theoretical models by Berg (1962), McClintock (1968); and Rice and Tracey (1969). These models are based on a number of assumptions which do not necessarily apply to real materials. In particular, they assume that no interaction effect takes place between two neighbouring cavities. In the Rice and Tracey model the cavity growth rate at large stress triaxialities is given by:

$$D = \frac{1}{3f}\frac{df}{d\varepsilon_{eq}} = 0.283 \exp\left(\frac{3}{2}\frac{\sigma_m}{\sigma_{eq}}\right) \tag{6.8}$$

where R is the cavity size and $d\varepsilon_{eq}$ is the increment of von Mises plastic strain. There are relatively few experimental verifications of this theoretical expression. However, recently a comparison between the results of numerical calculations and experiments was made by Worswick and Pick (1991). This comparison, shown in Fig. 6.11, indicates that the low triaxiality void growth measurements from Worswick and Pick (1991), Bourcier et al. (1986), and Barnby et al. (1984) are in general agreement with the analytical expression of Equation 6.8, and finite element predictions, although there is considerable scatter. The straight line results at intermediate triaxiality in Fig. 6.11 are best fit curves to experimental void growth measurements from notched tensile tests of Beremin (1981) and Marini et al. (1985). These results show that the slope of the exponential dependence of D on σ_m/σ_{eq} predicted by Eq. 6.8 is correct, but that the pre-exponential coefficient is larger than 0.283 and is an increasing function of initial volume fraction, f_o. This was attributed to void interaction effects and to the presence of a second population of cavities initiated from carbides. However,

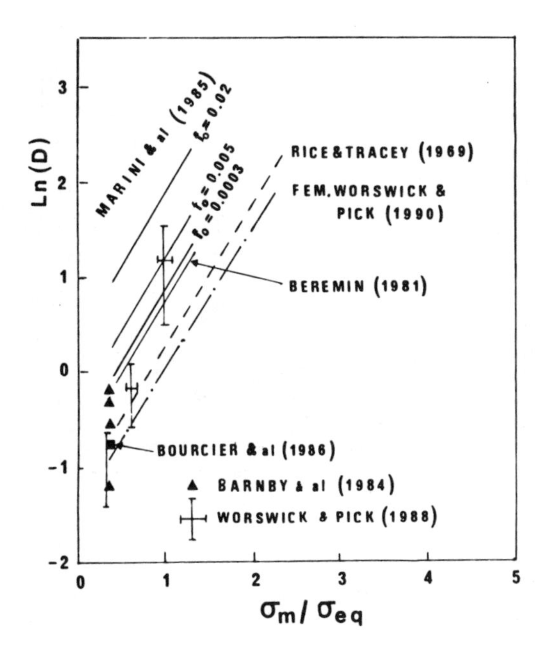

FIGURE 6.11. Cavity growth as a function of stress triaxiality (Worwick and Pick, 1990).

recent analytical work of Huang et al. (1991) indicates that the widely used high triaxiality approximation (Eq. 6.8) underestimates the dilatation rate of the voids by more than 50% at all levels of triaxiality above $\sigma_m/\sigma_{eq} = 1$. Whether the discrepancy between Eq. 6.8 and experimental results is due to the assumptions of the model or due to void distribution is not yet clear. More recently, analyses of the ductile fracture process have been based on phenomenological constitutive relations for porous plastic solids, such as the one proposed by Gurson (1975) which is briefly introduced later. In particular, results for clustered periodic arrays have been presented by Needleman and Kushner (1990). These numerical models are still far from actual distributions of voids, similar to those shown in Fig. 6.10. However, they can serve as ingredients in a homogenized material theory which needs to be developed.

Cavity coalescence and criteria for ductile rupture: Cavity coalescence is a phenomenon which is still very poorly understood and which requires a large research effort and a large number of approximations. As stated by McClintock (1971), "for most practical applications, it will be necessary to simplify the criterion by focusing attention on those phases of fracture that are most critical for the particular material and conditions at hand".

When the strain for cavity nucleation represents only a small fraction of

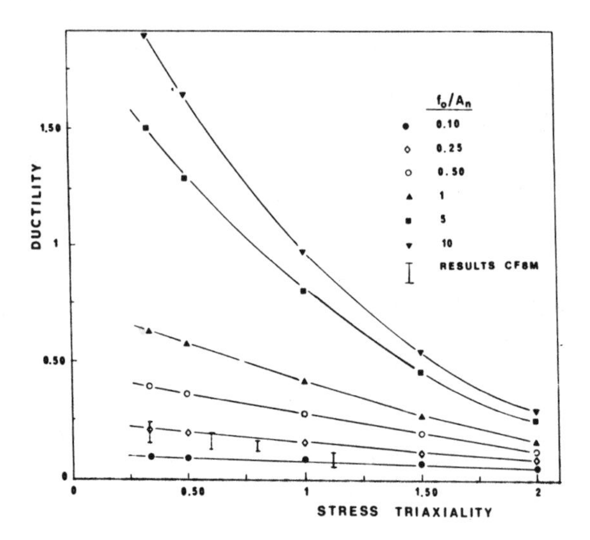

FIGURE 6.12. Ductility as a function of stress-triaxiality (Eq. 6.12). Results on a cast duplex stainless steel.

the ductility of the material, the simplest criterion, already introduced by McClintock is obtained by integrating Eq. 6.8. For a stress history during which the stress triaxiality is kept constant, this leads to:

$$\ln\left(\frac{R}{R_o}\right)_c = \frac{1}{3}\ln\left(\frac{f}{f_o}\right)_c = 0.283\varepsilon_R \exp\left(\frac{3}{2}\frac{\sigma_m}{\sigma_{eq}}\right) . \qquad (6.9)$$

It was experimentally shown that Eq. 6.9 is relatively well satisfied, and, in particular, that the ductility, ε_R is an exponentially decreasing function of stress triaxiality in various heats of a C-Mn-Ni-Mo steel containing different amounts of Mn S inclusions (Pineau, 1981). In these materials, it was found that the critical volume fraction of cavities inferred from Eq. 6.9 was small, since typically $1.20 \leq (R/R_o)_c \leq 1.80$. This represents only a mean value and simple reasoning can be made to show that "local volume fraction" must be as high as 10 or 20 times this "mean" volume fraction in order to obtain values for $(R/R_o)_c$ consistent with the experimental results of Pineau (1981).

In many situations, continuous cavity nucleation must be considered directly. This should produce different variations of ductility with stress triaxiality. Under these conditions, the increment in volume fraction of a cavity component can be expressed as:

$$df_N \simeq d^2 N_a d\varepsilon_{eq} = A_n d\varepsilon_{eq} \qquad (6.10)$$

where d is the size of the nucleation sites projected onto the fracture surface and N_a is the number of voids per unit volume at unit strain. These expressions lead after integration to

$$f = \frac{A_n}{K}\left[\exp(K\varepsilon - 1)\right] \tag{6.11}$$

with $K = 3 \times 0.283 \exp\left(\frac{3}{2}\frac{\sigma_m}{\sigma_{eq}}\right)$. If it is assumed that ductile failure occurs for a critical volume fraction of cavities, f_c, the variation of ductility with stress triaxiality can simply be expressed as:

$$\varepsilon_R = \frac{1}{K}\ln\left[1 + K\left(\frac{f}{A_n}\right)_c\right]. \tag{6.12}$$

This expression is plotted in Fig. 6.12 for various values of $(f/A_n)_c$. It is clear from the figure that, for large values of nucleation rate, A_n, i.e., small values of the ratio $(f/A_n)_c$, the ductility ε_R is no longer strongly dependent on stress triaxiality. In Fig. 6.12, we have also included experimental results obtained on the cast duplex stainless steel which was mentioned earlier since, in this material, the importance of continuous cavity nucleation is well established. These experimental results show that, in this material, the ductility is only slightly dependent on stress triaxiality which was not the situation found in steels where cavities are easily initiated from Mn S inclusions.

If critical cavity growth or critical volume fraction of voids is an empirical but useful criterion to model initiation of fracture from sharp cracks, but it does not give an insight into the final stage of ductile rupture where there exists a coupling effect between growing cavities and the constitutive equation of the material. A large research effort has been devoted, over the past decade, to model the strain softening effect produced by growing cavities using the mechanics of plastic porous materials. In these models, the plastic flow potential is dependent on cavity volume fraction. It is beyond the scope of this chapter to review all the existing theories but, instead, we show how these theories could be used to model not only numerically but also analytically the ductile rupture process. Here we refer only to the Gurson and the Rousselier models.

In the Gurson-Tvergaard potential, the yield criterion is written as (Tvergaard, 1981):

$$\sigma_{eq}^2/Y^2 + 2q_1 f \cosh(3q_2\sigma_m/2Y) - 1 - q_3 f^2 = 0 \tag{6.13}$$

where Y is the flow stress of the matrix, $q_1 \simeq 1.5, q_2 \simeq 1$ and $q_3 \simeq q_1^2$. For small volume fractions the Gurson expression can be written as:

$$\sqrt{(3/2)s_{ij}s_{ij}} = \sigma_{eq} = Y[1 - 0.50 f \exp(3\sigma_m/2Y)]\,. \tag{6.14}$$

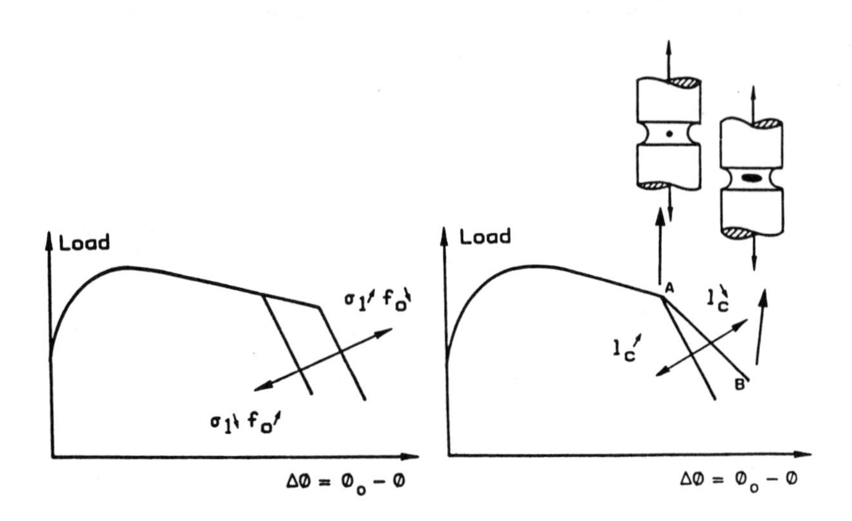

FIGURE 6.13. Sketch showing the softening effect associated with damage. Influence of σ_1 and f_o (Eq. 6.16) and effect of mesh size, l_c.

In this expression the softening effect due to the presence of cavities appears clearly. It is pronounced, especially at large stress triaxiality, as expected. The associated increment in cavity volume fraction is given by:

$$d\varepsilon_{ii} = \frac{df}{(1-f)} = 0.75 f \exp\left(\frac{3\sigma_m}{2Y}\right) d\varepsilon_{eq} \tag{6.15}$$

which is very similar to the Rice and Tracey expression (Eq. 6.8), except for the proportionality factor which is now equal to 3×0.28 instead of 0.75.

The application of the Gurson-Tvergaard potential has been reviewed recently by Needleman (1989), while Perrin and Leblond (1990) have used this type of potential to model the behavior of a material containing two populations of voids of different sizes. In the simulation of ductile fracture, it was observed that this model largely overestimated the observed ductility of structural materials. An accelerating function f^* was proposed by Tvergaard and Needleman (1984) to account for the effect of rapid void coalescence at failure. Intitially $f^* = f$ as proposed by Gurson but, at some critical void fraction, f_c the dependence of f^* on f is accelerated in order to simulate the effect of void coalescence. This model, therefore, bears a strong resemblance to the critical void growth criterion introduced previously.

The yield function introduced by Rousselier (1979, 1987) was derived from the application of thermodynamic theory of continuous plastic solids.

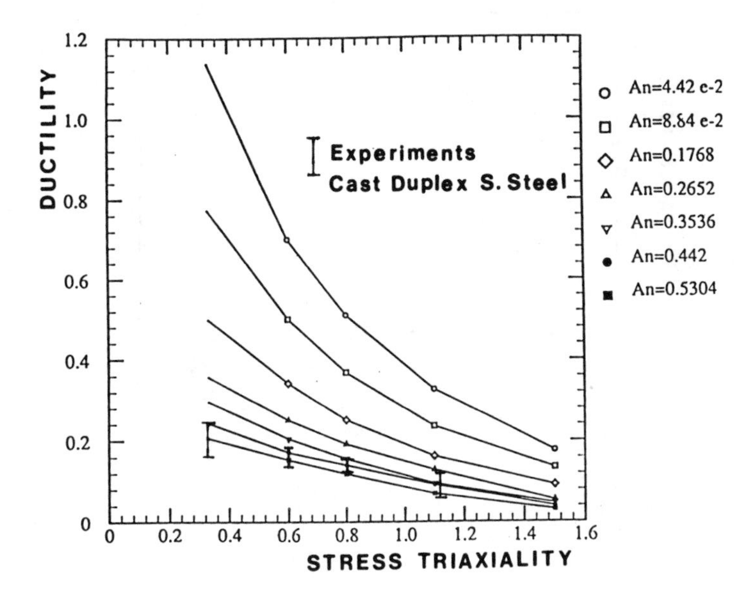

FIGURE 6.14. Application of the Gurson potential to the prediction of the variation of ductility with stress triaxiality in a cast duplex stainless steel. Effect of crack nucleation rate, A_n. Comparison with experimental results (Joly, 1991).

It can be written as:

$$\sigma_{eq}/(1-f)Y = 1 - (DB(\beta)/Y)\exp[\sigma_m/(1-f)\sigma_1] \tag{6.16}$$

with

$$d\beta = df/f(1-f) = D\exp[\sigma_m/(1-f)\sigma_1]d\varepsilon_{eq} \tag{6.17a}$$

and

$$B(\beta) = \sigma_1 f_o \exp\beta/(1 - f_o + f_o \exp\beta) \; . \tag{6.17b}$$

This criterion also bears a strong analogy to the Gurson potential and to the Rice and Tracey expression, provided it is assumed that $D = 3\times 0.283$ and $\sigma_1 = 2Y/3$. Rousselier introduced his criterion in a FEM code to simulate the failure of notched bars, as indicated schematically in Fig. 6.13. In this figure the effect of σ_1 and f_o (the initial volume fraction) parameters, and that of the mesh size l_c located in the center of the specimens are indicated. In particular, the mesh size l_c is used to simulate the slope of the branch AB of the loading curve. It can therefore be considered that, as in the Gurson-Tvergaard potential, it is necessary to introduce another parameter (f_c or l_c) to simulate cavity coalescence and final fracture.

Mudry (1982) has also attempted to use the Gurson potential without the accelerating effect in f introduced by Tvergaard and Needleman (1984) to simulate the ductility of various heats of a C-Mn-Ni-Mo steel. For a homogeneous distribution of inclusions, he confirmed also that the calculated

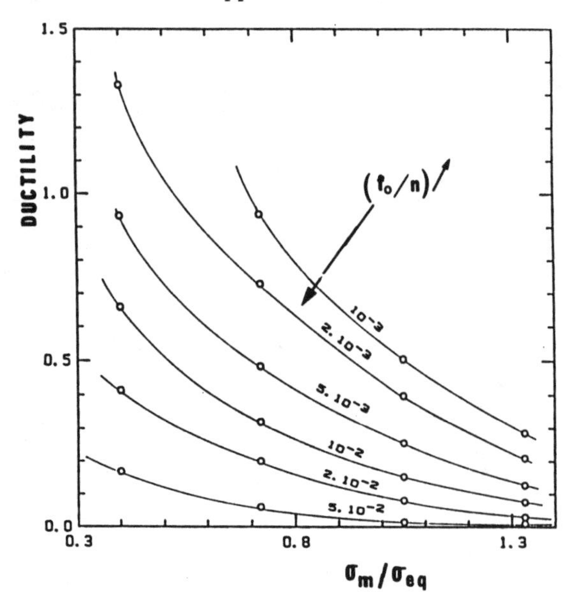

FIGURE 6.15. Theoretical variation of ductility with stress-triaxiality ratio. The numbers indicated on the curves correspond to various values of (f_o/n) ratio (Mudry, 1982).

ductilities were larger than those measured. More recently, a similar behavior was shown when the Gurson model was applied to cast duplex stainless steel, as illustrated in Fig. 6.14 (Joly, 1991). In this material, the comparison of the observed strains to failure and those calculated with various values of nucleation rate, A_n (Eq. 6.10) indicates that the "local" nucleation rate should be much larger, by more than one order of magnitude, to lead to the fracture of a volume element.

The Gurson potential can be used, in conjunction with a statistical analysis, to model ductile fracture as a function of stress triaxiality. The material of volume V is divided into a set of elementary cells of volume V_o which have either different nucleation rates or different initial volume fractions of cavities. One has to calculate, for a given strain, the probability of finding a cell which gives rise to a sufficiently large nucleation rate or growth rate to lead to fracture. All the cells are assumed to be submitted to the same equivalent stress. The fracture criterion is the achievement of a maximum for the "local" equivalent stress of that cell (Weakest link concept). The probability to failure can then be written as:

$$P_R = 1 - \exp\left(-P\frac{V}{V_o}\right) \tag{6.18}$$

where P is the probability of finding one elementary cell with either a nucleation rate larger than A_c^n or a void volume fraction larger than f_o^c.

This analysis was applied to model ductile rupture of various heats of a C-Mn-Ni-Mo steel by Mudry (1982). It was assumed that the inclusions were distributed according to a Poisson law. The results of these calculations are shown in Fig. 6.15 where it is observed that, within a first approximation, the variation of the ductility, ε_R with stress triaxiality ratio is only dependent on the ratio between the initial volume fraction of inclusions, f_o and the work-hardening exponent, n of the stress-strain curve written as:

$$\sigma = K\varepsilon^n . \tag{6.19}$$

In Fig. 6.15, only the curves corresponding to a probability of failure of 50 percent are drawn. However, with this specific inclusion distribution, it was shown that the difference between the curves corresponding to $P_R = 0.10$ and $P_R = 0.90$ was small. It was also observed that the local volume fraction in the cell which gives rise to the failure of the specimen can be as large as 5 or 10 times the mean volume fraction.

One difficulty with the application of Eq. 6.18 should be indicated. If, in the case of cavity nucleation from inclusions, the dimension V_o of the elementary volume is known to be of the order of $1/N_v$, where N_v is the mean number of inclusions per unit volume, the situation is far more complex when nucleation occurs continuously from sites which are not known *a priori*. In this case, it might be necessary to consider the volume V_o of the elementary cell as an adjustable parameter, as it is actually in many applications of the Weibull theory.

6.3 Fracture Toughness Measurements — Transferability of Laboratory Test Results to Components

In the standards a number of size requirements must be fullfilled to determine "valid" fracture toughness, as already discussed in the introduction. These requirements are determined without paying attention to the micromechanisms of failure, although there is no reason to believe that these conditions are exactly the same for a strain or a stress-controlled fracture mode. This is therefore an area where the application of local criteria can be useful. Another area is that of complex loading conditions which is also briefly discussed.

6.3.1 Characteristic Distances and Local Criteria

At the crack tip, as in many models based on local criteria, it is necessary to introduce a characteristic distance, λ, the famous "process zone" introduced in many research papers written by McClintock in this field. The strains and the stresses are averaged over this distance. In FEM calculations, this size is used for the first element located at the crack tip. The choice of this size is not too critical for a stress-controlled failure mode due to crack blunting effect, but it is much more critical because of the same reason when dealing with a strain-controlled fracture mode, as in ductile rupture. Theoretically, the introduction of constitutive equations in which a coupling with damage is introduced, such as the Gurson potential, could help in the solution of this difficult problem.

For brittle growth-controlled cleavage fracture, it was indicated previously that the distance, λ , or the unit volume, V_u in the Weibull statistics is determined from valid fracture toughness tests carried out at low temperature. For ductile rupture initiated from inclusions, the characteristic distance is related to the mean inclusion spacing, for instance the mean spacing in a plane perpendicular to the crack front when ductile tearing is modelled as a two-dimensional process. In order to take into account the effect of distribution of inclusions, a mean value for the cavity growth ahead of a crack tip can be calculated as:

$$< R/R_o > = \int_o^{R_c} [R/R_o(x/\mathrm{CTOD}, \theta) \cdot P(x, \theta)] dV(x) \qquad (6.20)$$

where $P(x, \theta)$ is the probability of finding an inclusion in a small volume $dV(x)$ located at a distance x from the crack tip. The cavity growth rate (R/R_o) at a given position $(x/\mathrm{CTOD}, \theta)$ can be obtained using the results of FEM calculations in which the crack tip blunting effect was taken into account (Rice and Johnson, 1970; McMeeking, 1977). Actually, fracture initiation from a sharp crack is a three-dimensional process. The choice of the third characteristic dimension along the crack front may be problematic, in particular in materials exhibiting a strong clustering effect similar to that shown in Fig. 6.10.

6.3.2 Theoretical Expressions between J_{IC} or K_{IC} and Local Criteria under Small-Scale Yielding Conditions

Analytical solutions for the crack tip stress or strain fields, the so called HRR field (Hutchinson, 1968; Rice and Rosengren, 1968), in conjunction with FEM plane strain small-scale yielding results, and modelling the crack tip blunting effect, have all been used to derive theoretical relationships between the fracture toughness expressed in terms of J_{IC} or K_{IC} and relate these to the local criteria, consisting of Weibull statistics for brittle cleavage fracture and critical cavity growth, (Eq. 6.9) for ductile rupture (Pineau and Joly, 1991; Mudry et al., 1989).

The probability of cleavage fracture can thus be expressed as:

$$P_R = 1 - \exp\left[-\frac{J_{IC}^2 E^2 B \sigma_o^{m-4} C_m}{(1-v^2)^2 V_u \sigma_u^m}\right] \quad (6.21)$$

where B is the specimen thickness, σ_o, the yield strength and C_m, a numerical factor. As shown earlier by Beremin (1983), and as confirmed more recently by Wallin (1991), the predicted variations of K_{IC} or J_{IC} with specimen size and temperature via the temperature dependence of the yield strength are in good agreement with the experimental results. Recently, Wallin (1991) introduced in Eq. 6.21 a threshold parameter and a simple correction function to account for ductile crack growth before cleavage initiation in the ductile-to-brittle transition region.

The validity of this approach applied to brittle intergranular fracture has not yet been established satisfactorily. As stated in the part devoted to the micromechanisms, it might be possible that this expression which is essentially based on a critical stress concept, similar to the RKR model (Ritchie et al., 1973) does not strictly apply under these conditions.

For the ductile rupture of materials in which cavity nucleation occurs easily and discontinuously, it was shown similarly that J_{IC} at crack initiation is related to the critical void growth, $(R/R_o)_c$, by the following expression (Mudry et al., 1989):

$$J_{IC} = \alpha \lambda \sigma_o \ln (R/R_o)_c \quad (6.22)$$

where α is a numerical factor.

Similar approaches have to be developed to model the fracture toughness of materials in which cavity nucleation occurs continuously and in which the ductility at failure is only slightly dependent on stress-triaxiality. A good example was shown previously with the embrittled cast duplex stainless steels. In those materials, a simple criterion in terms of a critical strain could be more appropriate than a critical void growth criterion but the effect of statistical scatter should also be included in the model.

Another domain where these approaches may be useful is that of the development of materials exhibiting a good compromise between their strength and their fracture toughness. As a rule, the fracture toughness is observed to be a decreasing function of the yield strength of the materials. Fully pearlitic steels with a small interlamellar spacing, Sp might provide an exception to this rule. This effect might be simply related to the increase of the cleavage stress, σ_c, for small Sp and for fine pearlitic microstructures, as discussed in the previous part devoted to the micromechanisms of brittle cleavage fracture.

6.3.3 Fracture Toughness under Large-Scale Yielding Conditions

Laboratory testing of fracture toughness specimens to measure elastic- plastic fracture toughness generally focusses on two objectives: (i) to rank potential structural materials; (ii) to provide fracture toughness values for failure studies and to insure a level of quality during construction. For the first objective, specimens containing two dimensional deep cracks ($a/W \simeq 0.50$) are largely used. It is thought that these specimens provide the most severe crack-tip conditions, provided a number of size requirements related to the thickness and the in-plane dimensions are fullfilled. For the second objective, it frequently becomes necessary to test short ($a/W \simeq 0.10$) three dimensional cracks which are more relevant to the service conditions.

For long 2D cracks, the effect of specimen geometry (tensile versus bend specimen) and specimen size (thickness and in-plane dimensions) on ductile tearing resistance, characterized in terms of $J - \Delta a$ curves is now being well documented. A number of experimental results show that the $J - \Delta a$ curves are dependent on specimen thickness, in-plane dimensions and, in particular, on the mode of loading, tension versus bending (see e.g. Garwood, 1982; Roos et al., 1987; and Marandet et al., 1984). It is frequently observed that the CT specimens, compared to the center-cracked panel (CCP) tensile specimens lead to lower J (or CTOD) values and to lower values of the slope of the $J - \Delta a$ curves. This influence of specimen geometry is mainly related to the plastic constraint effect in the plane of the specimens. In order to overcome this difficulty, there is an effort in the test standardization to impose a size requirement, such that the specimen ligament size, b is larger than $N(J/\sigma_o)$ where $N \simeq 25$ for CT specimens and 200 for CCP specimens. Rather surprisingly it is frequently believed that these requirements are independent on the failure micromechanisms. We will show below that this is not necessarily true.

More recently the effect of crack depth on brittle fracture toughness has been more thoroughly examined (see e.g. Sorem et al., 1991; and Dodds et al., 1991). In a structural steel (A 36), the critical CTOD values at brittle fracture for the short crack specimens ($a/W \simeq 0.15$) are reported to be 2-3 times the values for the deep crack specimens ($a/W \simeq 0.50$). Moreover the transition from brittle-to-ductile crack initiation occurs at a lower temperature for the short crack specimens. This crack size effect is related to the fact that, for short cracks, the plastic zone at the crack tip extends "backward" to the traction free surface, prior to the development of a plastic hinge. The J-dominance of the stress-strain field for a short crack in tension and bending has been investigated by FEM calculations (Al-Ani and Hancock, 1991; Sumpter and Hancock, 1991; Parks, 1991). The loss of J-dominance and therefore of single parameter characterization has been attributed to compressive T-stresses, parallel to the crack flanks.

In both modes of failure, predictive models which take account of the

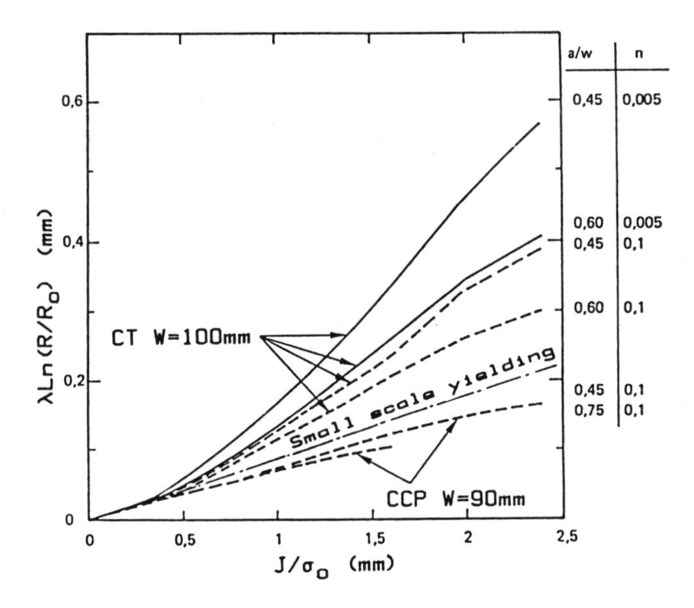

FIGURE 6.16. Variation of the ductile fracture criterion as a function of the loading parameter (J/σ_o). Influence of specimen geometry and work-hardening exponent (Mudry et al., 1989).

local crack-tip conditions and the micro-mechanisms of failure can be useful to satisfy the two objectives of laboratory testing of fracture toughness specimens and thus to provide guidelines for the transferability of the test results to components. This is clearly a research field which has not yet been investigated enough. We now present the results which can be derived from this approach to investigate the ligament size requirements for fracture toughness testing of 2D cracks.

Ligament size requirements in relation with the micromechanisms of failure: A numerical study was devoted to this specific topic by Mudry et al. (1989). In this study two local failure criteria, one for cleavage fracture (Eq. 6.3), the other one for ductile rupture (Eq. 6.20) were used to simulate numerically crack initiation under plane-strain conditions. Two widely different values for the work-hardening exponent were used, $n = 0.005$ and $n = 0.10$. Only deeply cracked CT and CCP specimens were simulated with different values of a/W $(0.45 < a/W < 0.75)$. The results corresponding to ductile rupture are reported in Fig. 6.16 where the product $\lambda \ln (R/R_o)_c$ introduced in Eq. 6.22 is plotted as a function of J/σ_o. Under small-scale yielding, there is no effect of specimen size or specimen geometry, as expected. On the other hand, it is observed that the relation between $\lambda \ln (R/R_o)_c$ and J/σ_o is largely specimen dependent when large-scale yielding conditions are reached. These results show that, for a material in which ductile rupture is essentially controlled by cavity growth, the val-

TABLE 6.1. Minimum size requirements for J_{IC} measurements

			Minimum $b\sigma_o/J$ for valid J_{IC}	
Specimen No. and Type	a/W	Strain Hardening Exponent	Local Ductile Fracture Criterion	Local Cleavage Fracture Criterion
53-CT	0.45	0.1	140	20
54-CT	0.60	0.1	85	15
55-CT	0.45	0.005	155	28
56-CT	0.60	0.005	90	40
258-CCP	0.75	0.1	25	185
259-CCP	0.45	0.1	30	990
268-CCP	0.75	0.005	45	560
269-CCP	0.45	0.005	50	1560

ues of J_{IC} are larger in CCP specimens than in CT specimens. Moreover the values of J_{IC} increase with the work-hardening exponent.

An attempt was made to use these results to provide an assessment of specimen size requirements. The values of the N coefficient introduced previously were calculated by using a deviation of 10 percent from the small scale-yielding solution. The results reported in Table 6.1 show that it is difficult to define a universal value for the minimum size requirement since it is dependent on the micro-mechanism of failure. For cleavage fracture the values of N derived from the analysis based on Weibull statistic (Eq. 6.4) are close to those recommended for fracture toughness tests. Larger values of N for CCP specimens reflect the fact that cleavage fracture is essentially stress-controlled. On the other hand, for ductile rupture which is essentially strain and stress-triaxiality controlled the size requirements are less stringent for CCP specimens than for CP specimens. This conclusion applies more specifically when ductile rupture is essentially strain-controlled, which is the situation corresponding to continuous strain-controlled nucleation, as discussed previously.

6.3.4 Other Applications and Developments

Further applications using criteria calculated from the local crack tip field are briefly presented before discussing the methods which could be developed, different from those based on a deliberate choice of finite element parameters to represent an appropriate averaging over the microstructure.

Three dimensional cracks

The eventual loss of J-dominance in two-dimensional plane-strain crack configurations, including important effects of degree of strain-hardening, crack geometry, and loading type is now well understood, and can be modelled using local criteria. However three-dimensional aspects of this problem are less well documented. Three dimensional FEM calculations have been presented recently by Parks (1991). Similar calculations coupled with the introduction of local damage criteria could be performed to determine not only the fracture toughness, but also the position along the crack front where the fracture process is initiated. Encouraging results along this line were reported at the EGF 9 conference for the ductile failure of an aluminum alloy (Greco et al., 1989).

Brittle fracture initiated from 3D cracks provides also a good example to illustrate the application of the local approach to fracture. Cleavage has already been investigated by Mudry (1987). More recently a study was also devoted to the brittle intergranular fracture of a temper-embrittled steel (Tigges et al., 1991). We concentrate only on cleavage fracture since, in both cases, a similar conclusion was reached.

Bending tests were carried out at low temperature on semi-elliptical cracks in a low alloy C-Mn-Ni-Mo steel. This produces a well-known variation of the stress-intensity factor, K, along the crack front, with a maximum K_b at the free surface and a minimum, K_a, at the deepest position of the ellipse. The usual fracture theory, based on strain energy release rate, predicts that failure is reached when the maximum value of K along the crack front reaches K_{IC}, i.e., $\max K(s) = K_b = K_{IC}$, where s is the curvilinian abscissa. This leads to a large overestimate of the fracture toughness. On the other hand, the application of the Weibull statistical theory (Eq. 6.21) leads to the following expression:

$$\ell \cdot K_{IC}^4 = \int K_I^4(s)ds \tag{6.23}$$

where ℓ is the length of the crack front and the thickness of the specimen used to measure the fracture toughness. It was observed that the results were in good agreement with Eq. 6.23.

Ductile-to-brittle-transition behavior

The ductile-to-brittle transition of fracture in ferritic steels is an important problem, both on a theoretical and a practical viewpoint which, in spite of many studies, is not yet fully understood. On pre-cracked specimens, it is frequently observed that this transition occurs either before or after some ductile crack initiation. Unstable cleavage fracture after large ductile crack extension takes place preferentially when the test temperature is increased. As a rule, a large scatter in the results is observed. Three physical explanations are frequently advanced to account for ductile-to-brittle-transition: (i) the increase in strain-rate due to crack extension produces an elevation of the maximum stress ahead of the crack tip; (ii) the crack extension produces an elevation of the load in the remaining ligament, and perhaps, more importantly (iii) crack growth produces an increase of the amount of

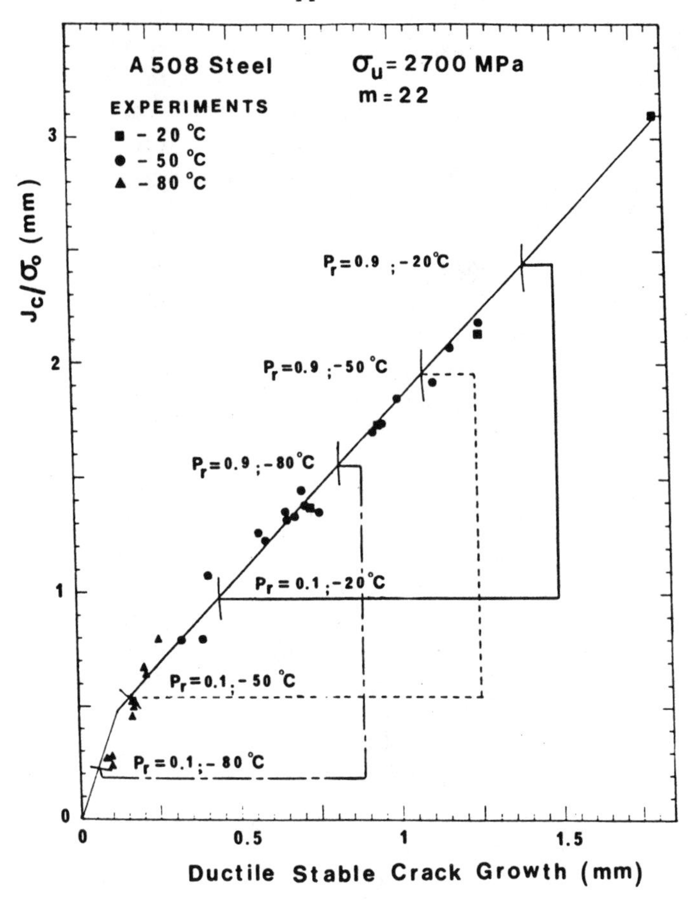

FIGURE 6.17. A-508 steel. Tests on axisymmetrically cracked specimens at various temperatures (-80°C; -50°C; -20°C). Ductile crack growth before unstable cleavage fracture. Comparison between measurements and a model based on Weibull statistics.

material which is sampled during crack propagation.

A number of results obtained on the ductile-to-brittle transition behavior of a C-Mn-Ni-Mo steel are given in Fig. 6.17 (Amar and Pineau, 1987). These results were obtained on axisymmetrically cracked tensile bars tested at three temperatures, -80°C, -50°C and -20°C. It is observed that, within a first approximation, a unique curve is obtained to characterize the ductile tearing resistance curve, $J - \Delta a$, provided that the value of J is divided by the yield strength, as suggested by Eq. 6.22. Finite element calculations were carried out to obtain the probability for failure at a given temperature, using the expression given in Eq. 6.3 with a strain correction factor which was introduced in order to account for the enhancement of the cleavage stress due to plastic strain. In this approach the statistical effects of

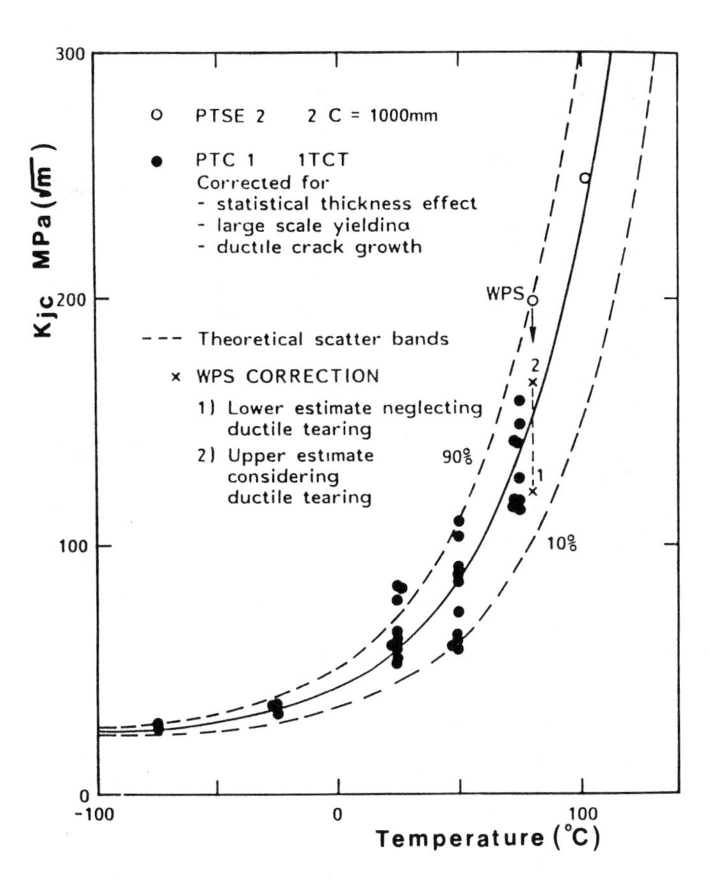

FIGURE 6.18. Analysis of ORNL pressurized thermal shock experiment 2 material, based on Weibull statistics (Wallin, 1991).

crack propagation were not taken into account. Figure 6.17 shows a broad agreement between the experimental results and those derived from the application of the Weibull theory. In particular, the large scatter in the results is well accounted for. Perhaps, a small correction for crack propagation effect, similar to that proposed by Wallin (1991) would have provided a better agreement. These results strongly suggest that, among the three factors which are advanced to explain the ductile-to-brittle transition, the two last ones, i.e. the increase in load or J, and perhaps the statistical effect associated with crack propagation are predominant. However further studies are necessary to reinforce this conclusion.

Thermal shock and size effects

The fact that Eq. 6.21 or those derived by Wallin (1991) to account eventually for a threshold effect and a crack propagation statistical effect yield a correct description of the brittle fracture probability, has widespread applications regarding fracture toughness testing in the ductile-to-brittle transition temperature regime. It means that it is possible to use quite small specimens for the laboratory tests, the results of which can be transformed to represent the behavior of a larger specimen or a structural detail which is not necessarily tested under isothermal conditions. As an example of its applicability this approach was successfully used to predict the event corresponding to brittle crack initiation during a thermal shock applied to a large cylinder (ϕ_{inner} = 920 mm, thickness = 230 mm) containing a long crack (1,000 mm), (Saillard et al., 1987). More recently, thermal shock experiments applied to thick (50 mm) and long (220 mm) cylinders containing a longitudinal sharp notch showed that it was also necessary to include the thickness effect predicted by Eq. 6.23 to explain the results for crack initiation (Di Fant et al., 1990). The analysis of two results of the ORNL pressurized thermal shock experiments have been presented by Wallin (1991) (Fig. 6.18). This author introduced a threshold effect and a correction for crack propagation in Eq. 6.21, as indicated previously. The results obtained on small specimens (25 mm CT) were therefore corrected to correspond to the actual (1,000 mm) wide crack and to include the effect of large-scale yielding and ductile tearing. Figure 6.18 shows that the results of these large scale thermal shock experiments are included within the calculated scatter band.

Scaling effects - Damage theory and the local approach

A large research effort is now being devoted to continuum damage mechanics and the local approach to fracture, especially in Europe. It is beyond the scope of the present chapter to review these approaches but it would be appropriate to comment concerning their developments. In the damage mechanics, the constitutive equations are modified to model the progressive damage which develops. A typical example is provided by the Gurson potential. In the local approach presented in this chapter the stress and strain histories are calculated from conventional constitutive equations to model both ductile and brittle fracture related to the micro-structure. Perhaps "one of the more urgent questions which should receive attention is the proper treatment of the scalings involved", as stated by Bilby (1991). The rapid development of image analysis to recognize the microstructural details can largely contribute to the knowledge of these scaling effects. However the post-mortem analysis of the fracture surfaces includes essentially the details associated with the third stage of ductile fracture, the coalescence stage which can be considered as an image of the "post-bifurcation". Perhaps, in this respect, more attention should be given to the observation of longitudinal sections ahead of the tip of propagating cracks in spite of the fact that these observations provide only a 2D analysis of the failure process.

Numerical methods involving softening theories have not yet been largely used to model crack initiation and crack growth. These methods, *a priori*, could be useful, in particular, to give solutions which are less mesh size dependent. However recent results tend to indicate that these approaches lead also to some strain localization and that they give results highly dependent of the mesh refinement (Saanouni and Lesne, 1989). This is the reason why perhaps still for a long period of time, some judicious choice of parameters reflecting the microstructural scale seems to be required. However, the rapid development of numerical methods and of the models for the microscopic details would make possible to incorporate in the "process zone", in the overall analysis, some specific modelling on the microstructural scale.

6.4 Conclusions

This chapter has shown that the approach largely developed by McClintock and which is based on the analysis of the local crack tip stress-strain field in conjunction with a detailed analysis of the fracture micro-mechanisms has largely contributed to the understanding of the fracture behavior of metallic materials.

However, many research areas remain to be explored, related to both fracture mechanisms and the prediction of fracture toughness. For most practical applications, due to the complexity of the problem, it is necessary to simplify the analysis and, in particular, to develop and test simple fracture criteria by focussing attention on the critical phases of fracture. With this respect it is clear that the relative simplicity of the criteria established for brittle cleavage fracture — either the concept of a critical stress over a characteristic distance — or more recent approaches based on Weibull statistics — may make important contributions to the modelling of this mode of failure. Brittle intergranular fracture has not yet received enough attention. Quantitative metallography and the mechanics of porous plastic materials have largely contributed to the understanding of ductile fracture by introducing a softening effect in the constitutive equations. However, much more research remains to be done to incorporate in particular the statistical aspects of this mode of failure and to understand the final stage of ductile fracture, that of coalescence, which is not yet fully understood. At the present time it is felt that, for practical applications, only simple criteria, such as the criterion based on the concept of a critical void growth at failure, can be used to model ductile rupture.

Analytical and numerical solutions for the crack tip stres-strain field have contributed importantly to the development of simple expressions relating either the crack tip opening displacement (CTOD) or the J-integral to local criteria established either for ductile rupture or cleavage fracture. These expressions apply when the small-scale yielding conditions are prevailing. The methodology based on local criteria is also extremely useful to deal with fracture accompanying large scale yielding, when the conditions of J-dominant field are lost, in particular to determine the size requirements for long two-dimensional cracks in relation to the damage micro-mechanisms, to predict the short crack effect in the ductile-to-brittle transition, and to model the behavior of three-dimensional cracks.

Acknowledgments: The author would like to acknowledge all the former PhD students of his research group, including, J.C. Lautridou, F. Mudry, B. Marini, E. Amar, M. Di Fant and now, P. Joly and Th. Iung. Thanks also to J.C. Devaux, A. Pelissier-Tanon and Y. Meyzaud from Framatome.

References

Al-Ani, A. M. and Hancock, J. W. (1991). J. Mech. Phys. Solids, 39:23–43.

Al Mundheri, M., Soulat, P., and Pineau, A. (1989). Fatigue and Fracture of Engineering Materials and Structures, 12:19–30.

Amar, E. and Pineau, A. (1987). Nuclear Engineering and Design, 105:89–96.

Argon, A. S., Im, J., and Safoglu, R. (1975). Met. Trans, 6A:825–837.

Barnby, J. T., Shi, Y. W., and Nadkarni, A. S. (1984). Int. J. Fracture, 25:273–283.

Beremin, F. M. (1981). Met. Trans. 12A:723–731.

Beremin, F. M. (1983). Met. Trans., 14A:2277–2287.

Berg, C. A. (1962). Proc. 4th U.S. National Congress of Applied Mechanics. Univ. of California, June 18–21.

Berveiller, M. and Zaoui, A. (1979). J. Mech. Phys. Solids, 26:325–345.

Bilby, B. A. (1991). Fundamentals of ductile fracture. ESIS/EGF 9, pages 3–18. See Wallin (1991).

Bourcier, R. J., Koss, D. A., Smelser, R. E., and Richmond, O. (1986). Acta Metall., 34:2443–2452.

Bowen, P., Condor, R. J., and Knott, J. F. (1990). Micromechanisms of cleavage fracture in ferritic steels. In Firrao, D., editor, *ECF8/Fracture Behaviour and Design of Materials and Structures*, I:25–30.

Cottrell, A. H. (1989). Mat. Science and Technology, 5:1165–1167.

Cottrell, A. H. (1990a) Mat. Science and Technology, 6:121–123.

Cottrell, A. H. (1990b). Mat. Science and Technology, 6:325–345.

Curry, D. A. and Knott, J. F. (1976). Metal Science, 10:1–10.

Di Fant, M., Genty, A, and Pineau, A. (1990). High Temperature Technology, 8:105–114.

Di Fant, M., Fontaine, A., and Pineau, A. (1991). Dynamic crack propagation and crack arrest in a structural steel: Comparison between isothermal and thermal shock tests. Int. Seminar on Dynamic Failure of Materials, Vienna, Austria, January.

Dodds, R. H., Anderson, T. L., and Kirk, M. T. (1991). Int. J. Fracture, 48:1–22.

Du, Z. Z. and Hancock, J. W. (1991). J. Mech. Phys. Solids, 39:555–567.

Fontaine, A., Maas, E., and Toulou, J. (1987). Nuclear Engineering Design, 105:83–88.

Garwood, S. J. (1982). Int. J. Pressure Vessels Piping, 10:297–319.

Greco, E, Roelandt, J. M., and Abisror, A. (1989). Numerical analysis of ductile rupture local onset: semi elliptical cracks-Influence of crack shape. European Symp. on Elastic-Plastic Fracture Mechanics, Freiburg, October.

Gurson, A. L. (1975). J. Engineering Mat. Techn., 99:2–15.

Holzmann, M., Vlach, B. and Man, J. (1991). The ductile-to-brittle transition of a pressure vessel steel embrittled by step cooling heat-treatment. ESIS/ECF 9, pages 569–585. See Wallin (1991).

Huang, Y., Hutchinson, J. W., and Tvergaard, V. (1991). J. Mech. Phys. Solids., 39:223–241.

Hutchinson, J. W. (1968). J. Mech. Phys. Solids, 16:13–31.

Jeulin, D. (1992). Morphological models of change of scale in brittle fracture statistics, To appear in Strength and Fracture, N°5.

Joly, P., Cozar, R., and Pineau, A. (1990). Scripta Metall., 24:2235–2240.

Joly, P. (1991). Mécanismes et mécanique de la rupture d'un acier biphasé (austénite/ferrite) moulé. Thesis.

Kameda, J. and Mc Mahon, C. J. (1980). Met. Trans., 11A:91–101.

Kavishe, F. P. L. and Baker, T. J. (1986). Materials Science and Technology, 2:583–588.

Kwon, D. and Asaro, R. J. (1990). Met Trans., 21A:117–134.

Lewandowski, J. J. and Thompson, A. W. (1986). Met. Trans., 17A:1769–1786.

Machida, S., Miyata, T., Hagiwara, H., Yoshinari, H., and Suzuki, Y. (1991). A statistical study of the effect of local brittle zone on the fracture toughness (CTOD) of weldments. ESIS/EGF9, pages 633–658. See Wallin (1991).

Marandet, B., Phelippeau, G., De Roo, G., and Rousselier, G. (1984). Effect of specimen dimensions on J_{IC} at initiation of crack growth by ductile tearing, ASTM STP 833, pages 606–621.

Marini, B., Mudry, F., and Pineau, A. (1985). Eng. Fracture Mechanics, 22:989–996.

McClintock, F. A. (1963). On the plasticity of the growth of fatigue cracks, Fracture of Solids, John Wiley, New York, pages 65–102.

McClintock, F. A. (1967). ASTM STP 415, pages 170–180.

McClintock, F. A. (1968). J. App. Mech., 35:363–371.

McClintock, F. A. (1971). Plasticity aspects of fracture, In Liebowitz, H., editor, *Fracture*, 3:47–225.

McMeeking, R. M. (1977). J. Mech. Phys. Solids, 25:357–381.

Mudry, F. (1982). Etude de la rupture ductile et de la rupture par clivage d'aciers faiblement alliés. Thesis.

Mudry, F. (1983). Methodology and applications of local criteria for the prediction of ductile tearing. In Larsson, M. H., editor, *Elastic-Plastic Fracture Mech*, 4th ISPRA Conf, Ispra, Italy, page 263.

Mudry, F. (1987). Nuclear Engineering Design, 105:65–76.

Mudry, F., Di Rienzo, F., and Pineau, A. (1989). Numerical comparison of global and local criteria in compact tension and centre-cracked panel specimens, ASTM STP 995, pages 24–39.

Needleman, A. (1989). Computational micromechanics. In Germain, P. et al., editor, *Theoretical and Applied Mechanics*, Elsevier Science Publishers, pages 217–240.

Needleman, A. and Kushner, A. S. (1990). European J. Mech. A/Solids, 9:193–206.

Orowan, E. (1948). Fracture and strength of solids, Report on Progress in Physics, 12:185.

Parks, D. M. (1976). J. Engineering Materials Technology, Trans ASME, pages 30–36.

Parks, D. (1991). Three dimensional aspects of HRR-dominance. ESIS/EGF 9, pages 205–231. See Wallin (1991).

Perrin, G. and Leblond, J. B. (1990). International Journal of Plasticity, 6:677–699.

Pineau, A., and Joly, P. (1991). Local versus global approaches to elastic-plastic fracture mechanics. Application to ferritic steels and a cast duplex stainless steel. ESIS/EGF 9, pages 381–414. See Wallin (1991).

Pineau, A. Review of fracture micromechanisms and a local approach to predicting crack resistance in low strength steels, In François et al., editor, *Advances in Fracture Research*, IFC5 Conf., pages 553–577.

Rice, J. R. and Rosengren, G. F. (1968). J. Mech. Phys. Solids, 16:1–12.

Rice, J. R. and Tracey, D. M. (1969). J. Mech. Phys. Solids, 17:201–217.

Rice, J. R. and Johnson, M. A. (1970). The role of large crack tip geometry changes in plane strain fracture. In Kanninen, M. F. et al., editor, *Inelastic Behaviour of Solids*, McGraw-Hill, New York, page 641.

Ritchie, R. O., Knott, J. F., and Rice, J. R. (1973). J. Mech. Phys. Solids, 21:395–410.

Roos, E., Eisele, U., Silcher, H., and Kiessling, D. (1987). Determination of J curves by large scale specimens, 13th MPA Seminar, Stuttgart.

Rousselier, G. (1979) Contribution àl'étude de la rupture des métaux dans le domaine de l'élasto-plasticité. Thèse, Ecole Polytechnique.

Rousselier, G. (1987). Nuclear Engineering Design, 105:97–111.

Saanouni, K. and Lesne, P. M. (1989). On the creep crack-growth prediction by a non local damage formulation. Eur. J. Mech., A Solids, 8:437–459.

Saillard, P., Devaux, J. C., and Pellissier-Tanon, A. (1987). Nuclear Engineering and Design, 105:83–88.

Sorem, W. A., Dodds, R. H., and Rolfe, S. T. (1991). Int. J. Fracture, 47:105–106.

Sumpter, J. D. G. and Hancock, J. W. (1991). Int. J. Pressure Vessel Piping, 45:207–221.

Tigges, D, Piques, R., and Pineau, A. Unpublished results.

Tvergaard, V. (1981). Int. J. Fracture, 17:389–407.

Tvergaard, V. and Needleman, A. (1984). Acta Metall., 32:157–169.

Wallin, K., Saario, T., and Torronen, K. (1984). Metal Science, 18:13–16.

Wallin, K. (1991). Statistical modelling of fracture in the ductile-to-brittle transition region, Defect assessment in components, In Blauel, J. G. and Schwalbe, K. H., editors, *Fundamentals and Applications*, ESIS/EGF 9, MEP, London, pages 415–445.

Walsh, J. A., Jata, K. V., and Starke, E. A. (1989). Acta Metall., 37:2861–2871.

Worswick, M. J. and Pick, R. J. (1990). J. Mech. Phys. Solids, 38:610–625.

Worswick, M. J. and Pick, R. J. Submitted to J. Applied Mech.

7

Growth of Cracks By Intergranular Cavitation in Creep

A. S. Argon, K. J. Hsia and D. M. Parks

ABSTRACT The spread of intergranular creep damage around blunt notches and sharp cracks in ductile single phase alloys is modeled by a mechanism based continuum material damage model and a finite element approach. The details of the material damage have been derived from extensive earlier experimental results on an austenitic stainless steel. The finite element simulation of the evolution of intergranular damage in the form of accumulating densities of grain boundary facet cracks has indicated that while this damage spreads out preferentially along inclined planes around the tips of sharp cracks, it localized in the symmetry plane ahead of a blunt notch. These results are in excellent agreement with the experimental observations of Ozmat, et al. on the directions of early crack growth from sharp cracks and blunt notches in Type 304 stainless steel.

7.1 Introduction

From its very beginning two competing views have prevailed in the description of the fracture phenomenon. In the one initiated by Griffith (1920) and formalized later by Irwin (1948), fracture is described as an overall system instability of a crack in a stressed part where the crack begins to grow and accelerates when the free energy of the overall system of cracked part and external stressing agency reaches a maximum. Or alternatively stated, when the rate of energy release from the overall system equals the work rate of fracture. This point of view, widely preferred in the engineering field of structural design and reliability, requires no mechanistic detail relating to microstructure and processes of material separation. It prescribes, what are thought to be, conservative, fracture mechanics tests that candidate materials must satisfy. Some of its practitioners often take pride in never having touched a microscope to see how the material is actually coming apart. In the second view, formulated first by Orowan (1934), fracture is considered to be the result of a local process of material separation occurring at the tip of a crack where the crack only serves to concentrate and maintain the required "driving forces" of stress (or strain), etc. to permit the separation processes to develop when the local ultimate material properties are reached. The two points of view are coincident in the case of purely brittle fractures, involving the propagation of a tensile crack under mode I loading (see Orowan, 1949, McClintock and Argon, 1966) when no dissipative fracture work is involved. Even here, however, in the case of fracture in compression where pre-existing cracks are subjected to mode II loading, fracture occurs across a direction of pure tensile separation when the cohesive strength of the solid is reached locally on the surface of the crack (Griffith, 1924). In the cases of fracture accompanied by plastic deformation, and particularly in fractures involving the development of deformation-induced damage processes, the second point of view is the only one that provides real understanding of the material separation phenomena, and the means of affecting them by microstructural modification. While this point of view began to be appreciated in the late 40's by a number of investigators (Bridgman, 1952; Tipper, 1949; Orowan, 1949), the pioneer in this field has been McClintock with his elastic-plastic mode III crack tip field analysis (Hult and McClintock, 1957) and more importantly, with his micro-mechanical model of ductile fracture by hole growth (McClintock, 1968). In a major exercise in synthesis, McClintock (1971) further emphasized the "process" nature of ductile fracture by considering in great detail a variety of concentrated slip line fields, emanating from crack tips, within which the process of ductile separation could develop. These considerations have led to more specific applications of plastic cavity growth in shear (McClintock, Kaplan, Berg, 1967), cavity growth in the concentrated strain field of a blunt mode I crack (McClintock, 1969) and finally to considerations of local plastic growth of planar arrays of cavities (Nagpal et al., 1972). Most of the considerations of plastic cavitation that relate to ductile fracture have their counterparts in creep fracture, where one of the more insidious forms of fracture is intergranular cavitation by a combination of continuum creep flow and diffusional flow (Needleman and Rice, 1979) where the deformations remain highly localized to the vicinities of boundaries resulting in rather low overall ductilities. It is this form of creep fracture mechanism that we will discuss here and how it manifests itself in the surprising modes of crack growth in the more ductile creeping alloys. Here we will only summarize this work and discuss its general implications, its presentation in full can be found elsewhere (Argon, et al. 1985; Hsia, Argon and Parks 1991a; Hsia, Parks and Argon, 1991b).

The phenomenon that we wish to explain that requires considerable in-depth understanding of the mechanisms of creep damage is shown in Figs. 7.1(a) and 7.1(b) reproduced from the experimental study of Ozmat, Argon and Parks (1991). When a very sharp initial crack is present in a sample, as might result from a previous fatigue history, the subsequent growth of this crack by the accumulation of local creep damage will not be in the plane of the initial crack, but will be split along inclined planes roughly at 45° toward both sides of the initial crack plane as shown in Fig.

7.1(a). On the other hand if the initial crack is not sharp, but is blunted, or acts more like a blunt notch, then the subsequent creep crack will meander much less up and down, but will grow roughly in the extension of the initial plane of the crack as shown in Fig. 7.1(b). It is this phenomenon that we wish to explain in the light of the best available mechanisms of creep damage.

7.2 Intergranular Damage Processes

7.2.1 Chronology of Intergranular Creep Damage

Ashby and Dyson (1984) have presented a general catalogue of phenomena that in addition to genuine microstructural damage and fracture, considered also many other forms of accelerations of creep due to microstructural instabilities and overall rupture. The temperature and stress conditions where these processes are dominant, have also been catalogued by Ashby et al. (1979) in the form of fracture mechanism maps. These all give a useful perspective to the specific process of intergranular fracture that we will present here, occurring in polycrystalline alloys under relatively low stresses typical of service conditions. Under these conditions creep damage is by intergranular cavitation emanating from hard grain boundary (G-B) particles (Argon, 1983) and involving growth of such cavities under stress on boundaries nearly normal to the local maximum principal tensile stress. The regimes and modes of growth of intergranular cavities by a combination of diffusional flow and creep deformation at different levels of stress and temperature have been studied extensively by many investigators over the past decade and a half. These developments have been reviewed by Argon (1982). Experiments on different alloys with different microstructure have identified two limiting types of damage morphology. In one limit are ductile solid solution alloys such as Type 304 stainless steel with considerable variation in the density of G-B particles among different grain boundaries. The growth and coalescence of the resulting cavities in these materials is quite inhomogeneous, giving rise to substantial densities of relatively isolated G-B facet microcracks (Chen and Argon, 1981a; Don and Majumdar, 1986). These subsequently grow further by creep deformation, to interact with each other and eventually result in overall fracture. The other limiting damage morphology is represented by less ductile, but much more creep-resistant alloys with built-in heterogeneous phases, such as Nimonic 80A (Dyson and McLean, 1977) and Astroloy (Capano et al., 1989). The intergranular cavitation process in these alloys is more homogeneous on all G-Bs and leads to overall fracture without the intermediate formation of an appreciable density of G-B facet microcracks (Capano et al., 1989). These different modes of damage development and their dependences on the type

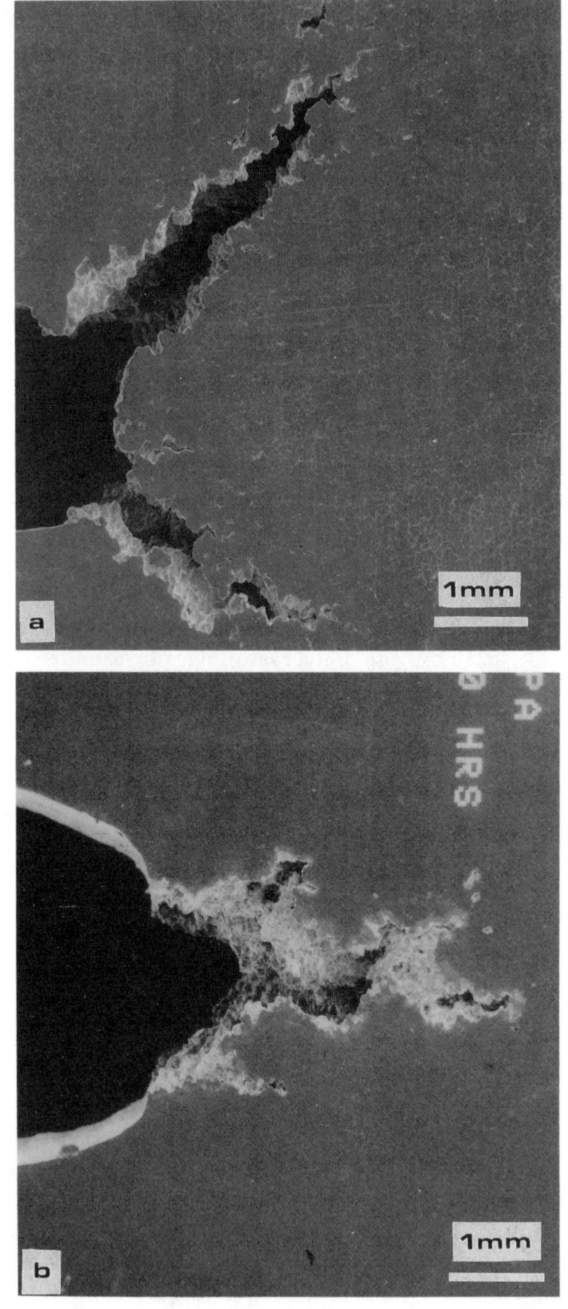

FIGURE 7.1. Two different modes of creep crack growth in Type 304 Stainless Steel: (a) bifurcation of crack front in sample with initially sharp crack; (b) quasi-planar crack growth from a blunt notch (Ozmat et al. 1991).

and level of local stress need to be described by mechanistic damage models. Such differences manifest themselves particularly in the accumulation of creep damage in notches and at the tips of macro-cracks in the form of crack plane meanderings reported by a number of investigators in the past (Davis and Manjoine, 1953; Leckie and Hayhurst, 1974; Hayhurst et al., 1978). Of particular interest to us here will be the experimental results on crack growth of Ozmat et al. (1991) in Type 304 stainless steel. These experiments have demonstrated that in this solid solution alloy, due to the non-uniformity of grain boundary cavitation, a significant volume density of traction-free G-B facet cracks emerge as a result of cavity coalescence before fracture happens at a macroscopic material point. It is the linkage of these facet cracks which then results in the final fracture. By assuming a random distribution of particle density among all the G-B's, Argon et al. (1985) developed a creep damage model in which only the contributions of traction-free G-B facet cracks to the creep rate were considered, even though the preparatory processes of intergranular cavitation were also considered in detail to account for the rates of appearance of the facet microcracks. In that model a mechanistic damage parameter was defined to be the area fraction of G-Bs which have become facet cracks, and the fracture mechanism was considered to be the eventual linkage of these cracks.

In reality, it must be recognized that both the traction-transmitting cavitating boundaries and the traction-free boundary facet microcracks will accelerate the overall creep rate, and will contribute to the damage at a material point. While the contribution of traction-transmitting cavitating G-Bs may dominate the damage process in relatively brittle heterogeneous superalloys, the contribution of traction-free facet cracks is likely to be much more significant in more ductile solid solution alloys such as Type 304 stainless steel.

Here we summarize the mechanistic damage model described in detail by Argon et al. (1985) that is intended for a ductile alloy in which the evolution of G-B facet cracks is the principal form of damage. The effects of G-B sliding on the damage evolution and creep constitutive law developed separately by Hsia et al. (1991) is incorporated into the model. Of primary interest is the application of the model to the spread of damage at the root of a British Standard Notch (BSN) and at the tip of a sharp crack. The numerical results of damage distribution and initial creep crack growth patterns from FEM computation will then be compared with experimental observations of Ozmat et al. (1991).

We note here that the model we are presenting, while independently developed, along the lines we have outlined above, has derived considerable stimulus from both phenomenological damage models of Hayhurst and co-workers (Hayhurst, Dimmer and Morrison, 1984; Hayhurst, Brown and Morrison, (1984), and mechanistic models of Tvergaard (1984; 1985a; 1985b; 1986). A more expanded discussion of these models, their strengths and weaknesses has been provided elsewhere by Hsia, et al. (1991a).

7.2.2 Effects of Grain Boundary Sliding

Grain boundary sliding plays a central role in intergranular fracture in creep. This role begins with the nucleation of cavities on grain boundary particles as the latter act as the chief source of stress concentration when they obstruct the otherwise free sliding of G-Bs. This phase of G-B sliding in a jerky and stochastic way, is integrally linked to the cavity nucleation rate that we will not discuss in any detail beyond recognizing that because the time average overall sliding rate of G-B's must be linked to the overall creep strain rate of the material, the cavity nucleation rate on the G-B particles must be proportional to the overall creep strain rate. An equally important second role of G-B sliding occurs in the accelerated rate of growth by creep flow of G-B size voids resulting from the continued deformation of G-B facet cracks. Such accelerated growth markedly affects the interaction of these separated G-B size voids and the associated tertiary creep rate. It is this second aspect of G-B sliding, which needs to be assessed separately, that was considered in considerable detail by Hsia, Parks and Argon (1991b) by a finite element method (FEM) that we will summarize in this section.

The effect of G-B sliding on the continuum creep rate had been studied in the past by many investigators using a variety of different approaches (Hart, 1967; Crossman and Ashby, 1975; Chen and Argon, 1979; and Ghahremani, 1980). Their principal finding is that when grain boundary sliding occurs freely, so that only normal stresses are transmitted across them, the effect on the overall creep rate of the solid is a simple stress enhancement (or a corresponding reduction of creep resistance). The 2-D FEM model of Hsia et al. (1991b) has verified these findings for both elastic and creeping polycrystalline grain assemblies. The subject of primary interest to Hsia et al. (1991b), however, has been the role of G-B sliding in the volumetric rate of growth of G-B facet microcracks and the qualitatively new dimension that it introduces to this response as a function of the applied stress state. This is shown in Fig. 7.2 where an assembly of regular hexagonal grains are subjected to a transverse compressive stress. If there is no G-B sliding and the assembly acts as a continuum, for a G-B facet crack with its plane parallel to the compression direction, there will be no opening of the crack, while very significant opening will occur when the surrounding G-B's are allowed to slide as the figure on the right demonstrates.

The growth rate of the cross sectional area of an ellipsoidal cavity, in 2-D plane-strain symmetry, in a power-law creeping solid, has been solved by a perturbation method by He and Hutchinson (1981) (H&H), who have also considered the effect of a dilute concentration (non-interactive behavior) of such cavities as G-B facet microcracks, on the overall creep rate. The FEM study of Hsia et al. (1991b) can be considered as a generalization of that of H&H to include G-B sliding.

The first significant finding of Hsia et al. (1991b) is illustrated in Fig. 7.3, which shows the normalized areal (cross-sectional) growth rate ΔA of facet

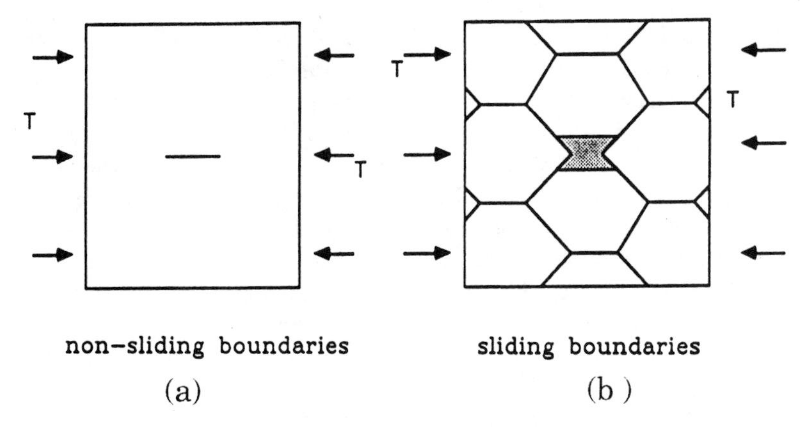

FIGURE 7.2. Role of grain boundary sliding in opening up of G-B facet cracks under transverse compression: (a) without G-B sliding; (b) with G-B sliding (Hsia, et al. 1991b).

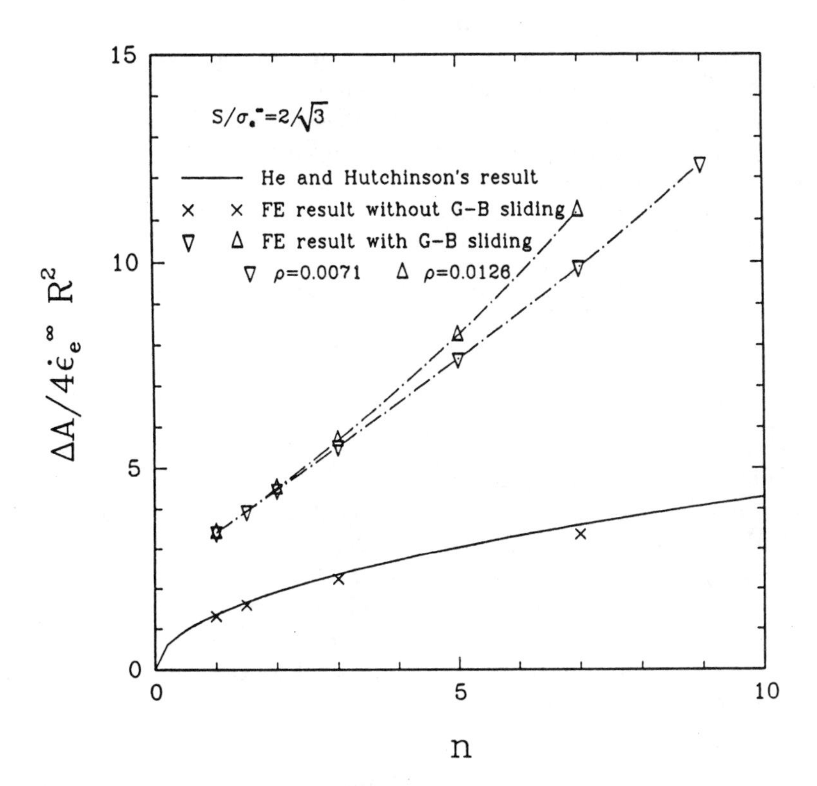

FIGURE 7.3. Dependence of facet crack opening rate on creep stress exponent for a given triaxiality (Hsia, et al. 1991b).

cracks of semi-major axis R in plane-strain tension with the crack plane being normal to the principal tensile stress direction. Here S is the far field maximum principal stress and σ_e^∞ is the far field Mises equivalent stress resulting in a far field equivalent creep strain rate of $\dot{\epsilon}_e^\infty$, and where n is the stress exponent in the power law creep rate expression. The figure shows that the FEM results of Hsia et al. compare very well with the results of H&H for the case of no G-B sliding. It also shows the result of a real growth rate of facet cracks for two crack densities ρ, defined as $\rho = \frac{3\pi}{2} N_A R^2$, where N_A is the number of facet cracks per unit projected area. [1] Thus, Fig. 7.3 shows the growth rate, for a given triaxiality ratio $S/\sigma_e^\infty = 2/\sqrt{3}$ (plane strain tension), as a function of the creep stress exponent. Figure 7.4 gives complementary information on the dependence of the growth rate on the triaxiality ratio for a stress exponent of 5 (characteristic of many pure metals.) The figures show that for any given condition the facet crack opening rate is very significantly higher when grain boundaries are allowed to slide, and that as the triaxiality ratio S/σ_e^∞ goes to zero, corresponding to zero maximum principal stress, the facet crack opening stops altogether if there is no grain boundary sliding, whereas non-zero facet crack opening rate results when there is grain boundary sliding, as is clear from Figure 7.4. Moreover, the figures show that the individual facet crack growth rates increase with increasing facet crack concentration which is a direct result of the interaction of facet cracks with each other. This facet crack areal growth rate has been parameterized by Hsia et al. (1991b) and can be given in analytical form as follows:

$$\frac{\Delta A}{\dot{\epsilon}_e^\infty R^2} = G_0(n) + G_1(n,\rho)\frac{S}{\sigma_e^\infty} \tag{7.1}$$

where

$$G_0(n) = 4.0 n^{0.65} \tag{7.2a}$$

$$G_1(n,\rho) = 4.0(2.05 + 15.5\rho)\sqrt{n}. \tag{7.2b}$$

Hsia et al. (1991b) have generalized their FEM findings to 3-D response by making use of the correspondence between 3-D and 2-D problems found by H&H. Through this comparison, based on the 3-D analysis of H&H, Hsia et al. (1991b) have proposed a modified creep potential function Φ from which all tensor creep rates can be obtained by differentiating with the appropriate elements of the stress tensor. We delay the introduction of this potential function to the section below where the mechanistic creep damage model is presented.

---

[1] A scaling error in Hsia, et al. (1991b), pointed out to us by Prof. V. Tvergaard has been corrected here. We are grateful to Prof. Tvergaard for this correction

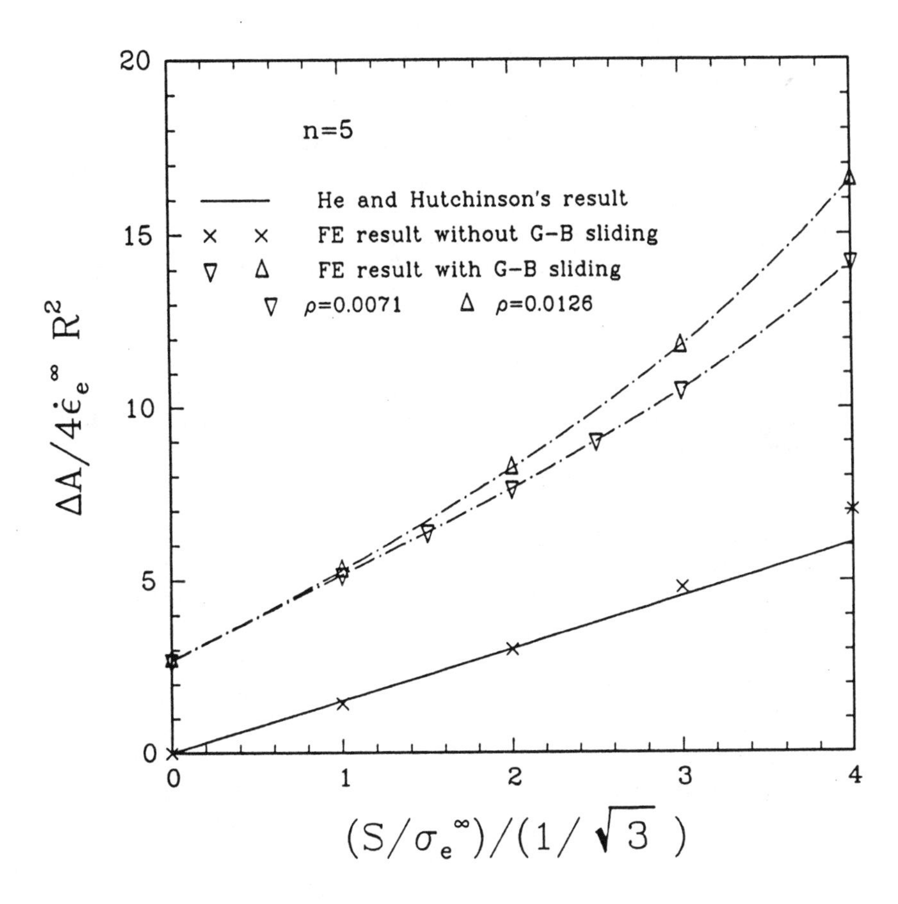

FIGURE 7.4. Dependence of facet crack opening rate on triaxiality ratio for a creep stress exponent of $n = 5$ (Hsia, et al. 1991b).

7.2.3 The Mechanistic Creep Damage Model

Damage Defining Parameters

In modeling creep damage problems by continuum damage mechanics, a set of internal state variables are introduced to describe the state of the material at each material point, and an evolution equation for each internal state variable is established, based on physical mechanistic considerations relying on experimental observations. In the simplest form, adopted here, a single scalar parameter α represents the state of damage. Thus, the creep constitutive relation incorporating material damage to accelerate the creep rate during tertiary creep will take the symbolic form of

$$\dot{\epsilon}^c = f(\sigma, \alpha, T), \tag{7.3}$$

where σ is the macroscopic stress, and T is the absolute temperature. The rate of evolution of damage parameter $\dot{\alpha}$ is then defined as a function of the current state of the material, i.e.,

$$\dot{\alpha} = g(\sigma, \alpha, T). \tag{7.4}$$

In the present model, the damage parameter α is defined as the area fraction of the grain boundaries which have become traction-free facet cracks, among all G-Bs. By this definition, in contrast to Tvergaard's (1984) model, only the contributions of the fully opened boundary facets to the overall creep rate are taken into account. The contributions from the partially cavitated boundaries are not specifically considered. This underestimates the creep rate acceleration due to the earlier forms of G-B cavitation. Since the contributions of facet microcracks continue to develop throughout creep life, they are deemed to be much more important than those of the partially cavitated boundaries for a ductile alloy.

We note that, in addition to the state variables discussed above, the evolution of damage could also depend on some other physical parameters such as microstructural changes. While these could be important in heterogeneous superalloys, in the solid solution alloy of 304 SS they are not present. Since G-B cavitation is a cumulative deformation-dependent phenomenon, it is expected to parallel closely the accumulated creep strain.

In the section below we first develop the creep damage model. In a later section we then couple the damage model to the creep constitutive law.

The Creep Damage Model

It has now been established conclusively that hard particles on G-Bs play a key role in cavity nucleation (Argon, 1982; Argon, 1983) during creep, but that to develop the required high interface stresses on the particles, G-B sliding is essential (Argon, 1982). Much evidence points in the direction that these high stresses are only produced during stochastic transients in boundary sliding when accommodation mechanisms are ineffective. Since such transients recur throughout creep life, nucleation of cavities is a continuous process and its rate scales directly with the creep strain rate. This was first established by Cane (1978) in experiments on polycrystalline iron, demonstrating that the cavity nucleation rate is given by

$$\dot{N} = \beta \dot{\epsilon}^c \qquad \text{or} \qquad N = \beta \epsilon^c, \tag{7.5}$$

where N is the number of cavities per unit G-B area, and β is the nucleation coefficient which relates directly to the G-B particle density (Argon et al., 1985). This continuous nature of grain boundary cavity nucleation has now been observed in many systems (iron (Cane, 1978), 304 stainless steel (Chen and Argon, 1981a), Nimonic 80A (Dyson and McLean, 1977), Astroloy (Capano et al., 1989)). The often observed saturation of nucleation is a direct result of exhaustion of cavitation sites when all potential effective particles have been used up (Chen and Argon, 1981a; Dyson and McLean, 1977). Since cavity growth dominates the damage evolution process in later stages, the neglect of the saturation of nucleation has little influence on the prediction of the area coverage of cavities on G-Bs. If the distribution function of β over all G-Bs is known, the overall cavity nucleation behavior can be obtained in terms of the accumulated creep strain from Equation 7.5.

In order to determine the distribution of the cavity nucleation coefficient experimentally, a large number of G-Bs must be sampled and statistical measurements of particle size, particle density, crystallographic misorientation, etc., must be made to correlate them with the distribution of cavity nucleation behavior. Such detailed measurements and correlations have not been made on any alloy. While Chen and Argon (1981a) reported important variations in the cavitation behavior of different G-Bs containing growth twins, and boundaries which had different levels of "roughness", no quantitative correlations were given. In another related experimental study of G-B cavitation and G-B structure of 304 stainless steel at elevated temperature, Don and Majumdar (1986) found that cavity size and distribution varied more significantly from one G-B to another than within the same G-B. They noted that since the random distribution of G-B particles differed only from boundary to boundary, this suggested that G-B structure must play an important role in G-B cavitation. Although a more definite relationship between cavitation and G-B structure still remains to be established, the above findings are in support of the assumption that as a first-order approximation β can be considered as constant on any one boundary but varies from boundary to boundary. On the basis of these observations, Argon et al. (1985) invoked the earlier measurements of Argon and Im (1975) on random distributions of carbides in spheroidized steels, and suggested that similarly a Poisson distribution of G-B carbides exists among different G-Bs given by

$$\eta(\xi) = 1 - (1 + \xi)e^{-\xi}. \tag{7.6}$$

In Equation 7.6 $\xi = \bar{\beta}/\beta$ is a dimensionless variable, $\bar{\beta}$ is the average value of the nucleation coefficient β over all boundaries, and η is the area fraction of boundaries having a local nucleation coefficient equal to or larger than β. In Equation 7.6 the implicit assumption is made that cavity nucleation occurs at G-B carbides and that, therefore, β represents the reciprocal G-B area per carbide. Note that those boundaries which have a larger nucleation coefficient will cavitate earlier, so that they will become facet microcracks first. Thus, this fraction η is directly related to the damage parameter α.

Because the formation of facet cracks is associated with the growth and coalescence of G-B cavities, the growth law of such cavities has been of great interest to many investigators (Chuang et al., 1979; Needleman and Rice, 1980; Sham and Needleman, 1983; Hull and Rimmer, 1959; Speight and Beere, 1975; Chen and Argon, 1981b; Pharr and Nix, 1979; Martinez and Nix, 1982). Wide ranging models developed by these investigators have included both quasi-equilibrium growth mechanisms and non-equilibrium mechanisms of crack-like growth of cavities. It was first noted by Dyson (1976) that often only isolated boundaries cavitate (as is the case in 304 stainless steel), and that the thickening rate of these boundaries due to cavity growth must be compatible with the creep rate of the surrounding material. In a specific model of this effect, Rice (1981) demonstrated that the isolated boundaries that cavitate must shed load to the surroundings to meet the compatibility conditions. Based on this condition, Rice provided expressions for the cavity growth rate on isolated traction-shedding boundaries constrained by the creep flow of the far field under a uniaxial tensile stress. Hsia et al. (1991b) have shown, however, that if G-B sliding is taken into account, the traction-shedding effect becomes more significant due to the introduction of additional kinematic degrees of freedom of boundary sliding. Then, the cavity growth rate is substantially higher than in the case without G-B sliding, and the effect of the transverse stresses must also be considered. An important additional qualitative difference between the cases without and with sliding boundaries is that in the case of no sliding, tensile stresses do not appear across a boundary that is not subjected to such a stress, S, at a distance, regardless of the nature of transverse stress, T, acting on the material. However, when boundaries slide, tensile stresses are produced across a boundary through the remote application of transverse compressive stresses, T, parallel to the boundary in question. This yields finite values of facet stress, and results in cavity growth caused by transverse compression. Comparative results of cavitation models without and with sliding grain boundaries have been given by Hsia, et al. (1991b). Here we will consider only the models including G-B sliding.

On a particular G-B, if a representative right-circular cylinder of radius b containing a cavity of radius a at its center is chosen to examine cavity growth rate, the area coverage of cavities on this comparatively isolated G-B becomes $A = a^2/b^2$, and the cylinder size b satisfies the expression $\pi b^2 N = 1$, where N represents the number of cavities per unit G-B area.

Utilizing Equation 7.5, the rate of change in this area coverage of cavities can be given as,

$$\dot{A} = \pi a^2 \beta \dot{\epsilon}^c + 2\pi a N \dot{a}. \tag{7.7}$$

Considering only the creep constrained cavity growth which is the case when the far field stresses and creep strain rate are sufficiently low and are kept constant, and following the development of Rice (1981), Hsia et al. (1991b) obtained the cavity growth rate $\dot{a}$ on a typical boundary, which for the case of sliding G-Bs in the surrounding material, subject to a tensile stress S and a transverse stress T is:

$$\dot{a} = \frac{D}{a^2 h(\psi)} \frac{\frac{G_0(n)}{G_1(n,\rho)} + \frac{S}{\sigma_e^\infty}}{\frac{4\pi D}{b^2 R} \frac{\sigma_e^\infty}{\dot{\epsilon}_e^\infty f_1(n) G_1(n,\rho)} + Q(a/b)} \sigma_e^\infty \tag{7.8}$$

In Equation 7.8, $D = D_b \delta_b \Omega / kT$ is the boundary diffusivity parameter, $D_b \delta_b$ is the boundary diffusive conductance, Ω is the atomic volume, R is the radius of a G-B facet, $h(\psi)$ is a function of the dihedral half angle ψ at the cavity tip, $h(\psi) = [1/(1+\cos\psi) - \frac{1}{2}\cos\psi]/\sin\psi$, $S, \sigma_e^\infty, \dot{\epsilon}_e^\infty$ are, in turn, the far field maximum principal stress, the Mises equivalent stress and the equivalent strain rate, respectively. The functions $f_1(n)$, $G_0(n)$, $G_1(n,\rho)$ in Equation 7.8 are given in the Appendix and are obtained from the numerical solutions of Hsia et al. (1991b) utilizing the same FEM approach discussed here. The function $Q(a/b)$ is a function of the ratio a/b, alone, and is

$$Q(a/b) = \ln[(a/b)^{-2}] - \frac{1}{2}[3-(a/b)^2][1-(a/b)^2] = \ln(A^{-1}) - \frac{1}{2}(3-A)(1-A). \tag{7.9}$$

Under fully constrained growth conditions, i.e., sufficiently low stress levels, it can be shown (Hsia, 1989) that $Q(a/b)$ in the denominator of Equation 7.8 is usually much smaller than the first term unless the ratio a/b is very small corresponding to the very early stages of cavitation. Therefore, the second term in the denominator can usually be neglected which results in a new and considerably simpler expression for the rate of change of cavitated area fraction,

$$\dot{A} = [\pi a^2 \beta + (\frac{G_0(n)}{G_1(n,\rho)} + \frac{S}{\sigma_e^\infty}) \frac{R f_2(n) G_1(n,\rho)}{2\pi h(\psi) a}] \dot{\epsilon}_e^c. \tag{7.10}$$

The first term in this expression is the contribution of nucleation to the cavitated area fraction while the second term is the contribution of growth. Generally the first term can be ignored because the cavity size at nucleation is very much smaller than the average cavity size that has been subject to a growth history. Therefore, the contribution of growth to overall damage always dominates over that of nucleation during most parts of the damage life. Then, through integration, the area coverage of cavities on a particular

G-B becomes

$$A = [(\frac{G_0}{G_1}+\frac{S}{\sigma_e^\infty})^2(\frac{Rf_1G_1}{2h(\psi)})^2\frac{\beta}{\pi}]^{1/3}\epsilon_e^c = [(\frac{G_0}{G_1}+\frac{S}{\sigma_e^\infty})^2(\frac{Rf_1G_1}{2h(\psi)})^2\frac{\bar{\beta}}{\pi}]^{1/3}\xi^{-1/3}\epsilon_e^c. \tag{7.11}$$

Comparing the predictions of different models with experimental results, Riedel (1987) showed that the constrained cavity growth model fits the experimental data much better than the unconstrained growth models under most typical test conditions for creeping ductile alloys such as 304 stainless steel. In the more brittle heterogeneous superalloys, however, the conditions are different (Capano et al., 1989). We exclude the discussion of these from our present model.

On a particular G-B, this cavity growth process becomes unstable at a critical value A_{crit} of this area coverage ($A_{\text{crit}} \approx 0.4$ (Argon, 1982)), and the cavities then rapidly coalesce to form a facet crack. This critical value can be approximately determined by considering the load-shedding behavior due to cavity growth using the relationship between the facet stress and the cavitated fraction A in the constrained cavity growth model of Rice (1981), when a cavitating G-B can no longer support load. The boundary can then be considered as having transformed into a facet crack. Since the load-shedding behavior is dependent on the stress level (Hsia, 1989), A_{crit} should be a function of the applied stress, but here for simplicity, we assume it is constant.

As creep strains increase, those G-Bs with higher values of the nucleation coefficient β will cavitate earlier, and become facet cracks earlier. For a given macroscopic creep strain, the minimum value of the nucleation coefficient (or maximum value of $\xi = \xi_{\text{crit}} = \bar{\beta}/\beta_{\text{crit}}$) for a particular G-B to reach the critical area coverage A_{crit} can be determined from Equation 7.11. Then, the boundaries with a nucleation coefficient higher than β_{crit} (or those with $\xi \leq \xi_{\text{crit}}$) have already become facet microcracks at this creep strain level. Thus, they represent material damage at a macroscopic material point. From this consideration, the damage parameter α for a given creep strain level at a material point can then be obtained, in an integrated form, as,

$$\alpha = 1 - (1 + \xi_{\text{crit}})e^{-\xi_{\text{crit}}} \tag{7.12}$$

And the rate of evolution of damage parameter is found to be,

$$\dot{\alpha} = 3\xi_{\text{crit}}^2(t)\exp(-\xi_{\text{crit}}(t))\frac{\dot{\epsilon}^c}{\epsilon^c(t)} \tag{7.13}$$

It should be noted that in this model, only the diffusional growth of G-B cavities is considered. Tvergaard (1984), on the other hand, considered both diffusional growth and growth due to coupled power law creep of the material between the cavities. If the applied stress is sufficiently low, however, as would be the case under typical service conditions, the contribution

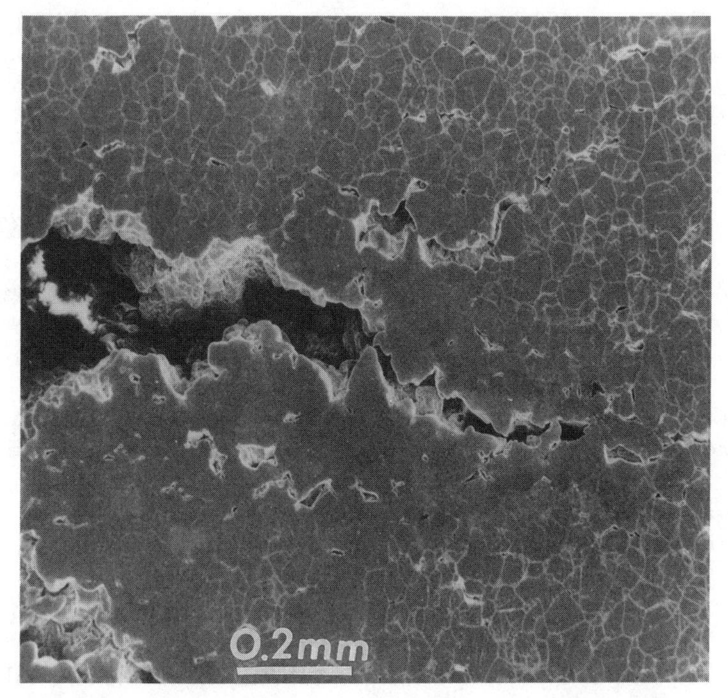

FIGURE 7.5. Distribution of G-B facet crack damage in 304 stainless steel near a macrocrack (Ozmat et al. 1991).

of power law creep to the cavity growth rate is unimportant compared to the part due to G-B diffusion, thus, as a first approximation, the power law creep contribution was neglected. Figure 7.5 shows a polished and etched region around an advancing macro-crack in Type 304 stainless steel where the etching has revealed many grain boundaries and has also accentuated the contrast of the relatively high density of grain boundary facet cracks that were described in the model above (Ozmat et al., 1991).

The Three-Dimensional Creep Constitutive Equation

Since damage is a measurement of the G-B facet microcrack density, the effect of damage on the overall creep constitutive behavior of the material can be investigated by considering a material containing distributed traction-free microcracks. Hutchinson (1983) studied a power law creeping material containing a dilute concentration of microcracks, normal to the maximum principal tensile stress, and obtained an expression of the stress potential function Φ using the solution of He and Hutchinson (1981) of a penny-shaped traction-free crack in a power law creeping material without G-B sliding. Following Hutchinson's derivation, and the considerations of Parks (1987) on the validity of this approach as discussed in the section above, Hsia et al. (1991b) have performed a detailed numerical study utilizing a modified stress potential function incorporating the effect of G-B sliding to model the acceleration of creep due to damage. The resulting potential function Φ is:

$$\Phi = \{1 + \frac{4}{3}\pi\rho_{3D} f_2(n)[H_0(n,\rho) + H_1(n,\rho)(\frac{S}{\sigma_e}) + H_2(n,\rho)(\frac{S}{\sigma_e})^2]\}^{(n+1)/2} \tag{7.14}$$

where the functions $f_2(n), H_0(n,\rho), H_1(n,\rho), H_2(n,\rho)$ are given in Appendix I, ρ_{3D} is a microcrack density parameter defined as $\rho_{3D} = N_v R^3$, where N_v is the number of cracks per unit volume, R is the average radius of grain boundary facets. It is shown in Appendix II that ρ_{3D} and its 2-D counterpart ρ satisfy the relationship $\rho_{3D} \approx 0.09877\rho$.

In Equation 7.14 the crack density ρ_{3D} is directly related to the damage parameter α, by Equation 7.15 below, which incorporates also the interaction of facet cracks by means of a factor $1/(1-\alpha^p)^{n/q}$ based on the properties of centrally cracked panels with increasing crack length to panel width ratios (Kumar et al., 1981),

$$\rho_{3D} = \frac{1}{\pi\sqrt{14}} \frac{\alpha}{(1-\alpha^p)^{\frac{n}{q}}} \tag{7.15}$$

where the constants p and q are obtained from the tabulated results of Kumar et al. (1981).

The three-dimensional creep strain rate tensor $\dot{\boldsymbol{\epsilon}}^c$ can then be obtained by differentiating the potential function with respect to the stress tensor $\boldsymbol{\sigma}$, i.e., $\dot{\boldsymbol{\epsilon}}^c = \partial\Phi/\partial\boldsymbol{\sigma}$, to obtain,

$$\begin{aligned} \dot{\boldsymbol{\epsilon}}^c = \; & \dot{\epsilon}_0(\tfrac{\sigma_e}{\sigma_0})^n[\quad]^{(n-1)/2}[\tfrac{3}{2}\tfrac{\boldsymbol{\sigma}'}{\sigma_e}(1 + \tfrac{4}{3}\pi\rho_{3D} f_2(H_0 + \tfrac{1}{2}H_1\tfrac{S}{\sigma_e}) \\ & + \tfrac{4}{3}\pi\rho_{3D} f_2(\tfrac{1}{2}H_1 + H_2(\tfrac{S}{\sigma_e}))\mathcal{M}], \end{aligned} \tag{7.16a}$$

where

$$[\quad] = 1 + \frac{4}{3}\pi\rho_{3D} f_2(H_0 + H_1(\frac{S}{\sigma_e}) + H_2(\frac{S}{\sigma_e})^2). \tag{7.16b}$$

In Equation 7.16a $\boldsymbol{\sigma}'$ is the deviatoric stress tensor, $\mathcal{M} = \partial S/\partial\boldsymbol{\sigma}$ is a tensor whose components in the principal stress coordinate system are zero except the normal component in the maximum principal stress direction which is unity.

Under uniaxial tensile stress σ, the creep rate constitutive equation takes the special form,

$$\dot{\epsilon}^c = \dot{\epsilon}_0(\frac{\sigma}{\sigma_0})^n[1 + \frac{4}{3}\pi\rho_{3D} f_2(H_0 + H_1 + H_2)]^{(n+1)/2}. \tag{7.17}$$

TABLE 7.1. Summary of Material Constants

Parameters	304 S.S.	Tough Pitch Copper
	(760^0C)	(250^0C)
E (MPa)	133.75×10^3	96.2×10^3
σ_o (MPa)	50	50
$\dot{\epsilon}_o$ (s^{-1})	1.0762×10^{-7}	2.1577×10^{-8}
n	6.84	4.84
p	6.84	4.84
q	6.84	4.84
$h(\psi)$	0.33	0.33
R (m)	2×10^{-5}	2×10^{-5}
$\bar{\beta}$ (m^{-2})	5.74×10^8	1.14×10^{11}
A_{crit}	0.4	0.4

Equation 7.17 is of a conventional power law form with an accelerator due to material damage. The more severe the accumulated damage is, the larger the value of ρ_{3D} will be, and the higher the creep strain rate will become.

Uniaxial Tension Behavior

To test the creep constitutive model with damage we compare its predictions against actual uniaxial tension results. Because the acceleration of creep rate during the tertiary stage could be caused not only by the damage mechanism we examined, but also by some other possibilities (Ashby and Dyson, 1984), and the creep constrained growth model is valid only for certain materials and within certain stress ranges, care is needed in applying this model. Therefore, it is important to check the damage mechanism against an experiment before comparing the prediction of our model with the experimental results.

The uniaxial tension behavior for two materials exhibiting the features of our model have been chosen for comparison. One is Type 304 stainless steel tested by Swindeman (1982) at $760°C(T/T_m \approx 0.58)$ within a stress range of $5 \times 10^{-4} < \sigma/E \leq 10^{-3}$; the other is a tough pitch copper tested by Dyson (1987) at $250°C(T/T_m = 0.4)$ within roughly the same normalized stress range. Although the homologous temperatures for 304 stainless steel and copper are considerably different, because the 304 stainless steel is a solid solution alloy, the creep rates of the two materials are comparable in the above stress range.

The values of the relevant parameters for these two materials are given in Table 7.1, based on data accumulated by Needleman and Rice (1980), Chen and Argon (1981a), and by fitting the model parameters to the experimental results. In the model predictions only elastic deformation and steady state power-law creep deformation was considered, but primary creep was not. Since primary creep for copper can be substantial, we determined the strains at the intercepts of the back extrapolated minimum creep rate line and the strain axis for each stress level, and simply considered these as instantaneous initial strains.

Figure 7.6(a) shows the uniaxial creep curves of 304 stainless steel under different levels of applied stress. The prediction of the uniaxial tensile creep behavior by the present model and Swindeman's (1982) experimental results for this material are in good agreement with each other. Figure 7.6(b) shows the uniaxial creep curves of tough pitch copper due to Dyson (1987). Although the minimum creep rates are well described by a power law relation, the agreement between the model prediction and the experimental results for the tertiary stage creep is not as good as for stainless steel. The experimental creep strain rates accelerate faster than the model predicts. This indicates that there may be additional processes in the copper that govern its behavior in the stress range considered. Furthermore, it can be shown that the stress range for tough pitch copper is a little beyond the limit of fully constrained creep cavity growth for this material (Hsia, 1989).

The sudden increases in strain rate near the end of some creep curves in the tests of 304 stainless steel and even more so in copper indicate that final fracture of the specimen is caused by some localization of deformation or other kind of strain rate enhancement such as, perhaps, surface oxidation which was not considered in our model.

7.3 Creep Damage Fields at Crack Tips

7.3.1 THE FINITE ELEMENT METHOD

The finite element method (FEM) was used to solve the boundary value problems of primary interest, on initial directions of creep crack growth presented in Section 7.1, using the creep damage model established in the previous section. Primary creep was not taken into account in the computation, as already mentioned above, because the transient response of the material covers a short period of time compared to the steady state stage and tertiary stage, and usually it does not contribute much to G-B cavitation. It can, however, produce plastic resistance inhomogeneities around the crack which the model can not represent. Both elastic and power law creep deformations were considered. These are taken to be additive, giving a total strain rate as

$$\dot{\epsilon} = \dot{\epsilon}^e + \dot{\epsilon}^c \tag{7.18}$$

where $\dot{\epsilon}^c$ is the creep strain rate tensor given by Equation 7.16a, and $\dot{\epsilon}^e$ is the elastic strain rate tensor which is related to the stress rate tensor

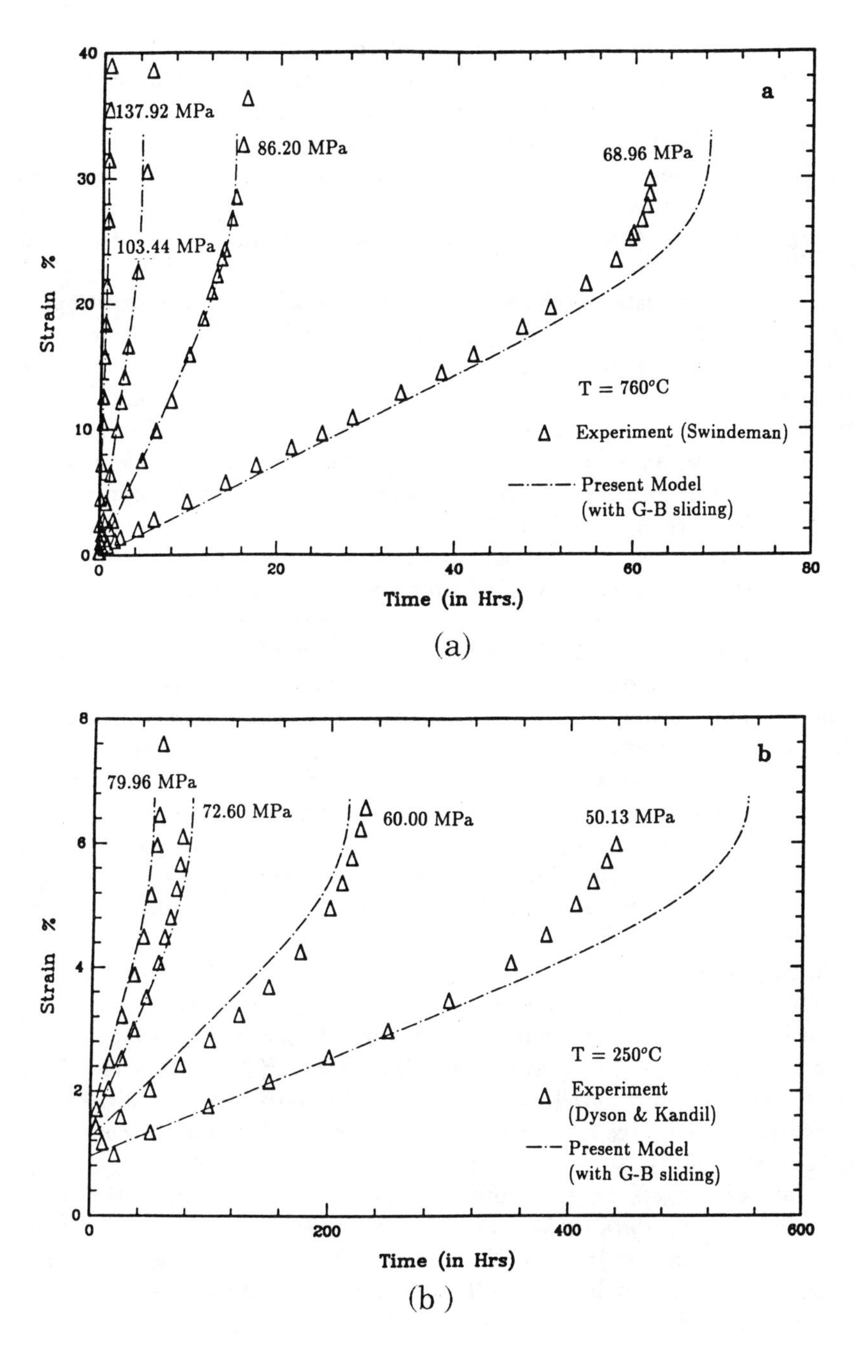

FIGURE 7.6. Computed uniaxial tensile creep curves compared with experimental data: (a) in Type 304 stainless steel; (b) in copper (Hsia, et al. 1991a).

$\dot{\sigma}$ by Hooke's law in the case of small deformations. In finite deformation, however, $\dot{\epsilon}^e$ and $\dot{\sigma}$ are replaced with the elastic stretch rate tensor and the Jauman derivative of the Kirchoff stress tensor, respectively.

The ABAQUS finite element code was used for the FEM computations incorporating nonlinear geometry effects to deal with finite deformations around the crack tip region. The details of the computation have been presented by Hsia (1989) and can also be found summarized in Hsia et al (1991a).

In order to simulate the creep fracture and initial creep crack growth process, a critical value of damage parameter α_{crit} was incorporated into the program. When the damage parameter α at a material point reaches this critical value α_{crit}, this material point is considered to be fractured. Since the damage parameter α represents the area fraction of cracked boundaries, this criterion means that when a substantial number of G-Bs ($\alpha \geq \alpha_{\text{crit}}$) become facet cracks, the material at this point can no longer support any load and fracture occurs. Then the stresses at this point are systematically reduced within a few time increments. It should be noted that the formation of boundary facet cracks is not isotropic, most of the facet cracks produced by boundary cavitation are approximately normal to the maximum principal tensile stress direction. Therefore, when the material at one point loses its strength in one direction (maximum principal stress direction), it can still have substantial strength in other directions. This tensorial feature of damage has been considered by several investigators in the past (Ashby and Dyson, 1984; Leckie and Hayhurst, 1977; Riedel, 1987), but has not yet been successfully modeled. Since the damage parameter used in our model is scalar and isotropic, all stress components were cut down to zero gradually when the critical damage was reached, without regard to residual strength in directions other than the principal extension direction just mentioned above.

In our program, the distribution and eventual propagation of damage was obtained by recording the values of damage parameter at each integration point in every element as an internal variable. To bring the damage action to a conclusion smoothly a stress reduction factor $f(t)$ (which varies from one to zero) was introduced into the program so that all stresses at the dead integration point were systematically decreased in a manner discussed by Hsia et al. (1991a).

As already stated above, the material constants in the computation were obtained from the data for 304 stainless steel given by Argon, et al. (1985), Chen and Argon (1981a), and Needleman and Rice (1980) and the curve fitting data from Swindeman's (1982) experimental results. These are summarized in Table 7.1.

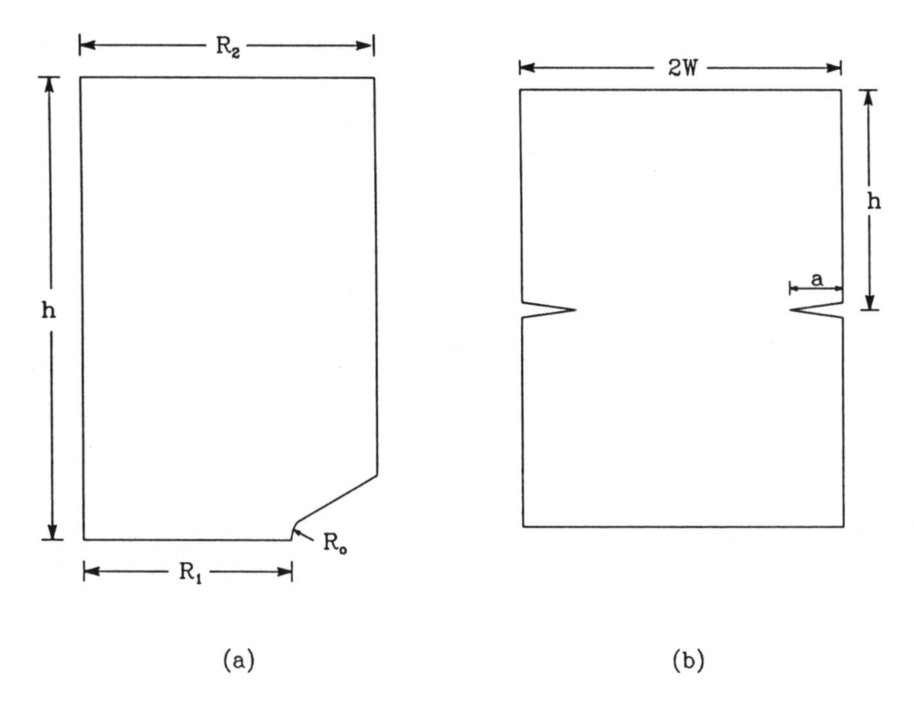

FIGURE 7.7. Geometries of the British Standard Notched bar and the double-edge cracked plate: (a) $R_1/R_0 = 18.2; R_2/R_0 = 25.7; h/R_0 = 40$; (b) $W/a = 3; h/a = 5$.

Computational Results

Finite element computational results have been obtained for two typical configurations: a cylindrical bar with a British Standard Notch (BSN), and a double-edge cracked plate in plane strain. The geometries of these two configurations are shown in Fig. 7.7(a) and Fig. 7.7(b). Creep times presented in the following figures were normalized by the characteristic relaxation time discussed by Bassani and McClintock (1981),

$$t_N = \frac{\sigma_{net}/E}{\dot{\epsilon_o}(\sigma_{net}/\sigma_o)^n} \tag{7.19}$$

where $\sigma_{\rm net}$ is the average net section stress acting on the ligament of the notched bar and the cracked plate.

The finite element mesh for the BSN problem around the notch tip region is shown in Fig. 7.8. The typical distributions of accumulated creep strain around the notch tip are shown in Fig. 7.9 for a normalized time of 516.6. Although the contours show that low levels of creep strain tends to concentrate in an inclined zone, making an angle of roughly 55° with the symmetry plane, the contours for higher values of creep strain still show

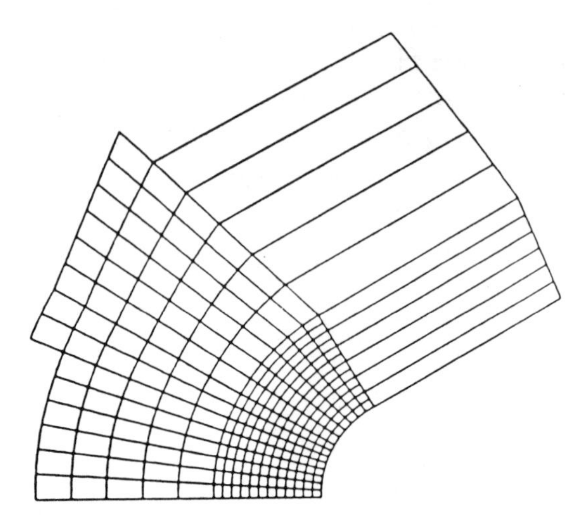

FIGURE 7.8. Finite element mesh of the BSN bar around the notch tip region (Hsia, et al. 1991a).

that the severely strained region centers around the symmetry plane for the BSN configuration. Figure 7.10 shows the contours of damage distribution at the notch tip at different creep times obtained from the FEM damage model. The shaded regions are where the damage exceeds its maximum critical value, thus the stresses in that region are zero. It is clear that the damage is highly concentrated at the notch tip region, but decreases very rapidly to some negligible levels away from the heavily damaged zone. It must be noted that the shaded region representing the damaged and unloaded material must be larger than the actual crack width. The latter can not be specifically identified by the present model.

The FEM mesh at the crack tip for the double-edge cracked plate is shown in Fig. 7.11. Singular elements were used at the crack tip by collapsing three nodes along one side of an element into one node and maintaining the remaining other side nodes at the middle of the sides. It can be shown (Barsoum, 1977) that these elements have a strain singularity of r^{-1} which is a good approximation to the strain singularity of $r^{-n/(n+1)}$ of the HRR field at the crack tip.

It should be noted that for this sharp crack problem, an instantaneous plastic deformation occurs in the vicinity of the crack tip due to the stress singularity when a step load is applied. Such plastic deformation contributes very little to the intergranular cavitation but produces blunting. In our computation, this plastic deformation was not considered. Therefore, the initial stress and strain singularity was of a linear elastic nature. This approximation, however, should not have an important influence on the final results, even though it should result in a faster initial creep rate

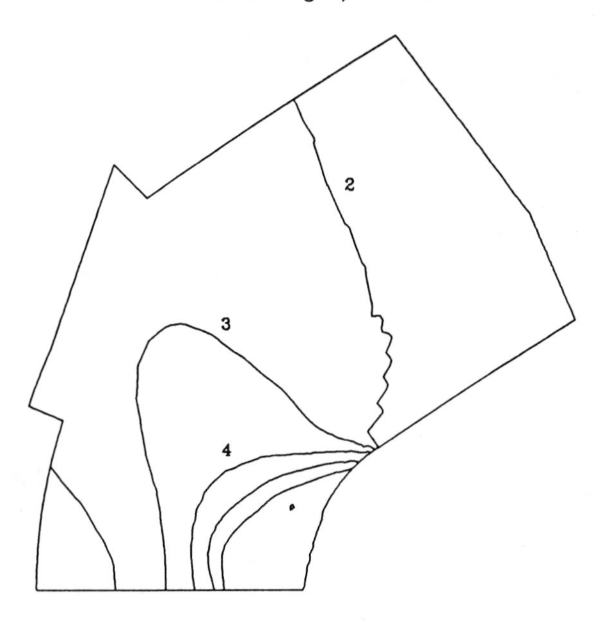

FIGURE 7.9. Creep strain contours around the notch tip for a BSN bar, at $t/t_N = 516.6$ and $T = 760°C$ (contour No./creep strain level: 1/0.05; 2/0.08; 3/0.11; 4/0.14; 5/0.17; 6/0.20) (Hsia, et al. 1991a).

that would lead to a shorter incubation time.

The creep strain distributions around the crack tip regions at different creep times are plotted in Fig. 7.12. For this geometry, the creep strains concentrate along inclined planes making angles between 60° and 65° with respect to the initial crack plane.

The resulting damage contours around the crack tip at increasing times are plotted in Fig. 7.13 with the shaded regions representing the heavily damaged zone. It is clear that the damaged zones are propagating out along inclined planes with respect to the initial crack plane. Clearly, in this case the creep macrocrack is growing along inclined planes in which the equivalent creep strain and the resulting damage is concentrated.

It should also be noted that, as the creep strains increase, the initial sharp crack tip becomes more and more blunted. Because of this effect, further increments of strain develop notch-like behavior and the heavily damaged zone begins to concentrate in later stages in the symmetry plane of the crack. The macrocrack then develops three branches. This tendency can be seen in Fig. 7.13 from the shapes of the damage contours at later stages of creep.

Hsia et al. (1991a) have also developed a model for the entire creep process and damage development discussed above, without any G-B sliding to compare with the results of the model using G-B sliding. This comparison

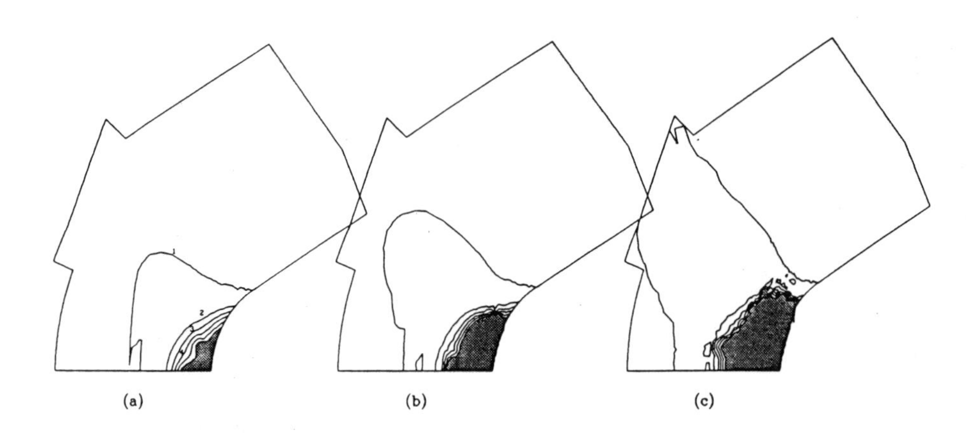

FIGURE 7.10. Creep damage evolution around notch tip region for a BSN bar at $T - 760°C$ after: (a) $t/t_N = 460$; (b) $t/t_N = 516.6$; (c) $t/t_N = 577.7$ (contour No./damage level: 1/0.04; 2/0.08; 3/0.12; 4/0.16 5/0.20) (Hsia, et al 1991a).

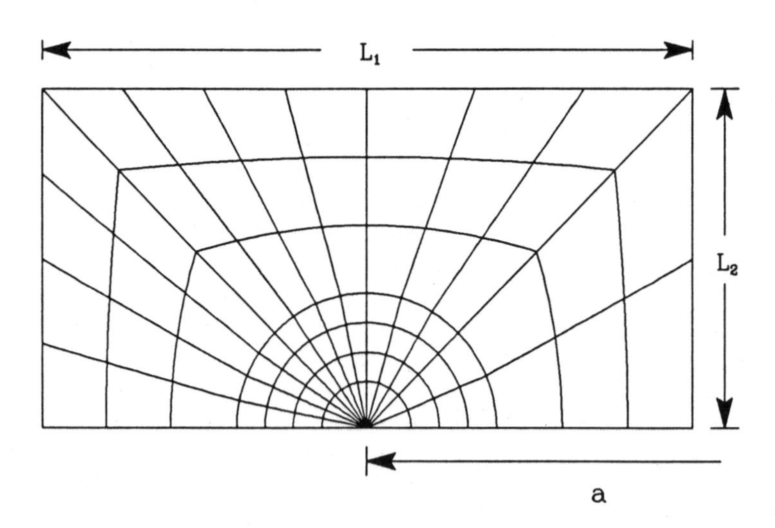

FIGURE 7.11. Finite element mesh of the plane strain cracked plate around the crack tip region. $L_1/a = 2.0/15.0; L_2/a = 1.0/15.0$ (Hsia, et al. 1991a).

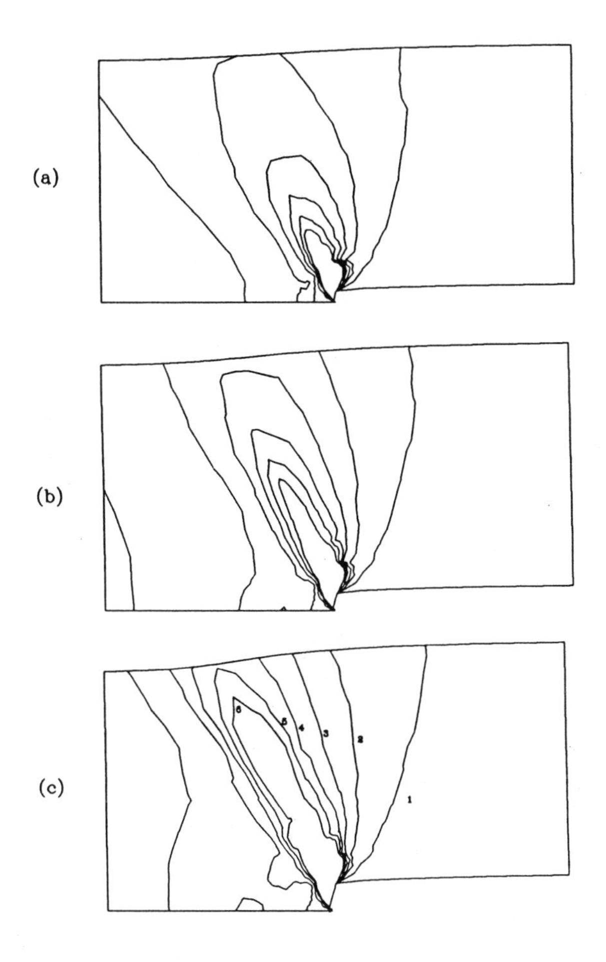

FIGURE 7.12. Creep strain contours around the crack tip region for a plate with an initially sharp crack, at $T = 760°C$ after: (a) $t/t_N = 28.3$; (b) $t/t_N = 41.9$; (c) $t/t_N = 61.2$ (contour No./creep strain level: 1/0.020; 2/0.056; 3/0.092; 4/0.128; 5/0.164; 6/0.200) (Hsia, et al. 1991a).

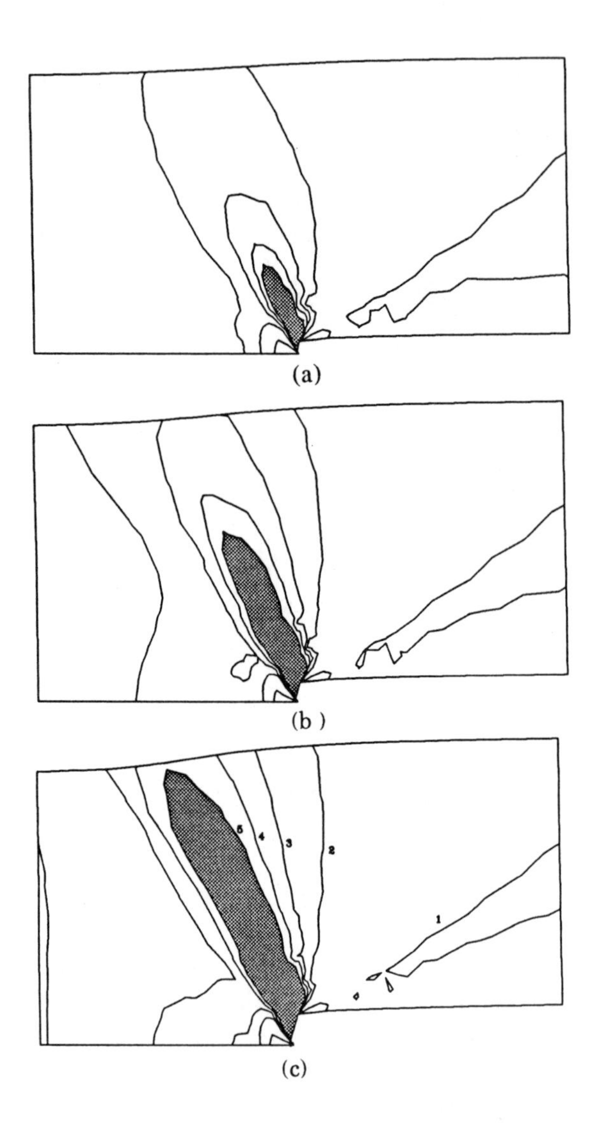

FIGURE 7.13. Creep damage evolution around the crack tip region for a plate with an initially sharp crack at T = 760°C after: (a) $t/t_N = 28.3$; (b) $t/t_N = 41.86$; (c) $t/t_N = 61.16$ (contour No./damage level: 1/0.002; 2/0.051; 3/0.101; 4/0.150; 5/0.200) (Hsia, et al. 1991a.)

showed that in all cases, under the same conditions and for the same overall times the concentration of strain and its associated damage is 2-3 times larger than in the cases without G-B sliding, as can be ascertained already from the results shown in Figs. 7.3 and 7.4.

7.4 Discussion

Here we have presented a creep damage model starting with the development of relatively randomly positioned G-B facet microcracks from the nucleation, growth, and linking of G-B cavities under stress, followed by the growth and linking of these microcracks by creep flow. In this scenario, our model differs significantly from the ones used by Riedel (1987) and Tvergaard (1985a, 1985b, 1986) both of whom consider creep damage only as the nucleation and growth of G-B cavities, i.e., only the preparatory phase of our damage model. The damage model advanced here is more appropriate for the ductile, relatively pure metals such as copper, and solid solution alloys such as Type 304 stainless steel with considerable variability in G-B particle coverage and having the capacity of accumulating substantial G-B facet crack densities prior to fracture. It has the basic feature of giving relatively prolonged tertiary stages of creep with very considerable creep rate enhancement. In comparison, the damage model which terminates with the nearly simultaneous separation of most G-Bs by only the nucleation and growth of G-B cavities is more appropriate for the brittle superalloys such as Nimonic 80A (Dyson and McLean, 1977) and Astroloy (Capano, et al., 1989), where nearly all G-Bs have the same particle coverage.

In our damage model we have used the known mechanistic details of the experimentally observed processes to provide a proper kinetics framework and a correct stress dependence. We have then homogenized the model for application on the continuum level which implies a representative volume element (RVE) having dimensions on the order of several grain sizes. The adequacy of this homogenization scale has been established in the experiments of Ozmat et al. (1991) which have indicated that in all cases the characteristic dimensions of the creep damage gradients around notches and even sharp cracks is of considerably longer range than the grain size of the material.

The principal application of our model has been to provide a quantitative explanation for the very different initial crack extension behavior of sharp cracks versus rounded notches.

The models for the relatively blunt notches represented by the BSN configuration of our results, plotted in Fig. 7.10, show that the creep damage in this case is more diffusely distributed in the symmetry plane of the notch and that macrocrack growth eventually starts and localizes in this region. This damage accumulation pattern is a direct result of the concentration of integrated creep strains which directly control the integrated damage.

Such concentration of creep strain is a characteristic of blunt notches or blunting cracks as had been noted first by Rice and Johnson (1970). This concentration is reflected in an earlier blunt crack deformation model studied by Argon et al. (1985) and is, of course, one of the experimental features observed by Ozmat et al. (1991) that the present model was developed to help explain.

In the sharp crack models the integrated creep strains and the associated damage tend to localize in inclined planes making an angle of 50-60° with the crack plane as is shown in Figs. 7.12 and 7.13 for the model. This bifurcation behavior of sharp cracks is reflected in the singular field model of creep cracks of Bassani and McClintock (1981), and has been observed experimentally by Ozmat et al. (1991) and some other investigators earlier (Davis and Manjoine, 1953; Leckie and Hayhurst, 1974; Hayhurst et al., 1978).

The qualitatively very different behavior of sharp cracks and blunt notches poses the question of what constitutes a sharp crack and what constitutes a blunt notch? Furthermore, since all initially sharp cracks will continue to blunt with increasing deformation, a related question is, "if fracture did not intervene, how much local strain is required to blunt the crack enough to redirect the further concentration of strain from the inclined planes to the median plane in front of the crack?" While the answers to these questions are not clear, and certainly must be influenced in the real material by primary creep and its associated elevation of plastic resistance in the regions starting to creep first, the experiments of Ozmat et al. (1991) show that in Type 304 stainless steel some initially sharp cracks appear to approach this transition in behavior. The key observation is an initial bifurcation of the sharp cracks on inclined planes followed by the extension of a crack along the center plane, giving rise to a three pronged crack front shown in Fig. 7.14 (Ozmat et al. 1991). A trend in this direction is also seen in the later stages of our damage accumulation model around a sharp crack shown in Fig. 7.13.

Our mechanistic damage accumulation model has successfully demonstrated the consistency of its predictions with the experimental observations of Ozmat et al. (1991) for the two qualitatively different types of behavior of sharp cracks and blunt notches. Of course, these same results have also been obtained by Tvergaard (1985a, 1985b, 1986) with a strain-controlled mechanistic damage model using only our preparatory intergranular cavity nucleation and growth phase of the damage, without considering the effect of the grain boundary facet cracks on the creep rate. Both models have been based on mechanistic detail and then have been applied in a homogenized sense as continuum processes, but are ultimately strain-controlled and therefore model well a strain-controlled damage process. Both models specifically include the effects of G-B sliding, albeit in somewhat different forms; this effect acts primarily as a strain rate enhancer but also results in qualitatively different response for G-B cavitation or facet crack opening

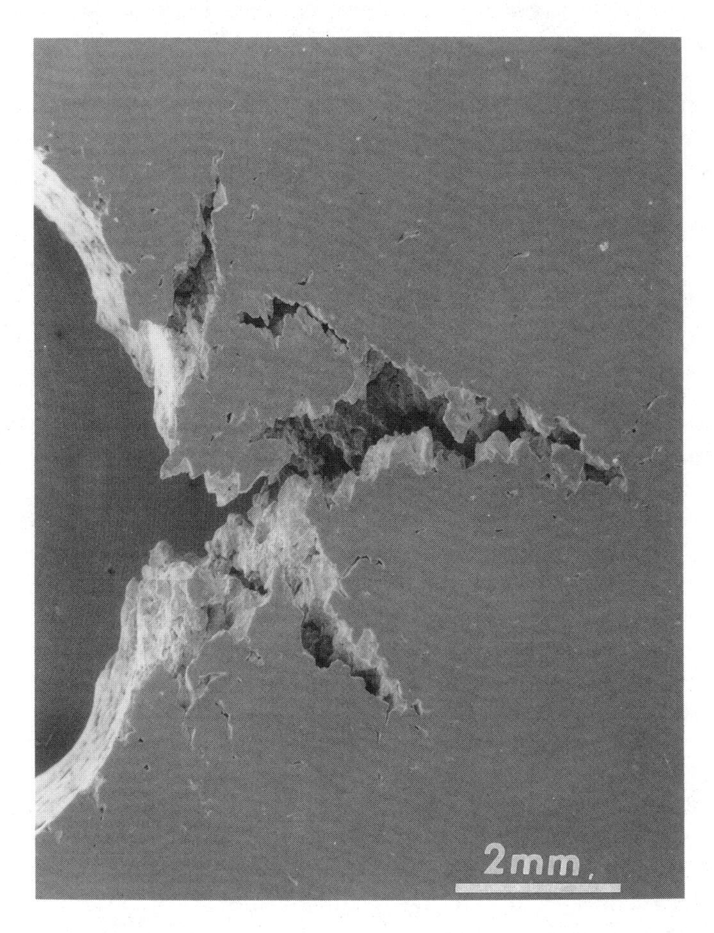

FIGURE 7.14. The tip of a sharp creep crack in 304 stainless steel breaks up into three separate cracks (Ozmat, et al. 1991).

with regard to the transverse stresses. While the Tvergaard model incorporates sensitivity to stress triaxiality this enters through the combined diffusional flow and power-law creep growth of G-B cavities as initially developed by Needleman and Rice (1980). The sensitivity to triaxiality of the stress state in our model is considerably stronger in the power law creep growth and eventual interaction of the G-B facet microcracks. These important differences, however, remain insufficiently explored since the crack tips or notch tip fields which have been primarily studied by all investigators have a relatively narrow range of high triaxiality. The purely phenomenological damage model of Hayhurst and co-workers (Leckie and Hayhurst, 1977; Hayhurst, Dimmer and Morrison, 1984; Hayhurst, Brown and Morrison, 1984) also has very similar features as the mechanistic models given by us and Tvergaard, and has the additional flexibility of response to maximum principal tensile stress incorporated in any desired arbitrary proportion to represent differing behavior of different alloys. Hayhurst's models have also been successful to nearly the same degree as ours in modeling the qualitatively different behavior of cracks and notches. Hayhurst's models, however, differ fundamentally from our mechanistic ones in attributing the entire effect of the damage to the acceleration of the deviatoric creep rate and considering no cavitational or dilatational strains and their effects. Clearly there must be important differences in relaxation of triaxial stress by the dilatation that it produces. While this important effect must manifest itself in any eventual fully predictive model making use of only the fundamental constitutive behavior of a creeping material, in the presently developed models in which parameters are adjusted by more macroscopic response patterns such as the uniaxial tension experiment, these differences do not stand out. Thus, we conclude that the present models developed by Hayhurst et al. (Leckie and Hayhurst, 1977; Hayhurst, Dimmer and Morrison, 1984; Hayhurst, Brown and Morrison, 1984), Tvergaard (1985a, 1985b, 1986) and that developed by us incorporate many redundancies and degrees of freedom that need to be explored further on problems with a wider range of stress triaxiality and using fewer adjustments of parameters to explore more fully their consequences.

In the experiments of Ozmat et al. (1991) it was observed that while initially sharp cracks bifurcate along inclined planes, the branch cracks do not continue to grow indefinitely but eventually rotate back into the plane normal to the opening mode direction. In a complementary manner quasi-planar cracks that emanate out from the notch tip along the symmetry plane of the blunt notch do not continue to grow in planar form indefinitely but eventually show a tendency of bifurcation. These two longer range responses result in considerable meandering of the plane of the creep crack. While modeling these effects should basically be within the reach of our model, they require much longer and more tedious computational runs which we have not performed. It is also likely that their full implementation may require incorporation of a primary creep response into the model which can account for developments of inhomogeneities of deformation resistance around cracks or notches.

Finally, we recognize that our detailed mechanistic descriptions of the developing damage and the complex effect it has on the creep rate lack the direct and compact approach of Kachanov (1958) or its somewhat generalized form of Hayhurst and coworkers (Hayhurst, Dimmer and Morrison, 1984; Hayhurst, Brown and Morrison, 1984). The latter phenomenological approaches with their several adjustable parameters can be readily adjusted to experimental data and therefore, they will remain to be more attractive to some practicing engineers. In that approach, however, there is little microstructural insight and no capacity to calculate increments of damage added to pre-existing damage, since the latter is not measurable and ill defined. Thus, the phenomenological approach lacks the flexibility needed for definitive assessments of remaining creep life in components that have already seen an unknown form of service life for which, however, the damage is physically measurable. Our approach, even though more tedious, has this capacity.

Acknowledgments: This research has been supported by the NSF/MRL through the M.I.T. Center for Materials Science and Engineering under Grant No. DMR-87-19217. The FEM computations have benefited markedly from the availability of an Alliant FX-8 computer made possible through a DARPA/ONR URI program on simulation of polymer properties under Contract No. N00014-86-K-0768. The ABAQUS finite element program was made available under academic license from Hibbitt, Karlson, and Sorensen, Inc., of Providence, Rhode Island.

REFERENCES

Argon, A. S. (1982). In Wilshire, B. and Owen, D. R. J., editors, *Recent Advances in Creep and Fracture of Engineering Materials and Structures*, Pineridge Press, Swansea, page 1.

Argon, A. S. (1983). Scripta Metall., 17:5.

Argon, A. S., and Im, J. (1975). Met. Trans., 6A:839.

Argon, A. S., Lau, C. W., Ozmat, B., and Parks, D. M. (1985). In Miller, K. J. et al., editors, *Fundamentals of Deformation and Fracture*, Cambridge Univ. Press, Cambridge, page 189.

Ashby, M. F. and Dyson, B. F. (1984). In Valluri, S. R. et al., editors, *Advances in Fracture Research '84* – Proceedings of ICF6, Pergamon Press, Oxford, 1:3.

Ashby, M. F., Gandhi, C., and Taplin, D. M. R. (1979). Acta Metall., 27:699.

Barsoum, R. S. (1977). Int. J. Numer. Met. Eng., 11:85.

Bassani, J. L. and McClintock, F. A. (1981). Int. J. Sol. Struct., 17:479.

Bridgman, P. W. (1952). *Studies in Large Plastic Flow and Fracture*, McGraw-Hill, New York.

Cane, B. J. (1978). Metal Sci., 12:102.

Capano, M., Argon, A. S., and Chen, I.-W. (1989). Acta. Metall., 37:3195.

Chen, I.-W. and Argon, A. S. (1979). Acta Metall., 27:749.

Chen, I.-W. and Argon, A. S. (1981a). Acta Metall., 29:1321.

Chen, I.-W. and Argon, A. S. (1981b) Acta. Metall., 29:1759.

Chuang, T.-J., Kagawa, K. I., Rice, J. R., and Sills, L. B. (1979). Acta Metall., 27:265.

Crossman, F. W. and Ashby, M. F. (1975). Acta. Metall., 23:425.

Davis, E. A. and Manjoine, M. J. (1953). In *Strength and Ductility of Metals at Elevated Temperatures*, STP-128, ASTM, Philadelphia, page 67.

Don, J. and Majumdar, S. (1986). Acta Metall., 34:961.

Dyson, B.F. (1976). Metal Sci., 10:349.

Dyson, B. F. (1987). private communication of data by Kandil and Dyson.

Dyson, B. F. and McLean, D. (1977). Metal Sci., 11:37.

Ghahremani, F. (1980). Int. J. Solids Struct., 16:847.

Griffith, A. A. (1920). Phil. Trans. Roy. Soc., (London), A221:163.

Griffith, A. A. (1924). Proc. First Intern. Conf. Appl. Mech., (Delft), page 55.

Hart, E. W. (1967). Acta Metall., 15:1545.

Hayhurst, D. R., Brown, P. R., Morrison, C. J. (1984). Phil Trans. R. Soc. (London), A311:131.

Hayhurst, D. R., Dimmer, P. R., and Morrison, C. J. (1984). Phil. Trans. R. Soc. (London), A311:103.

Hayhurst, D. R., Leckie, F. A., and Morrison, C. J. (1978). Proc. R. Soc., A360:243.

He, M. Y. and Hutchinson, J. W. (1981). J. Appl. Mech., 48:830.

Hsia, K. J. (1989). Modeling of intergranular creep damage and investigation of the role of creep crack growth in creep life prediction, Ph.D. Thesis in Mechanical Engineering, M.I.T., Cambridge, MA, U.S.A.

Hsia, K. J., Argon, A. S. and Parks, D. M. (1991a). Mech. Mater., 11:19.

Hsia, K. J., Parks, D. M. and Argon, A. S. (1991b). Mech. Mater., 11:43.

Hull, D., and Rimmer, D. E. (1959). Phil. Mag., 4:673.

Hult, J. A. H., and McClintock, F. A. (1957). IXe Congrés International de Mécanique Appliquée, Actes, 8:51.

Hutchinson, J. W. (1983). Acta Metall., 31:1079.

Irwin, G. R. (1948). In *Fracturing of Metals*, ASM: Metals Park, Ohio, page 147.

Kachanov, L. M. (1958). Izv. Akad. Nauk SSSR Otk. Teck. Nauk. 8:26.

Kumar, V., German, M. D. and Shih, C. F. (1981). An engineering approach for elastic—plastic fracture analysis, EPRI Topical Report NP-1931, Palo Alto, California.

Leckie, F. A. and Hayhurst, D. R. (1974), Proc. R. Soc., A340:323.

Leckie, F. A. and Hayhurst, D. R. (1977). Acta Metall., 25:1059.

Martinez, L. and Nix, W. D. (1982). Met. Trans., 13A:427.

McClintock, F. A. (1968). J. Appl. Mech., 35:363.

McClintock, F. A. (1969). In Argon, A. S., editor, *Physics of Strength and Plasticity*, MIT Press, Cambridge, MA page 307.

McClintock, F. A. (1971). In Leibowitz, H., editor, *Fracture: an Advanced Treatise*, Academic Press, New York, 3:47.

McClintock, F. A. and Argon, A. S. (1966). *Mechanical Behavior of Materials*, Addison Wesley, Reading, MA.

McClintock, F. A., Kaplan, S. M. and Berg, C. A. (1966). International J. Fract. Mech., 2:614.

Nagpal, V., McClintock, F. A., Berg, C. A. and Subudhi, M. (1973). In Sawczuk, A., editor, *Foundations of plasticity*, Noordhoff, Leyden, page 365.

Needleman, A. and Rice, J. R. (1980). Acta. Metall., 28:1315.

Orowan, E. (1934). Z. Kristallographie, 89:327.

Orowan, E. (1949). rep. Prog. Phys., 12:185.

Ozmat, B., Argon, A. S. and Parks, D. M. (1991). Mech. Mater., 11:1.

Parks, D. M. (1987). Nucl. Eng. Design, 105:11.

Pharr, G. M. and Nix, W. D. (1979). Acta Metall., 27:1615.

Rice, J. R. (1981). Acta Metall., 29:675.

Rice, J. R. and Johnson, M. A. (1970). In Kanninen, M. F. et al., editors, *Inelastic Behavior of Solids*, McGraw-Hill, New York, page 641.

Riedel, H. (1987). *Fracture at High Temperatures*, Springer, Berlin.

Sham, T.-L. and Needleman, A. (1983). Acta Metall., 31:919.

Speight, M. V. and Beere, W. (1975). Metal. Sci., 9:190.

Swindeman, R. (1982). private communication quoted by Argon et al. (1985).

Tipper, C. F. (1949). Metallurgia, 39:133.

Tvergaard, V. (1984). Acta Metall., 32:1977.

Tvergaard, V. (1985a). Mech. Mater., 4:181.

Tvergaard, V. (1985b). Int. J. Sol. Struct., 21:279.

Tvergaard, V. (1986). Int. J. Fracture, 31:183.

Appendix I

The functions obtained by curve fitting of the results of the finite element model of grain boundary sliding are

$$G_0(n) = 4.00n^{0.65}, \quad (A.1)$$

$$G_1(n,\rho) = 4.00(2.05 + 15.5\rho)\sqrt{n}, \quad (A.2)$$

$$H_0(n,\rho) = (1.399 + 2.909\rho)n^{(0.5838-3.636\rho)}, \quad (A.3)$$

$$H_1(n,\rho) = (2.699 + 32.80\rho) + (1.732 + 2.527\rho)n, \quad (A.4)$$

$$H_2(n,\rho) = (3.006 - 61.16\rho) + (1.054 + 38.94\rho)n, \quad (A.5)$$

and $\rho = N_v R^3$ is the volume fraction of facet cracked material, whereas N_v is the number density of facet cracks per unit volume and R is the average radius of cracked boundary facets.

The conversion factors from the two-dimensional plane strain solution to the three dimensional case are:

$$f_1 = \frac{16}{3\pi}\frac{1}{\sqrt{n}\sqrt{1+3/n}} \quad (A.6)$$

$$f_2 = \frac{8}{\pi^2}\frac{1}{\sqrt{n}\sqrt{1+3/n}} \quad (A.7)$$

Appendix II

For a two dimensional array of hexagonal grains, and a boundary facet crack density of one crack per grain, the microcrack density parameter ρ can be evaluated as follows,

$$\rho = \frac{3\pi}{2}N_A R^2 = 0.4535$$

where N_A in this case is the reciprocal of the area of one grain, R is the half length of the facet.

Consider for instance the corresponding three dimensional space filling configuration to be a Wigner-Seitz cell with one facet crack per grain, either on the hexagonal face, or on the square face. The volume of one grain

$$V = 8\sqrt{2}l^3$$

where l is the length of the edges. The definition of the microcrack density parameter in 3-D case is $\rho_{3D} = N_v R^3$, where N is the number of cracks per unit volume, R is the equivalent radius of the facet.

For a facet crack on the hexagonal face, the equivalent radius R can be obtained by equating the area of a circle to that of the hexagon, i.e., $\pi R^2 = 3\sqrt{3}l^2/2$. Thus

$$\rho_{3D,hex} = \frac{\left(\frac{3\sqrt{3}}{2\pi}\right)^{3/2} l^3}{8\sqrt{2}l^3} = 0.06647$$

Similarly, for a facet crack on the square face, the microcrack density parameter can be evaluated by,

$$\rho_{3D,sq} = \frac{l^3}{8\sqrt{2}\pi^{3/2}l^3} = 0.01587$$

Take the weighted average of these two cases, we obtain an estimate of the microcrack density parameter for a Wigner-Seitz cell, ρ_{3D}, as

$$\rho_{3D} = \frac{6}{14}\rho_{3D,sq} + \frac{8}{14}\rho_{3D,hex} = 0.04479$$

From the above calculation, the relationship between the ρ of the 2-D plane strain hexagon and the ρ_{3D} of the 3-D Wigner-Seitz cell can be given as,

$$\rho_{3D} = 0.09877\rho.$$

8

Cracking and Fatigue in Fiber-Reinforced Metal and Ceramic Matrix Composites

A. G. Evans and F. W. Zok

ABSTRACT The damage that occurs in unidirectional ceramic and metal matrix composites upon monotonic and cyclic loading involves coupled considerations of mechanics and stochastic processes. Some of the basic principles are described and models presented that both characterize damage evolution and govern mechanism changes. Comparisons are presented between predictions and experimental data for such phenomena as modulus degradation caused by matrix cracking, fatigue crack growth and tensile strength.

8.1 Introduction

The fracture and fatigue of fiber-reinforced metal, ceramic and intermetallic matrix composites involves coupled considerations of mechanics and weakest link statistics. Frank McClintock (1976) was among the first to address problems that required linkage between these two disciplines. These initial concepts and linkages have grown and become the foundation for understanding some of the mechanical properties of composite systems at the micromechanics level. To provide focus on these issues, a broad view of composite properties is not attempted in this chapter. Instead, emphasis is given to unidirectional materials, but with the appreciation that the associated mechanical responses provide a fundamental framework for addressing the properties of multi-directional laminated and woven systems.

The three fundamental constituents of fracture and fatigue models for unidirectional composites are schematically represented on Fig. 8.1. First, debonding occurs at fiber/matrix interfaces, requiring an understanding of interface fracture mechanics in mixed-mode (He and Hutchinson, 1989; Evans and Marshall, 1989). Second, fibers exert tractions on the crack surfaces, requiring a mechanics of large-scale bridging (Evans and Marshall, 1989; Aveston et al., 1971; Marshall et al, 1985; Zok and Hom, 1990; Bowling and Groves, (1979)). Third, fiber fracture may occur, usually at locations away from the matrix crack plane, as a result of weakest link

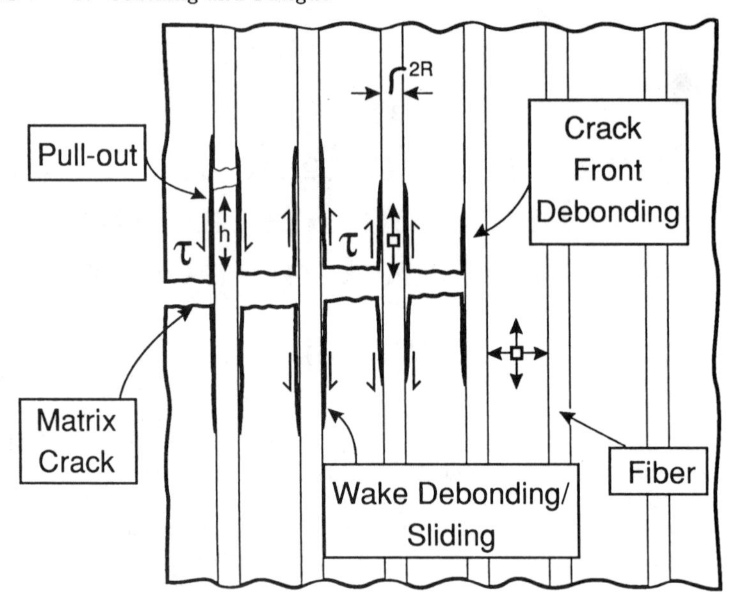

FIGURE 8.1. The various mechanisms that accompany mode I matrix crack propagation in unidirectional composites.

statistics (Thouless and Evans, 1980; Curtin, 1991a; Sutcu, 1989), resulting in pull-out. The dominant dissipation mechanism that allows fibers to enhance the fracture and fatigue resistance is caused by frictional sliding along previously debonded interfaces (Evans, 1990). Such dissipation occurs at both intact and failed fibers. However, the extent of the zone that provides dissipation is strongly influenced by the fiber failure site relative to the crack plane, which governs the pull-out length. This phenomenon arises from the stochastic nature of fiber failure. Large pull-out lengths relative to the crack opening also lead to large-scale bridging (LSB) (Zok et al., 1991), wherein the nominal crack growth resistance depends on crack size and specimen geometry. Consequently, there is an *intimate connection* between the probabilistic nature of fiber failure and the mechanics of crack growth.

A comprehensive understanding of crack growth and fatigue also recognizes that several crack growth mechanisms are possible, dependent upon the properties of the matrix, fiber and interfaces (Evans, (in press)) (Fig. 8.2); i) splitting can occur by either mixed-mode or mode II crack growth within the matrix and along the interfaces; ii) a single mode I crack may form, usually accompanied by fiber failure; iii) multiple mode I cracking may proceed with minimal fiber failure. The same range of behaviors occurs in the monotonic loading of brittle matrix composites and in the cyclic loading of metal matrix composites. This chapter deals with the two

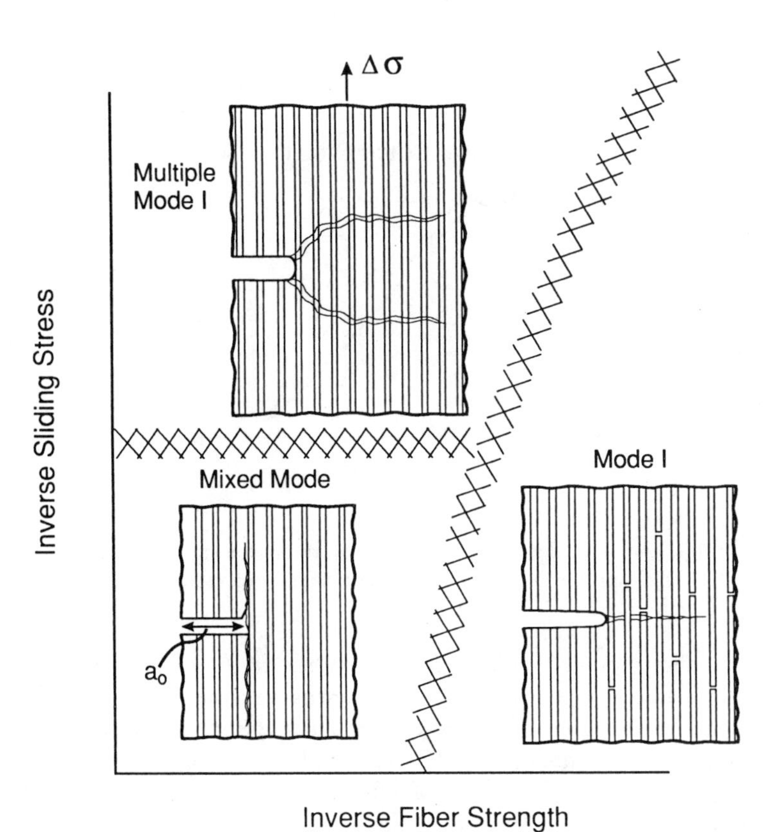

FIGURE 8.2. A mapping scheme that displays the various mechanisms of matrix cracking that can occur in notched composites subject to either monotonic or cyclic loading.

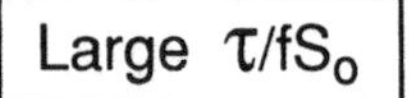

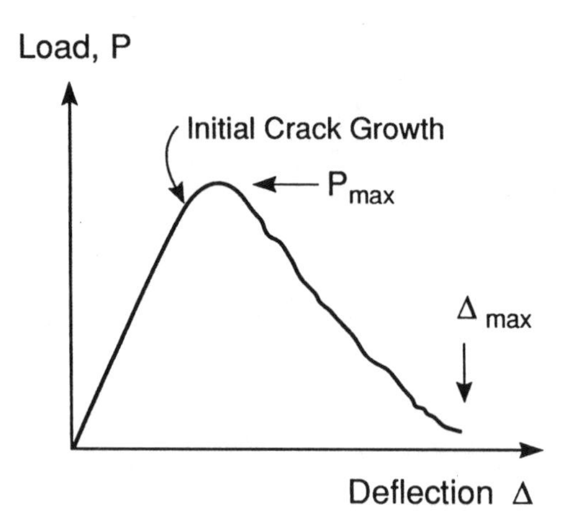

FIGURE 8.3. A typical load/deflection behavior for materials with large $\tau/fS_0(\tilde{>}0.5)$ that exhibit simultaneous matrix crack growth and fiber failure.

mode I mechanisms and the transition between them for both monotonic and cyclic loading. *The coupling of the stochastic processes of fiber failure with the mechanics of matrix crack growth is a central theme.* For discussion of basic aspects of debonding and mixed-mode cracking, the reader is referred to other reviews (Evans and Marshall, 1989; Evans, 1990; Hutchinson and Jensen, 1990; Evans, (in press); Evans et al., 1991).

The macroscopic load/deflection $[P(\Delta)]$ characteristics exhibited by unidirectional brittle composites connect with the above mechanisms of matrix and fiber cracking. There are two principal behaviors. First, when cracking occurs in mode I accompanied by the fiber failure and pull-out, the loading is essentially linear up to the point of first crack growth. Then, after a small non-linear increase in load, the load diminishes as the crack extends (Fig. 8.3). The rate of decrease of load, $-dP/d\Delta$, is governed by the extent of frictional dissipation associated with the fiber pull-out process. Second, when a crack extends across the matrix in mode I *without failing a significant fraction of the fibers*, the crack propagates at constant load (Aveston et al., 1971; Marshall et al., 1985). The material is also capable of sustaining larger loads and additional mode I matrix cracking occurs as the load increases (Marshall and Evans, 1985; Pryce and Smith, (in press); Beyerle et al., (in press)). The cracks reduce the elastic modulus and lead to a permanent strain (Fig. 8.4). The gradual failure of fibers occurs as the load increases. When a sufficient fraction of fibers has failed, the remaining fibers fracture abruptly, leading to a load drop. In both cases, the important parameters for design purposes are the peak load P_{max}, the failure displacement, Δ_{max}, and the change in compliance with load. The relationships between these composite properties and the *in situ* properties of the matrix, fibers and interfaces are addressed in this chapter.

8.2 Some Basic Mechanics

The approach used to simulate mode I cracking under monotonic loading is to define tractions σ_b acting on the crack faces, induced by the fibers (Fig. 8.1), and to determine their effect on the crack tip by using the J-integral (Marshal et al., 1985; Budiansky et al., 1986),

$$\mathcal{G}_{\text{tip}} = \mathcal{G}_{\infty} - \int_0^u \sigma_b du \tag{8.1}$$

where $\mathcal{G}$ is the energy release rate and u is the crack opening displacement[1]. Cracking is considered to proceed when $\mathcal{G}_{\text{tip}}$ attains the pertinent fracture energy. Since the fibers are not failing, the crack growth criterion involves matrix cracking only. A lower bound is given by (Budiansky et al., 1986; McCartney, 1987)

$$\mathcal{G}_{\text{tip}} = \Gamma_m(1-f) \tag{8.2}$$

with f being the fiber volume fraction and Γ_m the matrix toughness. Upon crack extension, $\mathcal{G}_{\infty}$ becomes the crack growth resistance Γ_R, whereupon

$$\Gamma_R = \Gamma_m(1-f) + \int_0^u \sigma_b du. \tag{8.3}$$

A traction law $\sigma_b(u)$ is now needed to predict Γ_R. A law based on frictional sliding along debonded interfaces has been used most extensively and appears to provide a reasonable description of many of the observed mechanical responses. Coulomb friction appears to have a key role in governing the sliding stress, τ, through a relation of the form (Hutchinson and Jensen, 1990; Mackin et al., (to be published));

$$\tau = \tau_0 - \mu p \tag{8.4}$$

where μ is the friction coefficient, p is the residual stress *normal* to the interface and τ_0 represents the sliding resistance afforded by asperities (or roughness). A constant sliding stress assumption, $\tau = \tau_0$, will be used to illustrate the basic mechanics. However, general results based on Eq. 8.4 are

[1]All symbols used in this chapter have been summarized at the end of the chapter for ready reference.

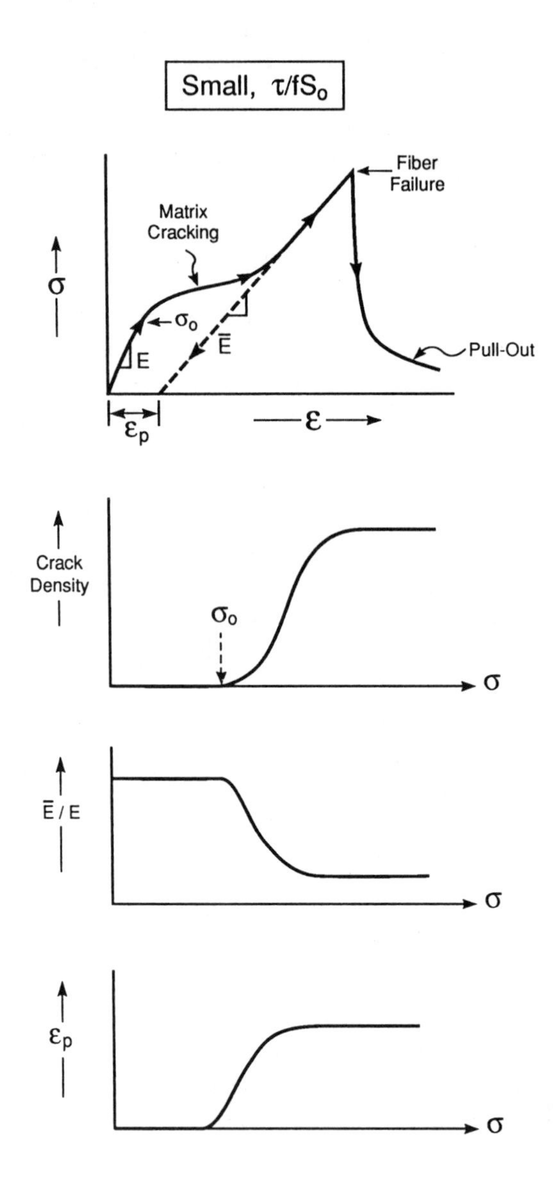

FIGURE 8.4. The stress/strain behavior exhibited by materials with small $\tau/fS_0(\tilde{<}0.1)$ subject to multiple matrix cracking. Also shown are schematics of changes in crack density, modulus and permanent strain.

available (Hutchinson and Jensen, 1990). In the absence of fiber failure, the sliding distance ℓ is related to the crack surface traction, σ_b, by (Aveston et al., 1971; Marshall et al., 1985)

$$\ell = \sigma_b R E_m (1 - f)/2\tau_0 E_f f \tag{8.5}$$

Here, R is the fiber radius, E is Young's modulus, and the subscripts f and m refer to matrix and fibers, respectively. The sliding length is, in turn, related to the crack opening displacement, such that the traction law becomes (Marshall et al., 1985; Budiansky et al., 1986)

$$\sigma_b = [2\xi\tau_0 E f u/R]^{1/2} - q(E/E_m) \tag{8.6}$$

where q is the residual axial stress in the matrix and ξ is defined in Table 8.1. When fiber failure occurs, statistical considerations are needed to determine $\sigma_b(u)$, as elaborated below (Section 8.3).

The matrix fracture behavior can also be described by using stress intensity factors, K. This approach is more convenient than the J-integral in some cases: particularly for short cracks and for fatigue (Marshall et al., 1985; McMeeking and Evans, 1990). To apply this approach, it is first necessary to specify the contribution to the crack opening induced by the applied stress, u_∞, as well as that provided by the bridging fibers, u_b. For a plane strain crack of length $2a$ in an infinite plate, these are given by (Tada et al., 1985)

$$u_\infty = (4/E)\sigma_\infty \sqrt{a^2 - x^2} \tag{8.7}$$

and

$$u_b = -(4/E)\int_0^a \sigma_b(\hat{x}) H(\hat{x}, x, a)\, d\hat{x} \tag{8.8}$$

with

$$H(\hat{x}, x, a) = \frac{1}{\pi}\log\left|\frac{\sqrt{a^2 - x^2} + \sqrt{a^2 - \hat{x}^2}}{\sqrt{a^2 - x^2} - \sqrt{a^2 - \hat{x}^2}}\right|$$

and x being the distance from the crack center. The net crack opening displacement is

$$u = u_\infty + u_b \tag{8.9}$$

The contribution to K from the bridging fibers is obtained using

$$K_b = -2\sqrt{\frac{2}{\pi}}\int_0^a \frac{\sigma_b(x)\,dx}{\sqrt{a^2 - x^2}}$$

which with σ_b given by Eq. 8.6 and $q = 0$ becomes

$$K_b = -\sqrt{\frac{a}{\pi}}\left[\frac{(1 - \nu^2) f a \xi \tau_0}{R}\int_0^1 \frac{\Sigma_b\, d\tilde{x}}{\sqrt{1 - \tilde{x}^2}}\right], \tag{8.10}$$

where Σ_b is a non-dimensional stress parameter defined in Table 8.1. The shielding associated with K_b leads to a tip stress intensity factor

$$K_{\text{tip}} = K_\infty + K_b \tag{8.11}$$

where K_∞ depends on the loading and specimen geometry.

A *criterion for matrix crack extension* based on K_{tip} is needed. For this purpose, to be consistent with many other *brittle cracking processes*, it has been assumed that fracture is governed by the *stress* in the matrix *ahead of the crack*. The corresponding crack growth criterion for the composite is (Marshall et al., 1985)

$$K_{\text{tip}} = (E/E_m)K_m \tag{8.12}$$

where K_m is the critical stress intensity factor for the matrix, given by

$$K_m = \sqrt{E_m \Gamma_m}, \tag{8.13}$$

whereupon the crack growth resistance is

$$K_R = (E/E_m)\sqrt{E_m \Gamma_m} - K_b. \tag{8.14}$$

This criterion would appear to differ from the energy criterion given in Eq. 8.2. Clearly Eq. 8.2 is a necessary condition for matrix crack extension, but it may not be sufficient. The criterion given by Eq. 8.13 is analogous to that used to successfully simulate brittle cracking in steels (Ritchie et al., 1973; Evans, 1983). As demonstrated below, the two criteria lead to essentially the same steady-state matrix cracking stress.

8.3 Basic Weakest Link Statistics

For fibers failing in a composite, the incidence of failure can be determined using straightforward weakest link statistics, provided that the interface is sufficiently "weak" that fiber fractures do not transmit stress concentrations to neighboring fibers: the so-called "global load sharing" condition (Thouless and Evans, 1980; Curtin, 1991a; Sutcu, 1989). This assumption appears to have validity for an appreciable number of brittle matrix composites (ceramic, carbon, glass, intermetallic) that use fiber coatings to impart fiber toughening (Evans et al., 1991). Conversely, this assumption is not normally valid for metal matrix composites with "strong" interfaces (Ritchie et al., 1973). For these cases, a "local load sharing" criterion is needed to understand failure.

When fibers do not interact, analysis begins by considering a fiber of length $2L$ divided into $2N$ elements each of length δz. The probability that an element will fail when the stress is less than σ is essentially the area under the probability density curve (Matthews et al., 1976; Freudenthal, 1967)

$$\delta\phi(\sigma) = \frac{\delta z}{L_0} \int_0^{\sigma} g(S) dS \tag{8.15}$$

where $g(S)dS/L_0$ represents the number of flaws per unit length of fiber having a "strength" between S and $S+dS$. The local stress, σ, is a function of both the distance along the fiber, z, and the reference stress, σ_b. The survival probability P_S *for all elements* in the fiber of length $2L$ is the product of the survival probabilities of each element,

$$P_S(\sigma_b, L) = \prod_{n=-N}^{N} [1 - \delta\phi(\sigma_b, z)] \tag{8.16}$$

where $z = n\delta z$, and $L = N\delta z$. Furthermore, the probability Φ_S that the element at z will fail when the peak stress is between σ_b and $\sigma_b + \delta\sigma_b$, *but not when the stress is less than* σ_b, is the change in $\delta\phi$ when the stress is increased by $\delta\sigma_b$ divided by the survival probability up to σ_b, given by (Matthews et al., 1976; Freudenthal, 1967; Oh and Finnie, 1970)

$$\Phi_S(\sigma_b, z) = [1 - \delta\phi(\sigma_b, z)]^{-1} \left[\frac{\partial\delta\phi(\sigma_b, z)}{\partial\sigma_b}\right] d\sigma_b. \tag{8.17}$$

Denoting the probability density function of fiber failure of $\Phi(\sigma_b, z)$, the probability that fracture occurs at a location z, when the peak stress is σ_b, is governed by the probability that all elements *survive up to a peak stress* σ_b, but that failure occurs at z when the stress reaches σ_b (Thouless and Evans, 1980; Oh and Finnie, 1970). It is given by the product of Eq. 8.16 with Eq. 8.17,

$$\Phi(\sigma_b, z)\delta\sigma_b\delta z = \frac{\prod_{n=-N}^{N}[1 - \delta\phi(\sigma_b, z)]}{[1 - \delta\phi(\sigma_b, z)]} \left[\frac{\partial\delta\phi(\sigma_b, z)}{\partial\sigma_b}\right] d\sigma_b. \tag{8.18}$$

While the above results are quite general, it is convenient to use a power law to represent $g(S)$,

$$\int_0^{\sigma} g(S) dS = (\sigma/S_0)^m. \tag{8.19}$$

where S_0 is the scale parameter and m the shape parameter for the fiber strength distribution. Alternative representations of $g(S)$ are not warranted at the present level of development. Using this assumption, Eq. 8.18 becomes (Thouless and Evans, 1980)

$$\Phi(\sigma_b, z) = \exp\left\{-2\int_0^L \left[\frac{\sigma(\sigma_b, z)}{S_0}\right]^m \frac{dz}{L_0}\right\} \left(\frac{1}{L_0}\right) \frac{\partial}{\partial\sigma_b} \left[\frac{\sigma(\sigma_b, z)}{S_0}\right]^m. \tag{8.20}$$

The above results may be applied to various fiber failure problems in composites. Two parameters repeatedly arise in the result (Table 8.1): a characteristic length (Curtin, 1991a)

$$\delta_c = L_0[S_0R/\tau_0L_0]^{m/(m+1)} \tag{8.21}$$

and a characteristic length (Curtin, 1991a)

$$\delta_c = S_0[\tau_0L_0/S_0R]^{1/(m+1)} \tag{8.22}$$

related by

$$S_c = \tau_0\delta_c/R \tag{8.23}$$

A number of important results for fiber bundle failure in brittle matrix composites have been derived using these *statistics*. All of the formulae have been derived by assuming that the interfaces have sufficiently low debond energy and sliding stress that the failed fibers do not concentrate stress (Thouless and Evans, 1980; Curtin, 1991a; Sutcu, 1989). In such a case, the matrix influences composite fracture only by transferring load from a failed fiber to *all* remaining intact fibers equally (so-called global load sharing) and by allowing load recovery from a fiber failure site, *along* a fiber, through the sliding stress. Since the load recovery length is related to τ, composite parameters such as the pull-out length and the *in situ* fiber strength have a direct connection with τ.

The nominal *in situ* fiber strength distribution, $G(S)$, may be obtained from measurements of fracture mirrors on fibers after composite tensile testing. The fiber strength S is given by the relationship (Jamet et al., 1984; Cao et al., 1990)

$$S \approx 3.5\sqrt{E_f\Gamma_f/a_m} \tag{8.24}$$

where Γ_f is the fracture energy of the fiber and a_m is the fracture mirror radius. It is found that $G(S)$ can be represented by empirically by (Cao et al., 1990)

$$G(S) = 1 - \exp\left[-(S/S_*)\right]^{m_*} \tag{8.25}$$

where the strengths, S, are subject to an implicit gauge length effect related to τ, through the load transfer length (Curtin, 1991a). The details of the preceding statistics depend on the fracture mechanism (multiple cracking or single cracks), as elaborated below.

8.3.1 Multiple Cracking

Analysis of fiber failure, based on Eqs. 8.20 and 8.25, has identified the relationship between the shape parameters m_* and m. It has also established that the scale parameters S_* is related to the characteristic *in situ* fiber strength, S_c (Curtin, 1991a) (See Fig. 8.5). Furthermore, S_c is related to the fiber scale parameters S_0 and L_0 by Eq. 8.22 (Curtin, 1991a).

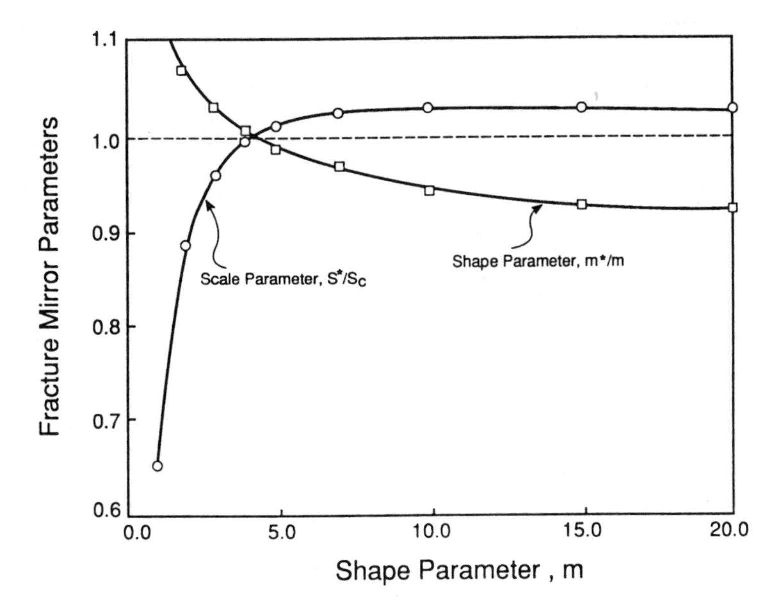

FIGURE 8.5. Relationships between fracture mirror parameters and the *in situ* fiber strength parameters (Curtin, 1991a).

In order to relate S_c determined from fracture mirrors to the strength S_b of the fibers measured in a bundle (without matrix), the difference in gauge lengths must be taken into account. In general, the *in situ* fiber bundle strength at gauge length L is given by (Daniels, 1945; Corten, 1967)

$$S_b \equiv S_0(L_0/Lme)^{1/m} \tag{8.26}$$

hence,

$$S_b = S_c^{(m+1)/m}(R/\tau_0 Lme)^{1/m} \tag{8.27}$$

where e is the base of natural logarithms. With S_c and m ascertained from fracture mirror data, S_b can be determined from Eq. 8.27, provided that τ_0 is known.

The sliding stress τ_0 can also be related to the mean pull-out length $\bar{h}$. To determine this relationship, the pull-out lengths are found by placing a randomly-located plane (the fracture plane) over the multiply fractured fibers and finding the fiber failure location relative to that plane by using Eq. 8.20. The results indicated that $\bar{h}$ and δ_c are related by (Curtin, 1991a)

$$\bar{h} = \lambda(m)\delta_c/4 \tag{8.28}$$

or

$$\bar{h}\tau_0/RS_c = \lambda(m)/4 \tag{8.29}$$

with $\lambda(m)$ plotted on Fig. 8.6. Consequently, either $\bar{h}$ can be predicted from independent measurement of τ_0 and S_c or $\bar{h}$ can be used to determine

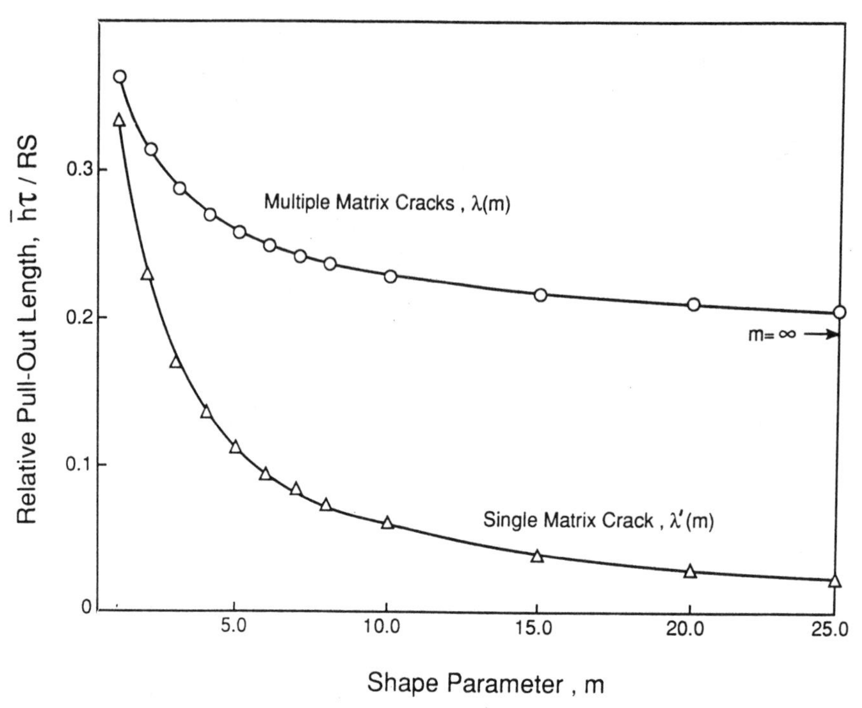

FIGURE 8.6. Pull-out parameters for single and multiple matrix cracking (Curtin, 1991a).

δ_c. Another result of importance concerns the ultimate strength of the composite, σ_u, which can be derived from Eq. 8.20 (Curtin, 1991a) as

$$\sigma_u = f S_c \left[\frac{2}{(m+2)}\right]^{1/(m+1)} \left[\frac{m+1}{m+2}\right] \tag{8.30}$$

It is of significance to appreciate that σ_u expressed in terms of S_c, through Eq. 8.30, has an implicit dependence on τ_0, as indicated by Eq. 8.22, but is independent of the composite gauge length.

The above formulae have been used as follows (Beyerle et al., (in press); Heredia et al., (to be published)): i) determine S_* and m_* based on fracture mirror measurements, and then obtain S_c and m from Fig. 8.5; ii) evaluate δ_c from $\bar{h}$ using Eq. 8.28, with m now known; iii) given S_c and δ_c, determine τ_0 from Eq. 8.22; with S_c, τ_0 and m known, the fiber bundle strength can then be obtained from Eq. 8.26; finally, calculate the ultimate strength σ_u from Eq. 8.30.

8.3.2 Single Cracks

When the fracture mechanism changes to that wherein a single mode I crack extends, accompanied by fiber failure, the fiber is subject to its *maximum stress* between the crack faces and the stress σ_z diminishes as (Thouless and Evans, 1980; Sutcu, 1989),

$$\sigma_z = \sigma_b(1 - z/\ell)/f. \tag{8.31}$$

Inserting this stress into Eq. 8.20 gives the probability density function, $\Phi(\sigma_b, z)$ (Thouless and Evans, 1980; Sutcu, 1989): the basic relation that governs several statistical parameters. Two results are of key interest. First, the effect of z on $\Phi(\sigma_b, z)$ dictates that, on average, the fibers *do not fail* at the location of highest stress on the fiber ($z = 0$, Eq. 8.31). Instead, fibers fail at an average distance, $z = \bar{h}$, leading to fiber pull-out and a related frictional contribution to crack growth. The average failure position for all fibers is given by an expression similar to Eq. 8.28 (Thouless and Evans, 1980; Curtin, 1991a; Sutcu, 1989),

$$\bar{h} = \delta_c \lambda'(m)/4 \tag{8.32}$$

where $\lambda'(m)$ is plotted on Fig. 8.6. Second, the peak stress $\sigma_{\max}$ on the composite when the fibers fail,[2] is given by (Thouless and Evans, 1980; Curtin, 1991a; Curtin, 1991b)

$$\sigma_{\max} = fS_c[(5m+1)/5m]\exp[-1/(m+1)] \tag{8.33}$$
$$\equiv fS_cG(m). \tag{8.34}$$

8.4 Matrix Cracking

Matrix crack evolution in brittle matrix composites subject to monotonic loading is connected to the incidence of accompanying fiber fracture (Fig. 8.2). Steady-state matrix cracking with intact fibers results in a reduced tensile modulus and permanent strain (Fig. 8.4) (Marshall et al., 1985; Beyerle et al., (in press); Kim and Pagano, 1991). This mode of cracking arises when the ratio τ/fS_0 is small (Evans, (in press)), as elaborated in Section 8.4.3. At larger τ/fS_0, fiber fracture may occur simultaneously with matrix crack growth. Then, a single crack occurs and extends subject to diminishing load (Fig. 8.3) (Zok et al., 1991). Further analysis of these mode I cracking phenomena is presented below, as well as considerations relevant to the transition criterion.

---

[2]This stress governs $P_{\max}$ in Fig. 8.3.

8.4.1 MULTIPLE CRACKING

The problem of matrix cracking with intact fibers has been comprehensively addressed, commencing with Aveston, Cooper and Kelly (1971) (ACK). More recent studies have both verified the ACK solution and introduced considerations additional to those envisaged by ACK (Marshall et al., 1985; Budiansky et al., 1986; McCartney, 1987; Zok and Spearing, (in press)). The present understanding involves the following factors. Because the fibers are intact, a steady-state condition exists wherein the tractions on the fibers in the crack wake balance the applied stress. This special case may be addressed by integrating Eq. 8.1 up to a limit $u = u_0$ (obtained from Eq. 8.6 by equating σ_b to σ_∞), giving (Budiansky et al., 1986)

$$\mathcal{G}_{\text{tip}}^{0} = \frac{(\sigma_\infty + qE/E_m)^3 E_m^2 (1-f)^2 R}{6\tau_0 f^2 E_f E^2} \tag{8.35}$$

A *lower bound to the matrix cracking stress,* σ_0, is then obtained by invoking Eq. 8.2 such that (Budiansky et al., 1986)

$$\sigma_0 = \sigma_* - qE/E_m{}^3 \tag{8.36}$$

where

$$\sigma_* = \left[\frac{6\tau_0 \Gamma_m f^2 E_f E^2}{(1-f)E_m^2 R}\right]^{1/3}.$$

Analogous results can be obtained using stress intensity factors (Marshall et al., 1985; McMeeking and Evans, 1990) (Eqs. 8.9 to 8.12). For the example of a center crack in a tensile specimen ($K_\infty = \sigma_\infty\sqrt{\pi a}$), Eqs. 8.10 and 8.11 give the steady-state result at large crack lengths (McMeeking and Evans, 1990),

$$K_{\text{tip}}^{0} = \sqrt{R}(\sqrt{6}T)^{-1}\sigma \tag{8.37}$$

where T is defined in Table 8.1 and $\sigma \equiv \sigma_\infty - qE/E_m$. When combined with the fracture criterion (Eq. 8.12), the matrix cracking stress is predicted to be the same as that given by Eq. 8.36 (Marshall et al., 1985; McMeeking and Evans, 1990), within 10% in the numerical coefficient.

In addition, this approach may be used to define a critical crack length a_c, above which, steady-state applies. This critical length occurs at $\Sigma \approx 4$ (Table 8.1), and is given by

$$a_c/R \approx E_m \left[\Gamma_m (1+\xi)^2 (1-f)^4 / \tau_0^2 f^4 E_f^2 R\right]^{1/3}. \tag{8.38}$$

Namely, when the initial flaw size $a_0 > a_c$, cracking occurs at $\sigma = \sigma_0$. Conversely, when the initial flaws are small, $a_0 < a_c$, Eqs. 8.9 and 8.10 give

---

[3]There is an interplay between the roles of τ and q in matrix cracking. For the case $\tau_0 = 0$, it has been shown that there is an *optimum* thermal expansion misfit at which σ_0 has the maximum possible value (Budiansky et al., 1986).

(McMeeking and Evans, 1990)

$$K_{\text{tip}} = \sigma\sqrt{\pi a}\left[1 - \frac{3.05}{\Sigma}\sqrt{\Sigma + 3.3} + \frac{5.5}{\Sigma}\right]. \tag{8.39}$$

This result, when combined with Eq. 8.12, gives the matrix cracking stress σ_c for $a_0 < a_c$. In this range, σ_c *exceeds* σ_0.

The evolution of additional cracks at stresses above σ_0 is less well understood, because two factors are involved: screening and statistics (Beyerle et al., (in press); Zok and Spearing, (in press)). When the sliding zones between neighboring cracks overlap, screening occurs and $\mathcal{G}_{\text{tip}}$ differs from $\mathcal{G}^0_{\text{tip}}$. The relationship is dictated by the location of the neighboring cracks. When a crack forms midway between two existing cracks with a separation $2d$, $\mathcal{G}_{\text{tip}}$ is related to $\mathcal{G}^0_{\text{tip}}$ by (Zok and and Spearing, in press.)

$$\mathcal{G}_{\text{tip}}/\mathcal{G}^0_{\text{tip}} = 4(d/2\ell)^3 \qquad (\text{for } 0 \le d/\ell \le 1) \tag{8.40}$$

and

$$\mathcal{G}_{\text{tip}}/\mathcal{G}^0_{\text{tip}} = 1 - 4(1 - d/2\ell)^3 \qquad (\text{for } 1 \le d/\ell \le 2) \tag{8.41}$$

Consequently, if matrix cracks develop in a strictly periodic manner, the evolution of the crack density with stress can be predicted by combining Eqs. 8.12, 8.35 and 8.41, leading to the result plotted on Fig. 8.7. In general, however, non-periodic crack locations exist, resulting in a different distribution of crack spacing. Computer simulations of *random* matrix cracking reveal similar trends (also shown in Fig. 8.7) and indicate that the average crack density reaches a *saturation* value, $\ell_*/\bar{d}_S = 0.575\ell_*$, when $\sigma/\sigma_0 > 1.3$.[4] An *upper bound* to the sliding stress can be inferred from $\bar{d}_S$, given by (Zok and Spearing, (in press))

$$\tau_0 \approx 2R(1 - f)(\Gamma_m E_f E_m / f E \bar{b}^3_S)^{1/2}. \tag{8.42}$$

The actual evolution of matrix cracks may deviate from the predictions, depending on the spatial and size distribution of matrix flaws. One statistical result of interest is derived by assuming that all matrix cracks initiate from flaws with $a_0 \le a_c$. The principal result is that significant matrix cracking is not predicted until stresses are appreciably above σ_0, depending on the magnitude of both the shape and scale parameters for matrix flaws (Beyerle et al., (in press); Curtin, 1991b). A schematic that combines the principal aspects of the mechanics and stochastics is presented in Fig. 8.8.

---

[4] ℓ_* is the sliding length at the onset of matrix cracking. It is given by Eq. 8.5 with $\sigma_b = \sigma_0$.

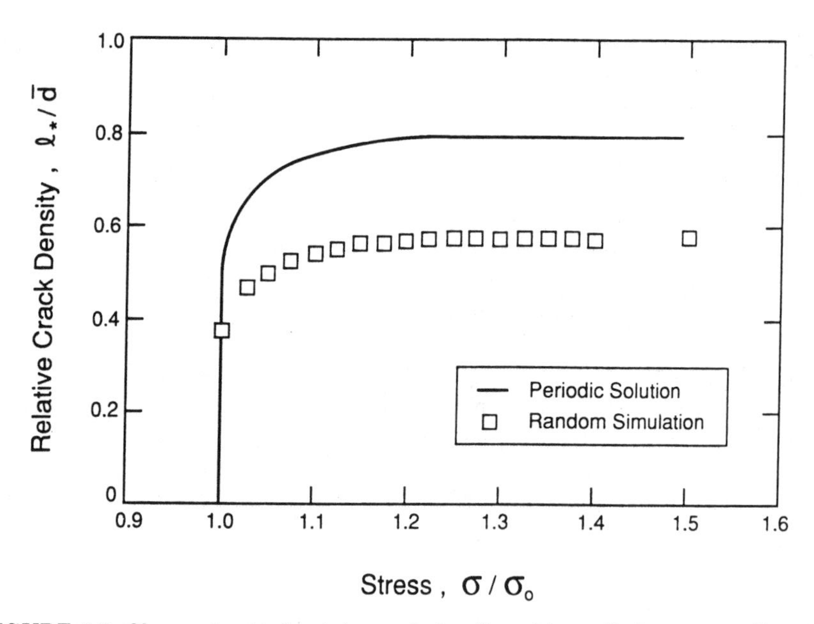

FIGURE 8.7. Change in steady-state crack density with applied stress predicted using the mechanics of matrix cracking (Zok and Spearing, (in press)).

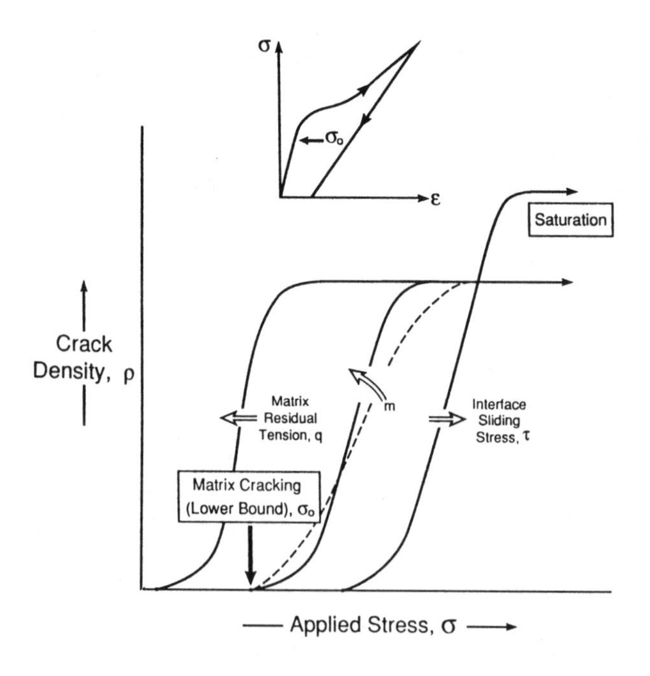

FIGURE 8.8. A schematic of general trends in crack density with stress.

The crack spacing relates to the change in Young's modulus and the permanent strain. Analysis of the sliding that occurs in the region between matrix cracks provides explicit predictions (Pryce and Smith, (in press)): the Young's modulus $\bar{E}$ is given by

$$E/\bar{E} = 1 + E_f R\bar{\sigma}/2\bar{d}\xi^2\tau E \tag{8.43}$$

and the permanent strain, ε_p, is

$$\varepsilon_p = \frac{R(1-f)^2q^2}{4\bar{d}f^2\tau_0 E_f}\left[\left(\frac{\bar{\sigma}E_m}{qE}\right)^2 - 4\left(\frac{\bar{\sigma}E_m}{qE}\right) + 2\right] \tag{8.44}$$

where $\bar{\sigma}$ is the stress reached upon loading the material.

8.4.2 Single Cracks

When fracture occurs by a single crack extending with failing filters, the "weaker" fibers fail in the immediate crack wake, near the crack plane. Upon additional crack extension, more fibers fail in the crack wake, at locations further from the crack plane. The statistics of the process that governs the associated traction laws are as follows (Thouless and Evans, 1980; Curtin, 1991a; Sutcu, 1989). For intact fibers, the bridging stress σ_b is given by Eq. 8.6, while for failed fibers,

$$\sigma_b = f(2\tau_0/R)(h-u). \tag{8.45}$$

The average stress on the fibers thus depends on the fraction of failed fibers f_x which is just the cumulative fiber failure probability at $\sigma_b = S$,

$$f_x = 1 - \exp[-(S/S_c)^{m+1}/(m+1)]. \tag{8.46}$$

These formulae may be combined to give the tractions. The result is unwieldy, but the dominant term is (Thouless and Evans, 1980; Curtin, 1991a; Sutcu, 1989)

$$\sigma_b/fS_c = (u/v)^{1/2}\exp[-(u/v)^{(m+1)/2}](m+1)^{1/(m+1)} \tag{8.47a}$$

where

$$v = S_c^2 R/4E_f\tau_0. \tag{8.47b}$$

This fundamental traction law may be used to compute the important fracture properties of the material.

A particularly useful simplification recognizes that the mean pull-length $\bar{h}$ is typically much larger than the crack opening displacement (Zok and Hom, 1990). Consequently, the traction exerted by the failed fibers is (Zok et al., 1991)

$$\sigma_b \approx 2\tau_0 f_x\bar{h}/R. \tag{8.48}$$

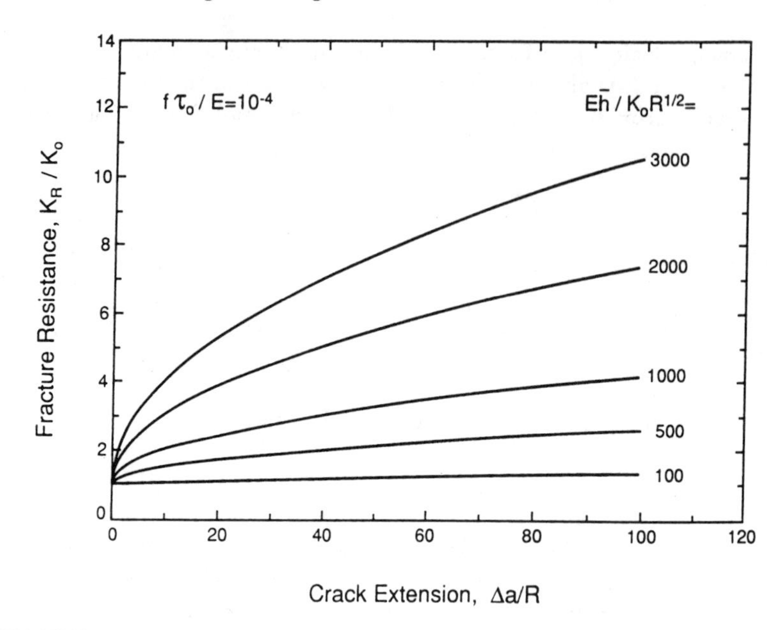

FIGURE 8.9. Predicted resistance curves for small-scale bridging (Zok et al., 1991).

With $\bar{h}$ given by Eq. 8.34 and $f_x \approx f$, the traction becomes

$$\sigma_b \approx 2fS_c \tag{8.49}$$

$$\approx 2f\left[\frac{\tau_0 L_0}{R} S_0^m\right]^{1/(m+1)}. \tag{8.50}$$

The traction is thus dominated by the *fiber strength*, S_c, and is insensitive to the sliding stress τ_0 and fiber radius, R. The basic traction law (Eq. 8.50) can be used with Eqs. 8.8–8.13 to calculate a resistance K_R given by (Zok et al., 1991)

$$\frac{K_R}{K_0} = 1 + 4\sqrt{2/\pi}\left(\frac{f\tau_0}{E}\right)\left(\frac{E\bar{h}}{K_0\sqrt{R}}\right)\left[\sqrt{\frac{\Delta a}{R}} - \sqrt{\frac{2}{\pi}}\left(\frac{K_0\sqrt{R}}{E\bar{h}}\right)\left(\frac{\Delta a}{R}\right)\right] \tag{8.51}$$

where Δa is the crack extension. A typical result is plotted on Fig. 8.9. The above formulation applies subject to small-scale bridging (SSB); wherein the bridging zone length is small compared with crack length and specimen dimensions.

Small-scale bridging is often violated and the crack growth resistance can only be deduced using a full, large-scale bridging (LSB) computation. Adequate prediction of the load/deflection characteristic also requires a mechanics of large-scale bridging. The basic approach, schematically illustrated in Fig. 8.10 (Zok and Hom, 1990; Cox, (in press); Cox and Lo, (in

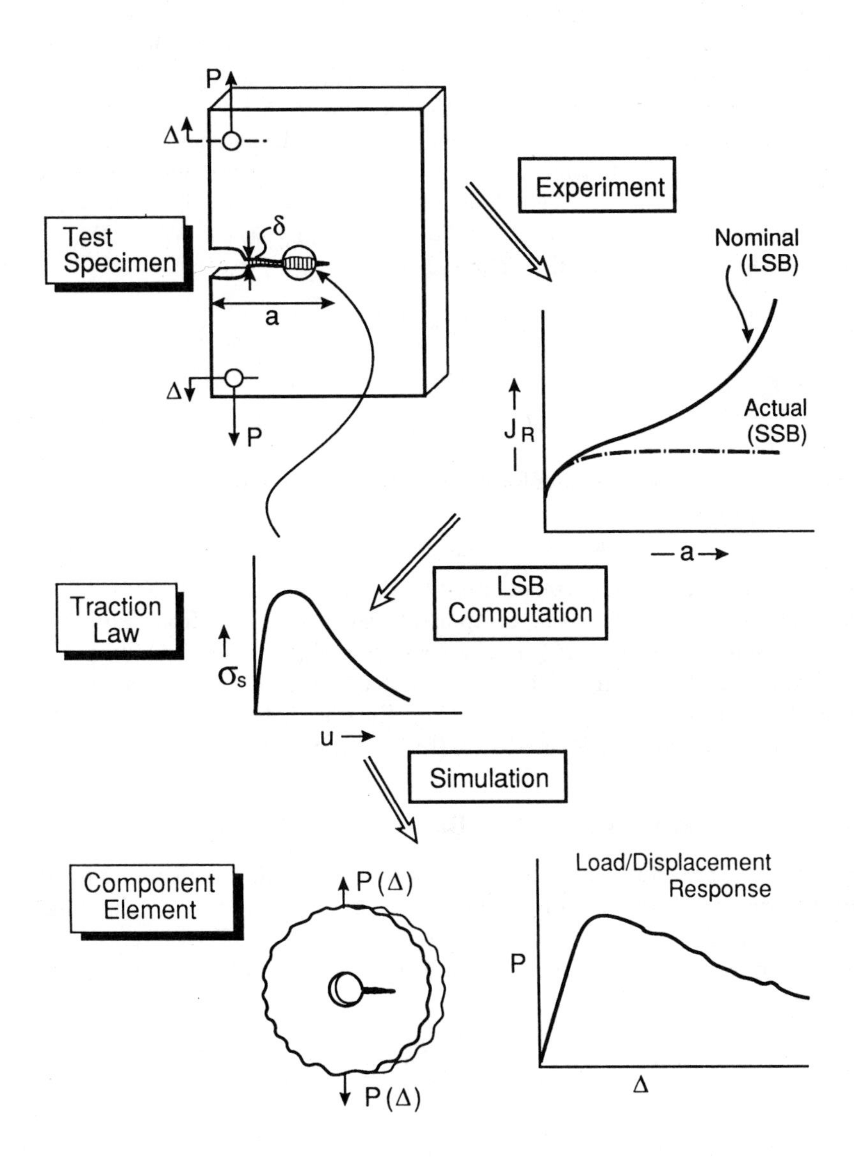

FIGURE 8.10. A scheme for calculating the load/deflection response in large-scale bridging.

press); Cox and Marshall, 1991), recognizes that the *fundamental property of the composite is the traction law*, $\sigma_b(u)$.[5] Once this law has been explicitly established, it may be used to predict either $P(\Delta)$ or $K_R(\Delta a)$. The procedure is essentially the same as that described in Eqs. 8.7 to 8.13, except that the loading function becomes explicit to the geometry of interest. One approach for facilitating this analysis is the use of canonical functions (Cox and Lo, (in press)). The choice of function is governed by the characteristic shape of curves of K_{tip}/K_∞ vs. bridging stress. When Eq. 8.6 applies, the preferred function is found to be

$$\begin{aligned} K_{\text{tip}}/K_\infty &= 1/2\Big\{1+\tanh\big[(1/4+\lambda_3 \\ &\quad \cdot\exp[\lambda_1(\omega-\omega_0)+\lambda_2(\omega-\omega_0)^2]) \\ &\quad \cdot(\omega-\omega_0)\big]\Big\} \end{aligned} \tag{8.52a}$$

where λ_i are fitting constants for each geometry (tabulated by Cox and Lo, (in press)), and

$$\omega = \log\left(\pi\sigma_\infty E_f R/16\xi E\tau w\right) \tag{8.52b}$$

with the parameter ω_0 corresponding to the value of ω when $K_{\text{tip}}/K_\infty = 0.5$, and w the width of the component. By using K_{tip} from Eqs. 8.12 and 8.13 and by equating K_∞ to the resistance K_R, the nominal composite resistance can be readily derived from Eq. 8.52a. More general numerical procedures have also been devised (Cox, (in press); Odette and Chao, (to be published); Bao and McMeeking, (to be published)).

8.4.3 The Mechanism Transition

A first order criterion for the mechanism transition from multiple to single matrix cracking is obtained by equating σ_0 (Eq. 8.56) and $\sigma_{\max}$ (Eq. 8.34), such that multiple cracking occurs when $\sigma_0 < \sigma_{\max}$ and vice versa. This result is unwieldy, but can be presented in an insightful form by neglecting elastic mismatch, leading to:

$$H(f)\left[\frac{E\Gamma_m\tau_c}{RS_0^3}\right]^{1/3} - \left(\frac{\tau_c \ell_0}{RS_0}\right)^{1/m} G(m) = \frac{q}{fS_0} \tag{8.53}$$

where $H(f) = [6/f(1-f)]^{1/3}$ and τ_c is the critical sliding stress, at the transition. For the case $q = 0$, the critical stress becomes

$$\frac{\tau_c}{fS_0} \approx \left[\left(\frac{S_0^2 R}{E\Gamma_m}\right)\left(\frac{\ell_0}{R}\right)^{3/m}\right]^{1-3/m} \frac{(1-f)}{6} f^{(m-3)/3} G^3(m). \tag{8.54}$$

---

[5]The law appropriate to a particular composite can either be obtained from basic principles of fiber sliding and failure (Eqs. 8.6 and 8.47a) or by direct measurement.

This result has the merit that it identifies τ_c/fS_0 as a key non-dimensional parameter in the transition and established the importance of a second non-dimensional parameter, $S_0^2R/E\Gamma_m$, in the sense that the larger this parameter is, the greater is the critical sliding stress. It is also evident from Eq. 8.53 that residual tensile stress in the matrix encourages the multiple cracking mode of behavior.

8.4.4 Cyclic Loading

Fatigue crack growth in either metal, intermetallic on ceramic matrix composites may be addressed with the same basic formalism used to describe monotonic crack growth (McMeeking and Evans, 1990; Bao and McMeeking, (to be published)). Once again, a key factor is the nature of the interface. When the interface is strong, there can be no contribution to the fatigue resistance from frictional dissipation and the composite behaves in approximate accordance with the rule-of-mixtures expected from fatigue properties of the matrix and reinforcement (Shang et al., 1987). Conversely, when the interfaces are "weak," fibers can remain intact in the crack wake and cyclic frictional dissipation can resist fatigue crack growth (McMeeking and Evans, 1990). The latter has been most extensively demonstrated on Ti matrix composites reinforced with SiC fibers (Walls et al., 1991; Sensmeier and Wright, 1989). The essential features of the "weak" interface behavior are as follows: intact, sliding fibers acting in the crack wake shield the crack tip, such that the stress intensity range at the crack tip, ΔK_{tip}, is less than that expected for the applied loads, ΔK; and subject to the reduced ΔK_{tip}, the fatigue crack grows in the matrix, in accordance with the Paris law[6],

$$da/dN = \beta(\Delta K_{\text{tip}}/E)^n. \tag{8.55}$$

Using this approach, it has been found that a simple transformation converts the monotonic crack growth parameters into cyclic parameters that can be used to interpret and simulate fatigue crack growth. The key transformation is based on the relationship between interface sliding during loading and unloading, which relates the monotonic result to the cyclic equivalent through (McMeeking and Evans, 1990)

$$\left(\frac{1}{2}\right) \Delta\sigma_b(x/a, \Delta\sigma) = \sigma_b(x/a, \Delta\sigma/2) \tag{8.56}$$

where $\Delta\sigma$ is the range in the applied stress. Notably, the amplitude of the *change* in fiber traction $\Delta\sigma_b$ caused by a change in applied stress, $\Delta\sigma$, is twice the fiber traction σ_b which would arise in the monotonic loading of

---

[6]The phenomenon has been addressed in terms of stress intensity factors (rather than energy release rates) because of the choice of a Paris law crack growth criterion.

a previously unopened crack, caused by an applied stress equal to half the stress change $\Delta\sigma$. *This result is fundamental to all subsequent developments* (McMeeking and Evans, 1990).

The stress intensity factor for bridging fibers subject to cyclic conditions is

$$\Delta K_b(\Delta\sigma) = -2\sqrt{\frac{a}{\pi}} \int_0^a \frac{\Delta\sigma_b(x, \Delta\sigma)}{\sqrt{a^2 - x^2}} dx \tag{8.57}$$

which, with the use of Eq. 8.56, becomes

$$\Delta K_b(\Delta\sigma) = 2K_b^{\max}(\Delta\sigma/2) \tag{8.58}$$

where the superscript "max" refers to the maximum values of the parameters achieved in the loading cycle and thus, $K_b^{\max}$ is the bridging contribution that would arise when the crack is loaded by an applied stress equal to $\Delta\sigma/2$. Furthermore, since ΔK is linear in $\Delta\sigma$, Eq. 8.58 is also valid for the tip stress intensity factor:

$$\Delta K_{\text{tip}} = 2K_{\text{tip}}(\Delta\sigma/2). \tag{8.59}$$

When the fibers remain intact, a cyclic *steady-state* (ΔK independent of crack length) is obtained when the cracks are long, given by the condition $\Delta\Sigma \leq 4$ (McMeeking and Evans, 1990) as

$$\Delta K_{\text{tip}} = \Delta\sigma\sqrt{R}(\sqrt{12}\Delta T)^{-1}. \tag{8.60}$$

For cyclic loading, the residual stress q does not affect ΔK_{tip}. The corresponding crack growth rate is determined from Eqs. 8.55 and 8.60 (McMeeking and Evans, 1990) as

$$\frac{da}{dN} = \beta\left[\frac{\Delta\sigma\sqrt{R}}{\sqrt{6}\Delta T E_m}\right]^n. \tag{8.61}$$

When short cracks are of relevance ($\Delta\Sigma > 4$),

$$\Delta K_{\text{tip}} = \Delta\sigma\sqrt{\pi a}\left[1 - \frac{4.31}{\Delta\Sigma}\sqrt{\Delta\Sigma + 6.6} + \frac{11}{\Delta\Sigma}\right]. \tag{8.62}$$

Consequently, at fixed $\Delta\sigma$, ΔK_{tip} increases as the crack extends, and the crack growth accelerates. However, the bridged matrix fatigue crack always grows at a *slower rate* than an unbridged crack of the same length.

To incorporate the effects of fiber breaking into the fatigue crack growth model, a deterministic criterion has been used (Bao and McMeeking, (to be published)): the statistical characteristics of fiber failure have yet to be incorporated. To conduct the calculation, once the fibers begin to fail, the unbridged crack length is continuously adjusted to maintain a stress at the unbridged crack tip equal to the fiber strength, fS. These conditions lead

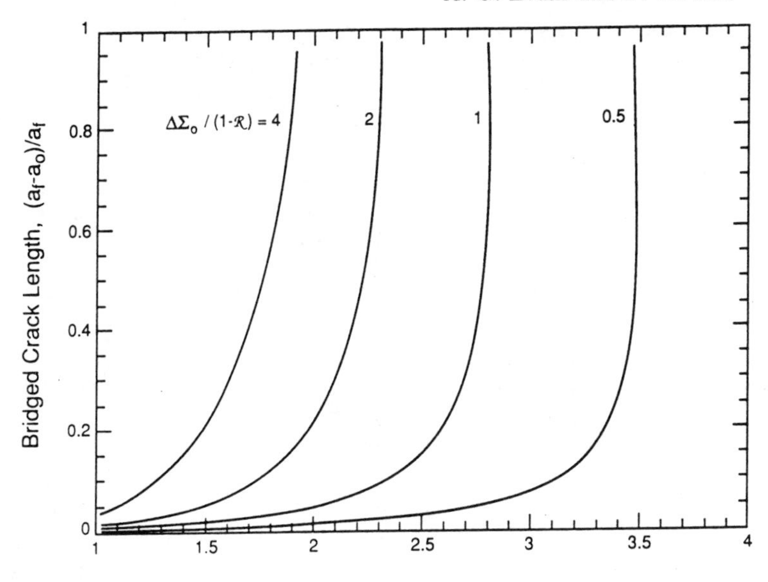

FIGURE 8.11. The length of the crack a_f at first fiber failure as a function of fiber strength for a range of stress amplitude, $\Delta\Sigma_0$.

to the determination of the crack length, a_f, when the first fibers fail, as a function of the fiber strength and the maximum applied load[7] (Fig. 8.11). As an illustration, for an initial condition characterized by $\Delta\Sigma_0/(1-\mathcal{R}) = 1$ and a strength parameter $S(1-\mathcal{R})f/\Delta\sigma = 2.5$, the onset of fiber failure occurs at $a_f = 1.18a_0$. Note that when either the fiber strength is high or the applied stress is low, no corresponding value of a_f can be identified and the fibers do not fail.

After the first fiber failure, fibers continue to break as the crack grows. Continuing fiber failure creates an unbridged segment larger than the original notch size. However, only the current unbridged length $2a_u$ and the current total crack length $2a$ are relevant, as determined from Fig. 8.11, with a_0/a_f replaced by a_u/a and $\Delta\Sigma_0$ replaced by $\Delta\Sigma_0/(a/a_0)$. This procedure has been used to compute a_u/a (Fig. 8.12) (Bao and McMeeking, (to be published)).

If the fibers are relatively weak and break close to the crack tip ($a_0/a \to 1$) the bridging zone is always a small fraction of the crack length. In this case, there is minimal shielding. If the fibers are moderately strong, the fibers remain intact at first. But when the first fibers fail, subsequent failure occurs quite rapidly as the crack grows. The unbridged crack length then increases more rapidly than the total crack length and ΔK_{tip} also

---

[7] $\sigma_{\max} = \Delta\sigma/(1-\mathcal{R})$, which defines R (the "R-ratio" in fatigue parlance).

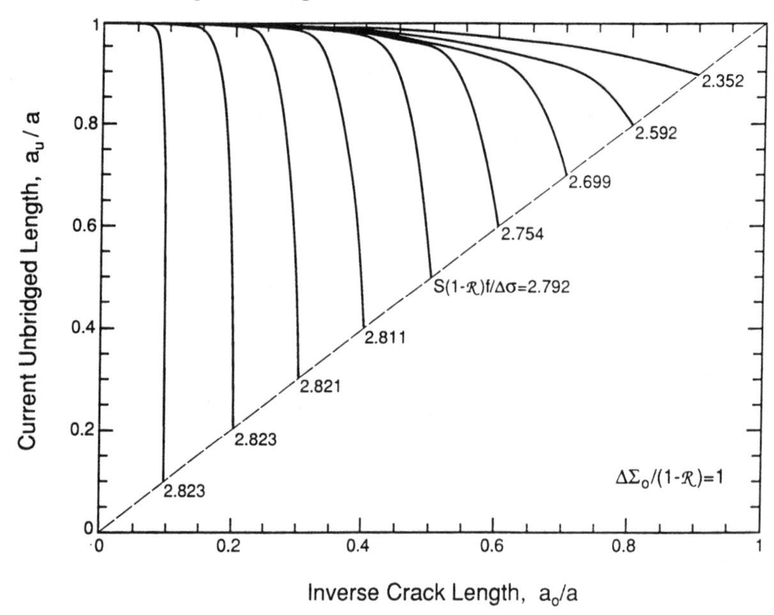

FIGURE 8.12. Fiber breaking rate, as manifest in the current unbridged length $2a_u$ as a function of total crack length, $2a$ ($2a_0$ is the initial notch length).

increases as the crack grows. When the fibers are even stronger, first fiber failure is delayed. But once such failure occurs, many fibers fail simultaneously and the unbridged length increases rapidly. This causes a sudden increase in the crack growth rate. Finally, when the fiber strength exceeds a critical value, they never break and the fatigue crack growth rate always diminishes as the crack grows. The sensitivity of these behaviors to fiber strength is quite marked (Fig. 8.12), with the different types of behavior occurring over a narrow range of fiber strength. Some typical crack growth curves predicted using this approach are plotted on Fig. 8.13. Finite geometry effects associated with LSB also exist (Bao and McMeeking, (to be published)).

The results of Fig. 8.11 can be used to develop a criterion for a "threshold" stress range, $\Delta\sigma_t$, below which fiber failure does not occur for *any* crack length. Within such a regime, the crack growth rate approaches the steady-state value given by Eq. 8.61, with all fibers in the crack wake remaining intact. The variation in the "threshold" stress range with fiber strength is plotted on Fig. 8.14. An upper bound to $\Delta\sigma_t$ is obtained when the notch length becomes exceedingly small ($\Delta\Sigma_0 \to \infty$), whereupon the threshold condition reduces to

$$\Delta\sigma_t = fS(1 - R) \tag{8.63}$$

A notable feature of the predictions pertains to the role of the stress ratio $\mathcal{R}$ in composite behavior. Prior to fiber failure, the crack growth rate

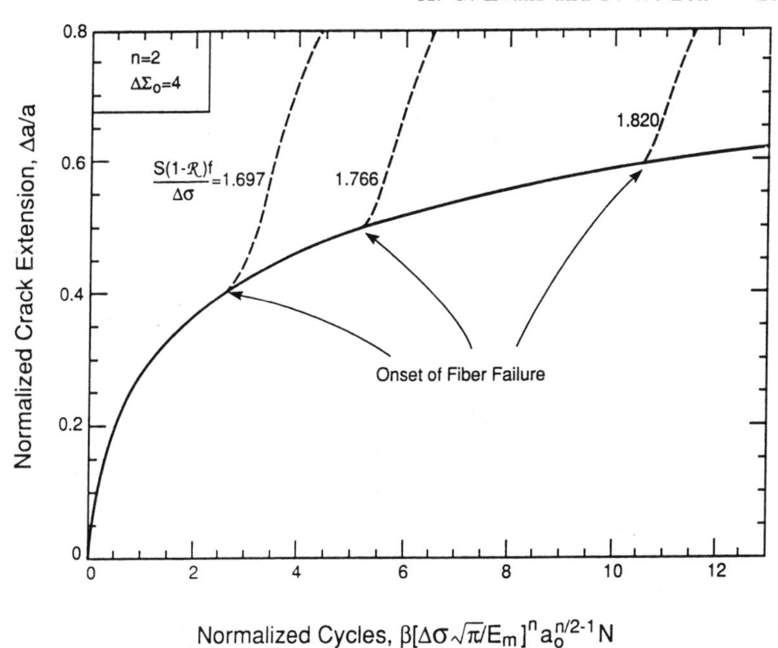

FIGURE 8.13. Predicted fatigue crack growth curves: normalized crack extension $\Delta a/a$ as a function of non-dimensional cycles.

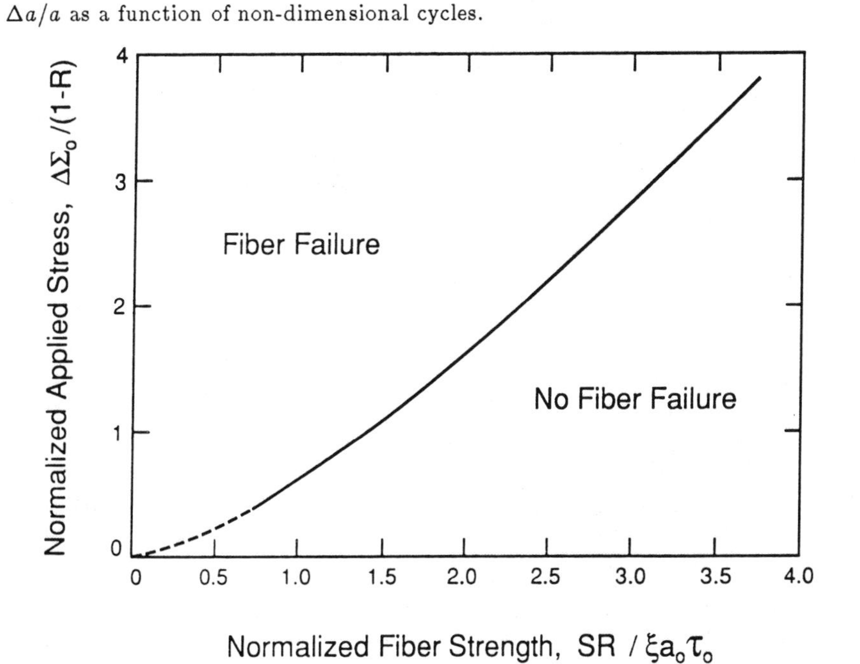

FIGURE 8.14. Prediction of the threshold stress range in terms of the fiber strength.

is independent of $\mathcal{R}$ (except for its effect on the fatigue properties of the matrix itself). However, $\mathcal{R}$ has a strong influence on the transition to fiber failure, as manifest in its effect on the maximum stress, and thus plays a dominant role in the fatigue lifetime. Such trends are evident in both the crack growth curves of Fig. 8.13 and the threshold stress range of Fig. 8.14.

8.5 Structural Performance

8.5.1 Monotonic Loading

The results in the preceding two sections may be combined to provide predictions of either load/deflection or stress/strain curves that may be compared with experiments and used to assess structural utility. The two features that have been most extensively addressed are the lower bound matrix cracking stress σ_0 (Eq. 8.36) and the ultimate strength σ_u (Eq. 8.31). To compare with experiments, a substantial range of independent measurements is needed to obtain τ, q, Γ_m, E, m, S_0, etc. The parameter most susceptible to measurement uncertainty is the sliding stress, τ. Consequently, it is judicious to plot predicted properties as a function of τ and compare these with experiments (Pryce and Smith, (in press)). There are relatively few comprehensive measurements of this type.

Subject to τ/fS_0 being sufficiently small ($\tilde{<}0.1$) that multiple matrix cracking occurs, the *matrix cracking* results (Marshall and Evans, 1985; Pryce and Smith, (in press); Beyerle et al., (in press); Kim and Pagano, 1991) (Fig. 8.15) are found to validate σ_0 as a lower bound. Results for small diameter Nicalon fibers ($R \sim 10\mu m$) also reveal that the first cracks usually form at stresses close to this bound, in accordance with the following features. The thermal expansion mismatch has a major influence on matrix cracking through its effect on both the residual stress q and the sliding stress, τ. In general, τ is larger in composites subject to positive misfit (residual compression normal to the interface), through the Coulomb term in Eq. 8.5. But, the effect of larger τ on σ_0 is offset by the influence of the residual tension in the matrix. These effects are evident in the comparison between results for LAS[8] and CAS[9] matrix composites reinforced with Nicalon fibers (Fig. 8.15) (Pryce and Smith, (in press)). The positive misfit for CAS leads to a larger τ, and a lower predicted matrix cracking stress.[10] Implicit in these findings is that $a_c < a_0$ for small R, because of the scaling of a_c with R (Eq. 8.38).

[8]lithium-alumine-silicate glass.

[9]calcium-alumine-silicate glass.

[10]Explicit measurements of τ by a variety of methods give values in the range 1–3 MPa for LAS and 15–25 MPa for CAS materials: the residual stress q ranges from -20 to -40 MPa and from 80 to 100 MPa for the two composite systems, respectively.

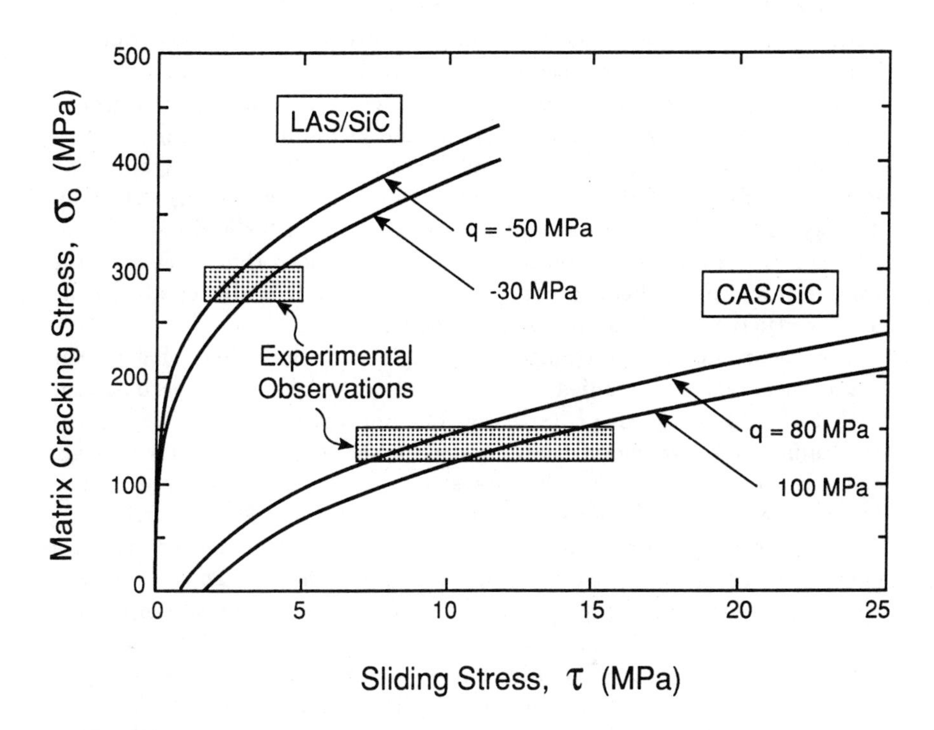

FIGURE 8.15. Comparison of predicted and measured lower bound matrix cracking stresses for two CMCs with small τ_0/fS_0: the independently measured values of τ are 1–3 MPa for LAS materials and 15–25 MPa for CAS materials. The residual stress range *measured* in these materials has governed the choices used to plot the predicted curves.

Initial cracking stresses appreciably above the lower bound are generally found for composites containing large diameter fibers ($R \approx 70 \mu m$) (Singh, 1989). In such materials, the relatively large matrix flaws needed to satisfy steady-state requirements, $a_0 > a_c$ (Eq. 8.38) would generally not occur during careful processing and machining. Statistical information about matrix flaws is needed to correlate the matrix cracking stress in such materials.

The *ultimate strengths* of brittle matrix composites having *small* τ_0/fS_0 also appear to satisfy global load sharing and yield ultimate strengths consistent with Eq. 8.30 (Curtin, 1991a; Evans et al., 1991; Cao et al., 1990; Prewo and Brennan, 1980). A typical example is shown in Fig. 8.16 (Budiansky et al., 1986). This analysis seemingly applies for unidirectional as well as 2-D (woven and laminated) materials, provided that f refers to the volume fraction of fibers in the loading direction (Heredia et al., (to be published); Beyerle et al., (to be published)).

The occurrence of the transition from global to local load sharing has not been extensively studied, but is found to be qualitatively consistent with Eq. 8.53: a rigorous comparison has not yet been possible, because of insufficient knowledge concerning residual stresses and matrix fracture energies. Nevertheless, the ultimate strengths at relatively large τ_0/fS_0 ($\gtrsim 0.5$), are found to be lower than predicted by Eq. 8.30 (Heredia et al., (to be published)). Furthermore, fracture involves a dominant matrix crack (Zok et al., 1991). In this case, a modelling approach based on crack growth with large-scale bridging, plus some premise concerning the size of the initial flaws, a_0, appears to predict the macroscopic response. A typical LSB, load/deflection curve predicted in this manner and comparison with experiment (Odette and Chao, (to be published)) is shown in Fig. 8.17. In this case, the initial flaw was introduced to give a well-defined a_0. Otherwise, independent estimates of a_0 are required.

The non-linearity and multiple cracking between σ_0 and σ_u in materials with $\tau_0 < \tau_c$ have yet to be comprehensively compared with models, primarily because of complexities associated with stress corrosion effects. Some typical experimental data (Pryce and Smith, (in press); Beyerle et al., (in press)) corresponding with the stress/strain curve of Fig. 8.16 are presented on Fig. 8.18.

8.5.2 Cyclic Loading

The important model predictions that require validation include the fatigue crack growth rates in the matrix both before and after the onset of fiber failure. For this purpose, experiments are conducted on a variety of specimen geometries (characterized by different notch lengths), and the measurements then compared with the model prediction for different values of τ. Following this approach, consistency between experiment and

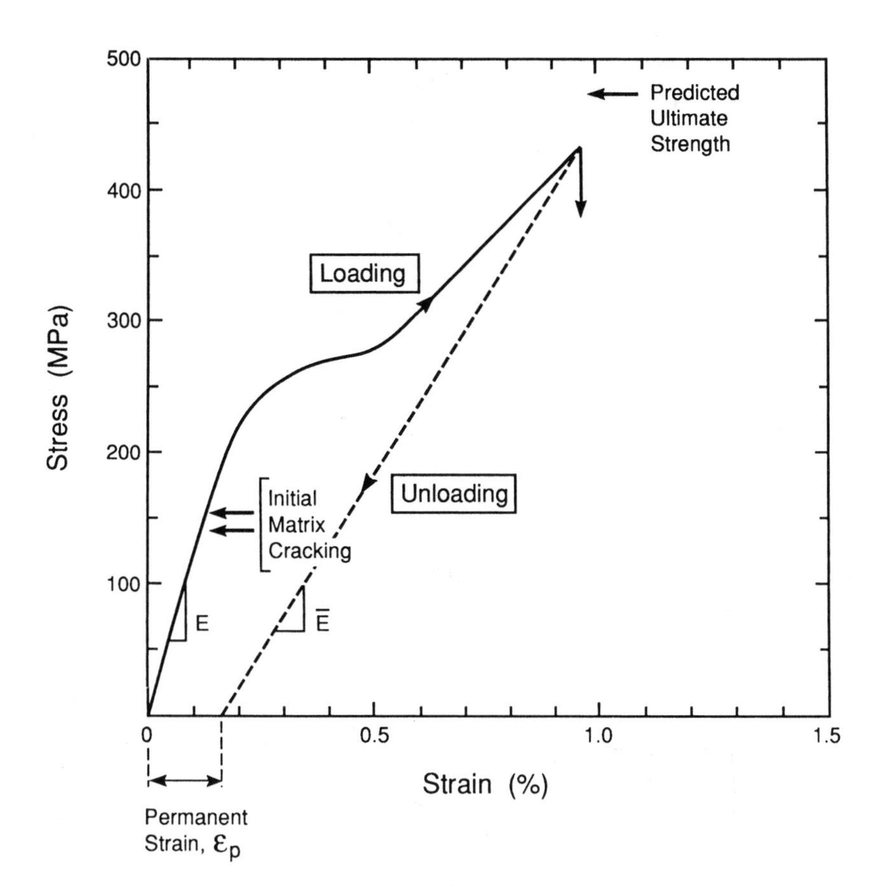

FIGURE 8.16. A tensile stress-strain curve for a 1-D CAS matrix composite showing the comparison between the measured ultimate strength and that predicted by the global load sharing analysis (Budiansky et al., 1986) ($\tau_0/fS_0 \approx 0.05$). Similar agreement (within 10%) has been shown for other CMCs with small τ_0/fS_0 (Curtin, 1991a; Heredia et al., (to be published)).

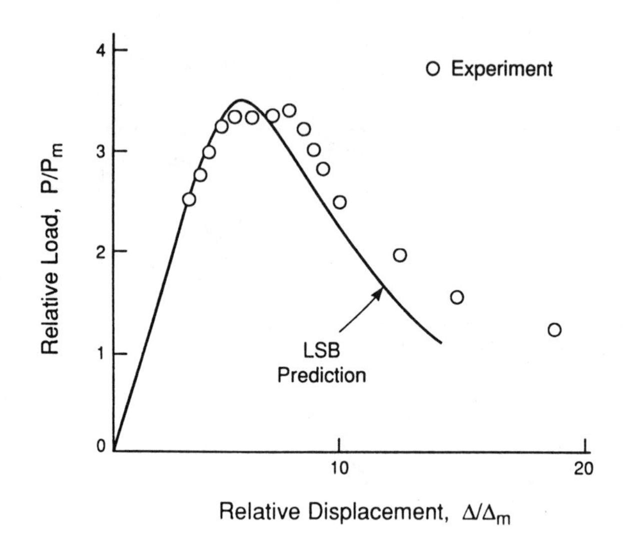

FIGURE 8.17. A load/deflection curve for a material for exhibits failure by the growth of a dominant crack. Also shown is a curve predicted from the known bridging traction law.

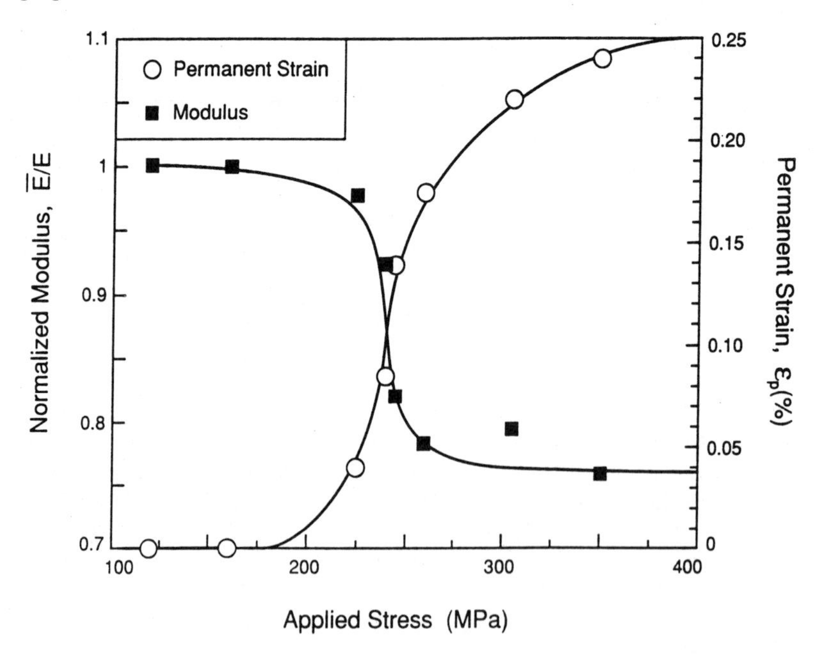

FIGURE 8.18. Changes in modulus and permanent strain measured for a 1-D CAS material.

theory can generally be obtained within a narrow range of τ, and the values of τ then compared with those obtained from alternate measurement techniques.

Most experimental studies have focused on Ti/SiC composites, as these exhibit sufficiently weak interfaces to allow matrix cracking to proceed with minimal fiber failure. Examples of crack growth curves in one such material are shown in Fig. 8.19. Comparisons with model predictions prior to fiber fracture suggest that τ is in the range of 30–60 MPa. Furthermore, the onset of fiber failure is consistent with a fiber strength of 3–4 GPa, which is in the range of values measured on individual fibers (Chawla, 1987). Transitions to fiber failure have been observed with increasing notch length and stress amplitude (Walls et al., 1991; Walls and Zok, 1991), and found to be in broad agreement with the predictions of Fig. 8.14.

A notable feature of the crack growth curves pertains to the assumption that τ remains constant throughout the entire loading history. During the early stages, the crack growth behavior is consistent with a relatively high value of τ ($\sim$ 50–60 GPa). As cracking proceeds, the growth rates in some instances exceed the predicted values (for the same value of τ), suggesting that the interface sliding stress is degraded and thus the shielding contribution ΔK_F is reduced. The data in this regime are consistent with a lower average value of τ, typically $\sim$ 30 MPa.

As a further check, the sliding stress τ has been measured using single fiber push-through tests (Warren et al., (in press)). Tests have been conducted on both pristine fibers and fibers located adjacent to a matrix fatigue crack. The results indicate that τ does indeed degrade with cyclic loading, from an initial value (obtained on pristine fibers) of $\sim$ 90 MPa to $\sim$ 20 MPa following $\sim 10^5$ loading cycles. These are consistent with the "average" values inferred from the fatigue tests (30–60 MPa) and provide confidence in the utility of the micromechanical model described in Section 8.4.4. It must be emphasized, however, that the process of interface degradation during cyclic loading is *not* yet understood, and thus extrapolation of the model predictions into other regimes (characterized by different values of $\Delta\sigma$, $\mathcal{R}$, a_0, etc.) may provide incorrect trends. It is anticipated that such changes in τ and the associated changes in slip length will also influence both the location of fiber failure relative to the matrix crack plane, as well as the values of σ_b at fiber failure. Additional study is required to establish the important connections between loading history, the statistics of fiber failure and the composite behavior.

8.6 Concluding Remarks

The majority of fiber-reinforced composites develop matrix cracks upon either monotonic or cyclic loading. These cracks have a major influence on the macroscopic properties of the composites. The modes of cracking are in-

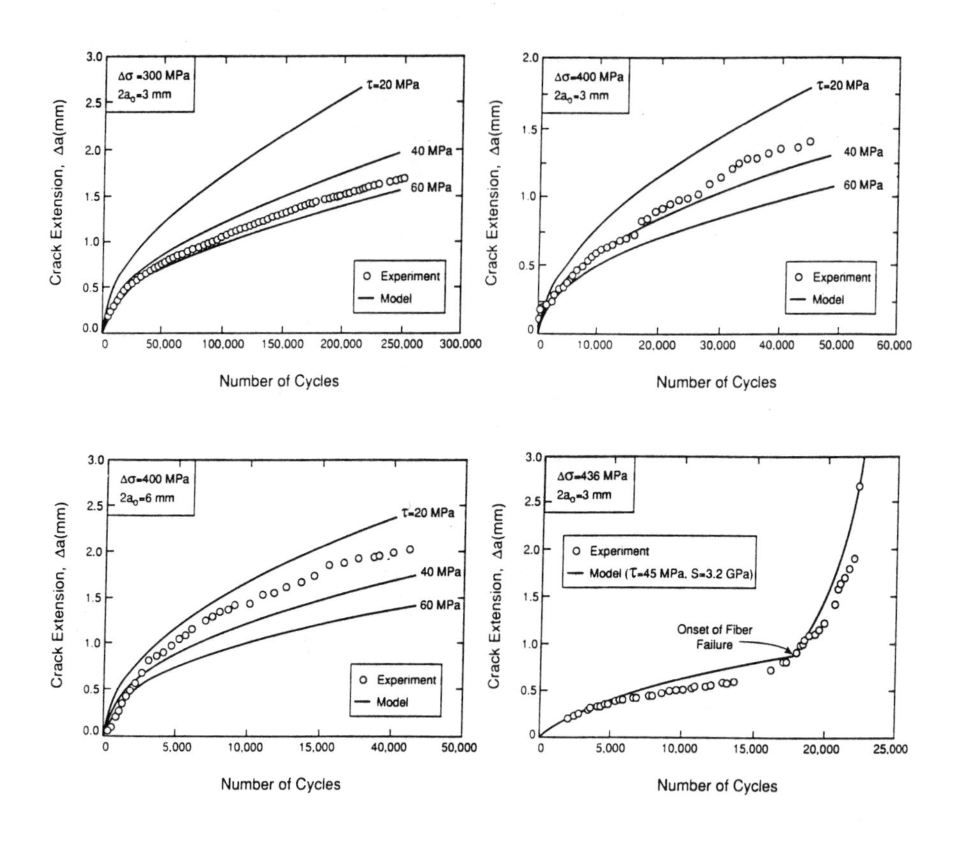

FIGURE 8.19. Comparisons of measured and predicted fatigue crack growth curves for a Ti-15V-3Cr-3Al-3Sn alloy reinforced with 35% unidirectional SiC (SCS-6) fibers: (a-c) no fiber failure in the crack wake; (d) with fiber failure.

TABLE 8.1. Non-Dimensional Parameters

Relative Stiffness	$\xi \rightarrow fE_f/(1-f)E_m$
Relative Sliding	$T \rightarrow \left[\frac{f\tau_0\xi}{\sigma}\right]^{1/2}$
Cyclic Sliding	$\Delta T \rightarrow \left[\frac{f\tau_0\xi}{\Delta\sigma}\right]^{1/2}$
Applied Stress	$\Sigma \rightarrow \left[\frac{2R\sigma}{f\xi a\tau_0}\right]$
Cyclic Stresses	$\Delta\Sigma \rightarrow \left[\frac{2R(\Delta\sigma)}{f\xi a\tau_0}\right]$
	$\Delta\Sigma_0 \rightarrow \left[\frac{2R(\Delta\sigma)}{f\xi a_0\tau_0}\right]$
Bridging	$\Sigma_b \rightarrow \left[\frac{2R\sigma_b}{f\xi a\tau_0}\right]$
Cyclic Bridging	$\Delta\Sigma_b \rightarrow \left[\frac{2R(\Delta\sigma_b)}{f\xi a\tau_0}\right]$
Characteristic Length[11]	$\delta_c \rightarrow \left[S_0 R L_0^{1/m}/\tau_0\right]^{m/(m+1)}$
Characteristic Strength[11]	$S_c \rightarrow [S_0^m \tau_0 L_0/R]^{1/(m+1)}$

[11] These parameters have dimensions but combine to form non-dimensional parameters

timately coupled with the statistical failure properties of the fibers and the nature of the interfaces. This chapter has summarized present understanding about the coupling between the mechanics and statistics, having origins in the pioneering work of Frank McClintock. A general conclusion is that problems related to the evolution of individual mode I matrix cracks (in either monotonic or cyclic loading) are rather well understood. Additionally, ultimate failure controlled by the fibers is well-comprehended when the interfaces are sufficiently "weak" by virtue of their debonding and frictional sliding characteristics that *global load sharing* applies. Phenomena that are not as well understood include: multiple mode I cracking (Zok and Spearing, (in press)), ultimate failure by local load sharing when the interfaces are relatively "strong" (Argon, 1972; Hu et al., (to be published)), and mixed-mode crack growth (Suo and Hutchinson, (to be published)).

Nomenclature

a	crack length
a_c	critical flaw size for steady-state
a_m	fracture mirror radius

a_0	initial flaw size or notch size
Δa	crack extension
d	crack spacing
$\bar{d}$	average crack spacing
$\bar{d}_S$	saturation crack spacing
E	Young's modulus of composite
E_m	Young's modulus of matrix
E_f	Young's modulus of fibers
$\bar{E}$	Young's modulus of composites with matrix cracks
f	fiber volume fraction
f_x	failed fiber fraction
h	fiber pull-out length
$\bar{h}$	mean pull-out length
K	stress intensity factor
K_f	critical stress intensity factors for fibers
K_m	critical stress intensity factor for matrix
K_0	initiation toughness
K_R	crack growth resistance
ΔK	stress intensity amplitude
n	fatigue growth rate exponent in Paris law
N	number of cycles
ℓ	slip length
L	gauge length
L_0	scale parameter
m	shape parameter for fiber strength
m_*	shape parameter for *in situ* strength
p	residual stresses normal to interface
P	load
P_S	survival probability
q	axial residual stress in matrix
R	fiber radius
$\mathcal{R}$	ratio of minimum to maximum applied stress
S	fiber strength

S_b	fiber bundle strength
S_c	*in situ* strength parameter
S_0	scale parameter for fiber strength
S_*	scale parameter for *in situ* strength
u	crack opening displacement
u_∞	crack opening caused by applied loads
w	width of component
$\mathcal{G}$	energy release rate
$\mathcal{G}_\infty$	applied energy release rate
$\mathcal{G}_{\mathrm{tip}}$	crack tip energy release rate
β	Paris law coefficient
σ	stress
σ_0	lower bound matrix cracking stress
σ_b	bridging stress
$\sigma_{\max}$	peak stress
σ_c	matrix cracking stress
σ_u	ultimate strength of composite
σ_∞	applied stress
$\Delta\sigma$	stress amplitude
$\Delta\sigma_b$	bridging stress amplitude
Δ	displacement
$\Delta_{\max}$	maximum displacement
ν	Poisson's ratio
δ_c	characteristic gauge length for fibers
Γ	fracture energy
Γ_f	fracture energy of fibers
Γ_m	matrix fracture energy
Γ_R	crack growth resistance
τ	sliding stress
τ_0	sliding stress caused by asperities
τ_c	transition sliding stress
ε_p	permanent strain
μ	friction coefficient
Φ	probability density function

References

Argon, A. S. (1972). *Treatise On Materials Science*, Academic Press.

Aveston, J., Cooper, G. A., and Kelly, A. (1971). *The Properties of Fiber Composites JPC.* pages 15–26.

Bao, G. and McMeeking, R. M. (to be published). *J. Mech. Phys. Solids.*

Beyerle, D., Spearing, S. M., and Evans, A. G. (to be published).

Beyerle, D., Spearing, S. M., Zok, F. W., and Evans, A. G. (in press). *J. Am. Ceram. Soc.*

Bowling, J. and Groves, G. W. (1979). *J. Mater. Sci.*, 14:43.

Budiansky, B., Hutchinson, J. W., and Evans, A. G. (1986). *J. Mech. Phys. Solids*, 34:164.

Cao, H. C., Bischoff, E., Sbaizero, O., Rühle, M., Evans, A. c., Marshall, D. B., and Brennan, J. J. (1990). *J. Am. Ceram. Soc.*, 73:1691.

Chawla, K. K. (1987). *Composite Materials Science and Engineering*, Springer-Verlag, NY.

Corten, H. T. (1967). *Modern Composite Materials* (Brontman, L. J. and Krock, R. H., editors), Addison, page 27.

Cox, B. N. (in press). *Acta Metall. Mater.*

Cox, B. N. and Lo, C. (in press). *Acta Metall. Mater.*

Cox, B. N. and Marshall, D. B. (1991). *Acta Metall. Mater.*, 39:579.

Curtin, W. A. (1991a). *J. Mater. Sci.*, 26.

Curtin, W. A. (1991b). *J. Am. Ceram. Soc.*, 74, 2837.

Daniels, H. E. (1945). *Proc. Roy. Soc.*, A183:405.

Evans, A. G. (1983). *Met. Trans.*, A14:1349.

Evans, A. G. and Marshall, D. B. (1989). *Acta Metall.*, 37:2567.

Evans, A. G. (1990). *J. Am. Ceram. Soc.*, 73:187.

Evans, A. G., Zok, F. W., and Davis, J. B. (1991). *Composite Science and Technology*, 42:3.

Evans, A. G. (in press). *Mat. Sci. Eng.*

Freudenthal, A. (1967). *Fracture* (Liebowitz, H., editor), Academic Press.

He, M.-Y. and Hutchinson, J. W. (1989). *International Journal Solids Structures*, 25:1053

Heredia, F. E., Spearing, S. M., Evans, A. G., Curtin, W. A., and Mosher, P. (to be published). *J. Am. Ceram. Soc.*

Hu, M. S., Cao, H. C., Yang, J. S., Mehrabian, R., and Evans, A. G. (to be published) *Met. Trans.*

Hutchinson, J. W. and Jensen, H. (1990). *Mech. of Mtls.*, 9:139.

Jamet, J. F., Lewis, D., and Luh, E. Y. (1984). *Ceram. Eng. Sci. Proc.*, 5:625.

Kim, R. Y. and Pagano, N. (1991). *J. Am. Ceram. Soc.*, 24:1082.

Mackin, T. J., Warren, P., and Evans, A. G. (to be published).

Marshall, D. B., Cox, B. N., and Evans, A. G. (1985). *Acta Metall.*, 33:2013

Marshall, D. B. and Evans, A. G. (1985). *J. Am. Ceram. Soc.*, 68:225.

Matthews, J. R., Shack, W. J., and McClintock, F. A. (1976). *J. Am. Ceram. Soc.*, 59:304.

McCartney, L. N. (1987). *Proc. Roy. Soc.*, A409:329.

McClintock, F. A. (1976). *Fracture Mechanics of Ceramics*, Bradt, R. C. et al., editor, Plenum, NY, 1:33.

McMeeking, R. M. and Evans, A. G. (1990). *Mech. of Mtls.*, 9:217.

Odette, G. R. and Chao, B. L. (to be published). *Acta Metall. Mater.*

Oh, H. L. and Finnie, I. (1970). *Intl. Jnl. Frac.*, 6:287.

Prewo, K. and Brennan, J. J. (1980). *J. Mater. Sci.*, 15:463.

Pryce, A. and Smith, P. (in press). *J. Mater. Sci.*

Ritchie, R. O., Knott, J., and Rice, J. R. (1973). *J. Mech. Phys. Solids*, 21:395.

Sensmeier, M. and Wright, K. (1989). *Proceedings TMS Fall Meeting*, Indianapolis, Law, P. K. and Gungor, M. N., editors, page 441.

Shang, J. K., Tzou, J. L., and Ritchie, R. V. (1987). *Met. Trans.*, A18:1613.

Singh, R. (1989). *J. Am. Ceram. Soc.*, 72:1764.

Suo, Z. and Hutchinson, J. W. (to be published). *Advances In Applied Mech.*

Sutcu, M. (1989). *Acta Metall.*, 37:651.

Tada, H., Paris, P. C., and Irwin, G. R. (1985). *The Stress Analysis of Cracks Handbook*, Del Research Corp., St. Louis.

Thouless, M. D. and Evans, A. G. (1980). *Acta Metall.*, 36:517.

Walls, D, Bao, G., and Zok, F. W. (1991). *Scripta Metall. Mater.*, 25:911.

Walls, D. and Zok, F. W. (1991). In Gungor, M. N., Lavernia, E. J., and Fishman, S. G., editors, *Advanced Metal Matrix Composites for Elevated Temperatures*, ASM, Metals Park, page 101.

Warren, P., Mackin, T. J., and Evans, A. G. (in press). *J. Am. Ceram. Soc.*

Zok, F. W. and Hom, C. L. (1990). *Acta Metall. Mater.*, 38:1895.

Zok, F. W., Sbaizero, O., Hom, C. L., and Evans, A. G. (1991). *J. Am. Ceram. Soc.*, 74:187.

Zok, F. W. and Spearing, S. M. (in press). *Acta Metall. Mater.*

9

Metal Fatigue — A New Perspective

K. J. Miller

ABSTRACT It is well understood that the fatigue limit behavior of a metal is a function of defect size within the Linear Elastic Fracture Mechanics (LEFM) regime. Conversely, it is not well understood how microstructural defects affect the fatigue limits of heterogeneous materials of apparently smooth specimens and engineering components. Consequently in the latter case investigations have concentrated on the cyclic stress-strain, or deformation, approach to fatigue fracture e.g. the Basquin and Coffin-Manson type studies.

By introducing micro-structural fracture mechanics and elastic-plastic fracture mechanics it is possible to link the deformation and the fracture approaches to metal fatigue investigations. This chapter considers these developments and their implications.

9.1 Introduction

The divide between death and everlasting life is the subject of this chapter. Under what conditions will a metal, component or structure survive indefinitely when subjected to cyclic forces, and what changes occur that introduce the possibility of failure?

Many trails have been followed to find answers to these questions. For example historically (perhaps based on the resemblance of fatigue fracture surfaces to brittle cleavage fracture in steel) it was incorrectly supposed that a material changed its structure due to the application of cyclic forces and became crystalline and brittle. In more recent times, studies of dislocations and the diffusion of small atoms through body-centered cubic structures led to theories and experiments to support ideas of dynamic strain ageing in ferrous materials.

At the present time investigations based on classical fatigue-endurance (S–N) curves are still hotly pursued in endeavors to find links between deformation behavior and fatigue failure. Sometimes the stress and strain range approaches have been combined to derive energy-based theories of fatigue, probably with the hope that below the fatigue limit a critical but as yet undefined state is attained in which all the available energy goes up

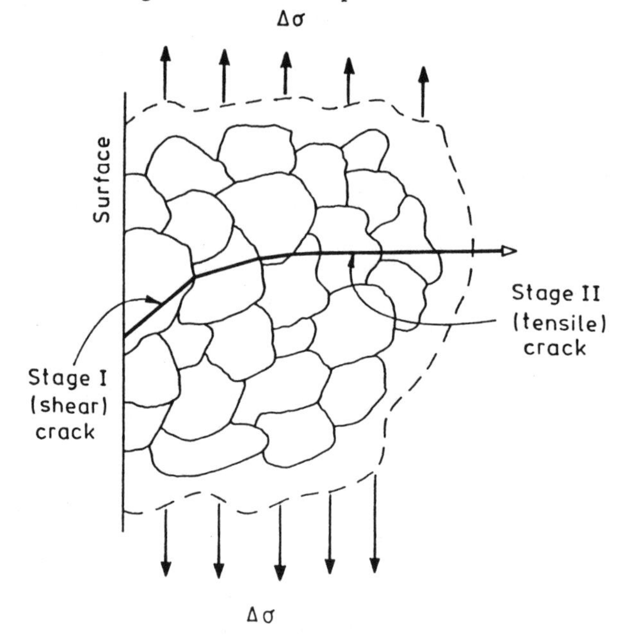

FIGURE 9.1. An initial shear (Stage I) crack developing into a tensile (Stage II) crack in a polycrystalline metal subjected to cyclic tensile stresses.

in smoke (heat) and none is available for the processes of fatigue.

Into this fiery arena came Frank McClintock (1956). In that 1956 conference in London at which world-wide experts met to discuss reasons for metal fatigue, only his paper out of a total 80 was related to the conditions required for the development and extension of a shear crack in a metal; an understanding of the Stage I crack had begun. Some five years later, Peter Forsyth (1961) presented his paper on Stage I and Stage II crack growth, see Fig 9.1.

Since metal fatigue failure under a cyclic tensile stress, $\Delta\sigma$, is caused by the growth of Stage I followed by Stage II cracking, both of the above quoted papers are central to this representation which concentrates on the behavior of cracks; first in relation to the orientation of cracks with respect to the applied stress or strain field, and second to the effect of crack size.

9.2 Orientation of Cracks

Figure 9.2 shows two identical stress or strain states. In particular Fig. 9.2(a) represents the critical zone of a component or structure while Fig. 9.2(b) is a laboratory simulation of that state. In both cases, the cyclic nature of the applied stresses is identical, as is the material, its processing route, the temperature, and the environment. Consequently the Tresca or

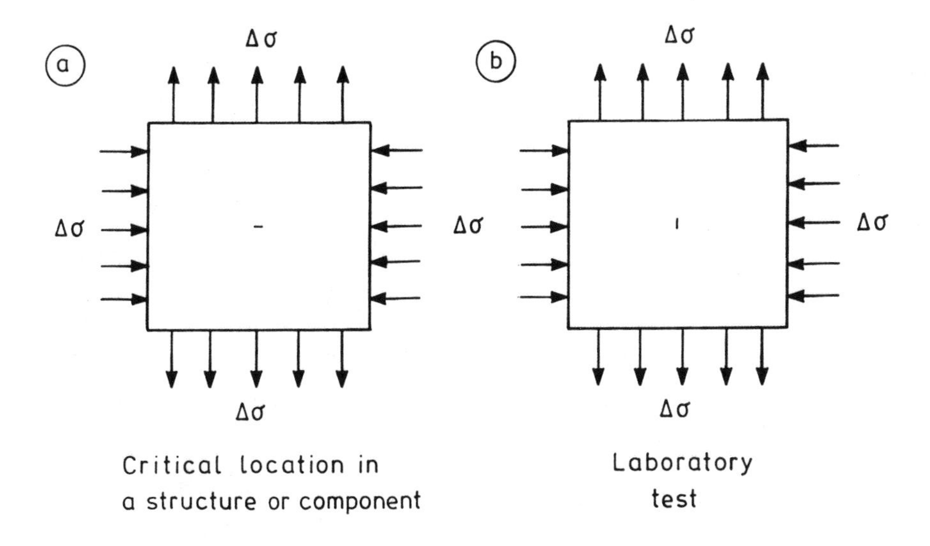

FIGURE 9.2. The importance of the orientation of the defect plane with respect to the direction of the applied stress field.

von Mises equivalent stresses are the same in both cases as are those calculated by any other theories. Also the cyclic energy input in both cases is identical and yet the component or structure depicted by Fig. 9.2(a) will fail whilst the laboratory simulation experiment will not fail. This indicates that von Mises, Tresca and energy theories etc., although possibly appropriate to the onset of yield deformation of a body subjected to complex stress, are not applicable to assessments related to fracture.

The difference between the two diagrams is that whilst both contain an unobserved central crack-like defect of say only 0.1 mm in length, the defect in Fig. 9.2(a) is horizontal whilst that in Fig. 9.2(b) is vertical, and very differently oriented with respect to the applied cyclic stress field. It follows that any fatigue fracture theory must recognize the orientation of the crack with respect to the orientation of the applied stress field.

However, even if this factor is taken into account, it is still not sufficient for a fatigue lifetime evaluation. For example in Fig. 9.3, which depicts two bodies suffering the same three-dimensional cyclically applied strain fields with $\Delta\varepsilon_1 > \Delta\varepsilon_2 > \Delta\varepsilon_3$, the orientation of both the Stage I and the Stage II cracks is identical, i.e. a Stage I crack develops on the plane of maximum shear which eventually turns into a Stage II crack that lies in a plane whose normal is in the direction of the maximum normal stress or strain; see Fig. 9.1. Nevertheless, despite their similarity, the sample shown in Fig. 9.3(a) will fail long before the sample illustrated in Fig. 9.3(b).

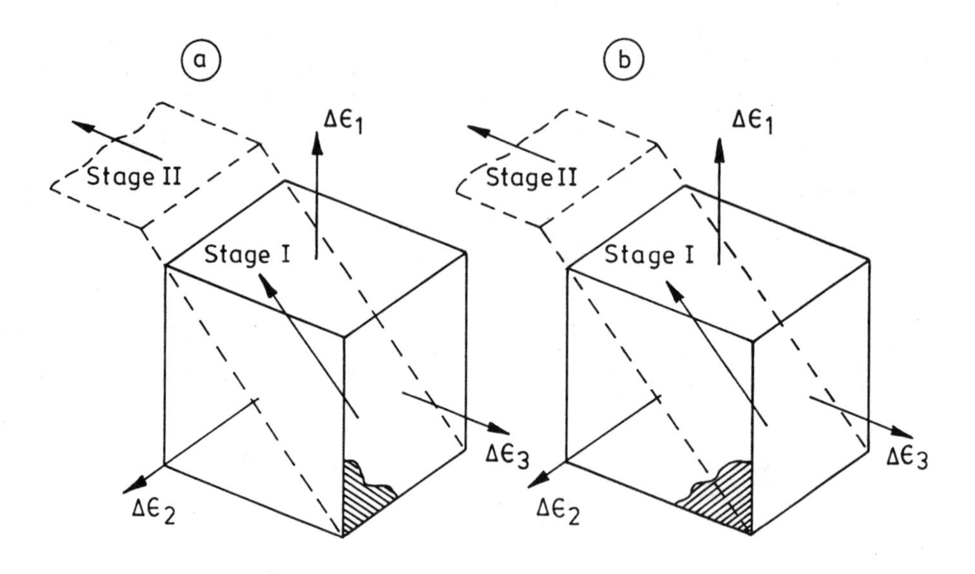

FIGURE 9.3. Identical cyclic stress-strain fields, but different crack growth directions with respect to the surface (shown shaded).

The difference here is that the free-surface position (indicated by the cross hatching) is not identical in these diagrams. Since most fatigue cracks start from or near a free surface, this is a critical consideration. In Fig. 9.3(a) the crack configuration is similar to an axial tension type of fatigue test while Fig. 9.3(b) is typical of a torsion test; the latter invariably lasting longer than a push pull test because the primary crack propagates in a direction that is parallel to the surface instead of into the depth of the material.

The first example quoted above indicates the importance of the orientation of the crack growth planes while the second indicates the importance of these planes with respect to the position of the free surface; both these aspects having to be considered with respect to the orientation of the three-dimensional applied cyclic stress or strain field.

For more information on multiaxial fatigue, readers are directed to the works of Miller and Brown (1985); Brown and Miller (1973, 1989); Kussmaul et al (1991).

9.3 Size of Fatigue Cracks

According to linear elastic fracture mechanics (LEFM) the longer the crack in a body, the weaker that body becomes, e.g. the lower the cyclic stress required to propagate it. For very long cracks, as used in conventional LEFM type tests, the stress levels are so low that it is difficult to observe and monitor crack-tip plasticity, even though such plasticity may possibly be extending across several grains. At the other end of the scale, when stress levels are very high and bulk plasticity is readily observed, it is difficult to observe cracks which may be growing very slowly across a single grain and measure only a few microns in length. In relation to the plasticity required to propagate such a crack, the question arises as to which comes first: the plastic deformation, say in the form of a persistent slip band (that is readily observed because of its size, e.g. $\sim 10^2$ microns in length); or the crack which may measure only 10^{-2} microns in length and be propagating very slowly, i.e. 10^{-3} microns per cycle. In my opinion it is impossible to categorically state that the persistent slip band (PSB) is generated before the crack. Indeed, I believe a crack-like defect is always present in polycrystalline metals, since they contain surface defects such as grain boundaries, triple points, inclusions/precipitates and surface machining marks, all of which act as micro-stress concentrators and from which micro-cracks can start propagating immediately and generate their own plastic zones. Thus the initial crack and the plasticity are generated immediately and simultaneously but only the plasticity is readily observable since it can be witnessed on any cross-section of a grain whilst the initial small crack cannot be so readily located and monitored.

Such a small crack can propagate if the shear stress is above the critical value to activate a favorably orientated slip system of a single grain on the metal surface. If the shear stress is at the critical value, however, the crack will stop propagating if a twin boundary is reached, since the resolved stress on the newly orientated slip system across the twin boundary will most likely be lower than the critical value for propagation.

The twin boundary thus represents a weaker barrier to crack growth and a fatigue limit condition, i.e. the crack growth rate reduces to zero. By increasing the applied stress, the crack will now continue propagation until it reaches the grain boundary where it will stop again thereby forming a new fatigue limit condition; see Fig. 9.4 (Jacquet 1956; Zhang 1991). The reason why the first grain boundary to be reached represents a strong barrier is because the crack front is now surrounded by several grains and is therefore required to propagate on several differently oriented planes and in several different directions under the action of a resolved shear stress that is equal to, or less than, the stress propagating the crack across the first grain. Furthermore, the size of the first grain to be cracked is invariably the largest grain, since this provides the best conditions for dislocation activity and a greater distance to the restraining barrier, hence the surrounding grains are usually much smaller and therefore provide a greater resistance to continued propagation.

The fatigue limit represented by the first grain boundary can be eliminated by increasing the stress level, thereby permitting the crack to grow across near neighbor grains, albeit in a somewhat irregular orientation; see

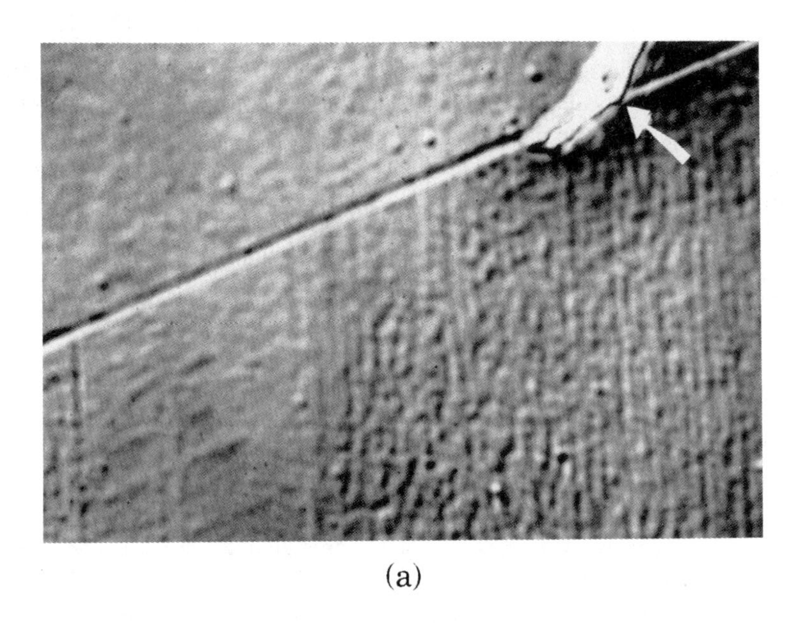

(a)

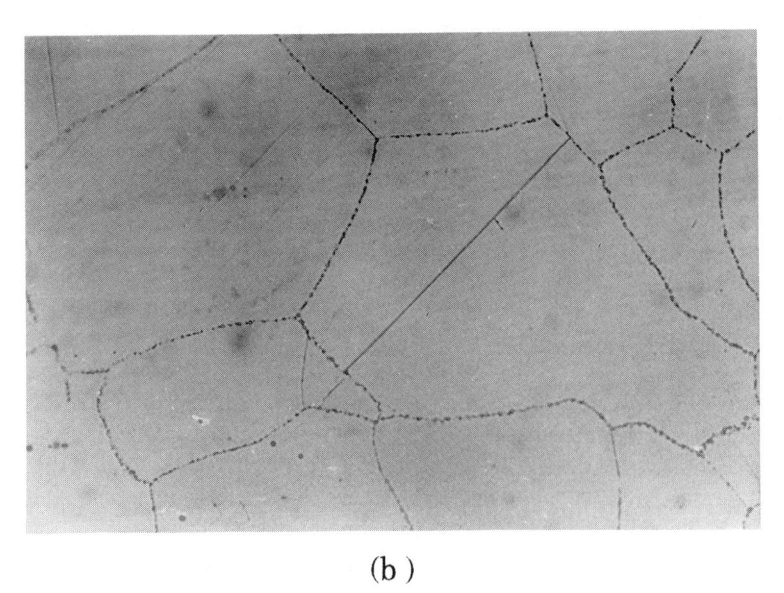

(b)

FIGURE 9.4. The influence of microstructure on fatigue crack growth: (a) a crack arrested at a twin boundary in Brass (Jacquet, 1956); (b) a crack arrested at a grain boundary in Waspaloy (Zhang, 1991).

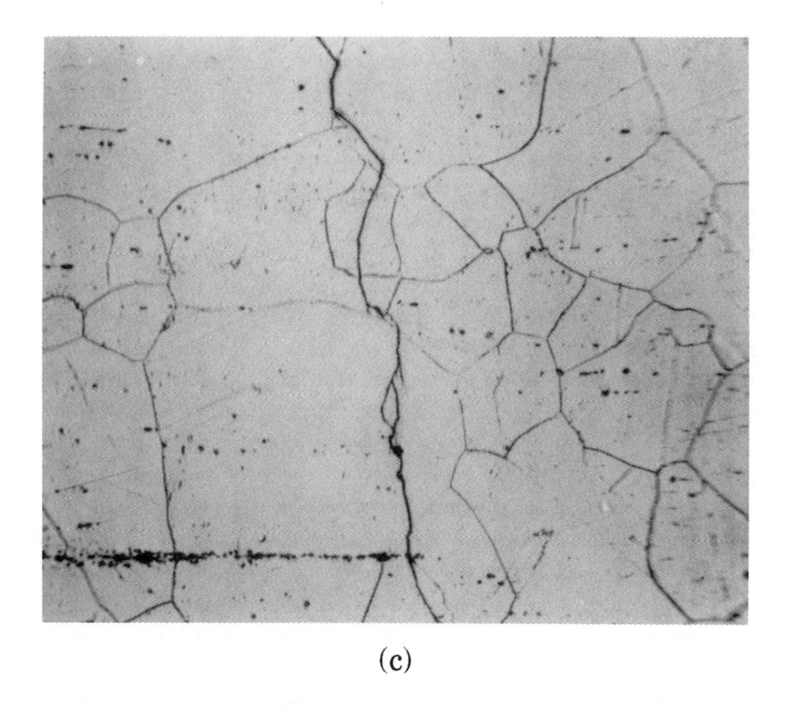

(c)

(d)

FIGURE 9.4. (Con't) (c) crack growth affected by several grain boundaries in Nimonic 90 (Allen and Forrest, 1956); and (d) a crack arrested at a pearlite zone in a medium carbon Steel (de los Rios et al., 1985).

Fig. 9.4(c), which also exhibits segments of grain boundary cracking plus Stage I transcrystalline facets (Allen and Forrest, 1956). Eventually a further and ultimate barrier may be reached that represents the fatigue limit of the microstructural system, e.g. a pearlite band, see Fig. 9.4(d) (de los Rios, et al, 1985). To breach such a strong barrier to crack propagation, it is required to once again increase the applied stress level. This new and higher stress level, together with the distance between pearlite bands, represents the fatigue limit conditions of this particular material.

The fatigue limit is therefore characterized by the maximum non-propagating crack size. This is easy to determine experimentally in large grained microstructures such as annealed mild and medium carbon steels, but is more difficult in small-grained heat-treated high strength steels. In recent experiments (Heyes, 1991) however, we have succeeded in isolating a 12μm sized crack in a high strength steel at a cyclic stress level just below the fatigue limit.

Very small Stage I cracks are difficult to distinguish because, as shear cracks, they remain closed, but with the advent of the acoustic microscope it is now possible to distinguish such cracks, even if they are just below the surface. Figure 9.5 presents two acoustic-micrographs. In Fig. 9.5(a), the crack is seen to be propagating along a twin boundary towards a grain boundary and with PSBs radiating from the crack tip at approximately 60 degrees to the crack plane. Figure 9.5(b) shows a different sample with the crack changing direction after crossing the grain boundary; it then follows the orientation of PSBs. It should be noted that the acoustic images can distinguish between PSBs and the crack, the latter giving a broader acoustic signal because of the sharp reflective nature of crack surfaces.

9.4 Fatigue Limits of Notches

The differences between notches and cracks and between the behavior of propagating and non-propagating cracks are essential features in understanding the fatigue limits of materials and the fatigue limits of components containing geometrical discontinuities. As stated previously, the fatigue limit of a metal is related to its microstructure while the fatigue limit of a component containing a notch can be dependent on the geometry of loading and shape. Additionally as the cyclic stress range increases, the maximum length of a non-propagating crack in an unnotched material decreases. Conversely, the maximum length of a non-propagating crack at a notch root increases as the cyclic stress range increases. These examples illustrate why the local-strain approach to the behavior of notches is fraught with difficulties. Perhaps the best way to study the apparently irreconcilable differences between the material cyclic deformation approach and the fracture mechanics approach to the behavior of notches is to examine the fatigue limit conditions of notched specimens.

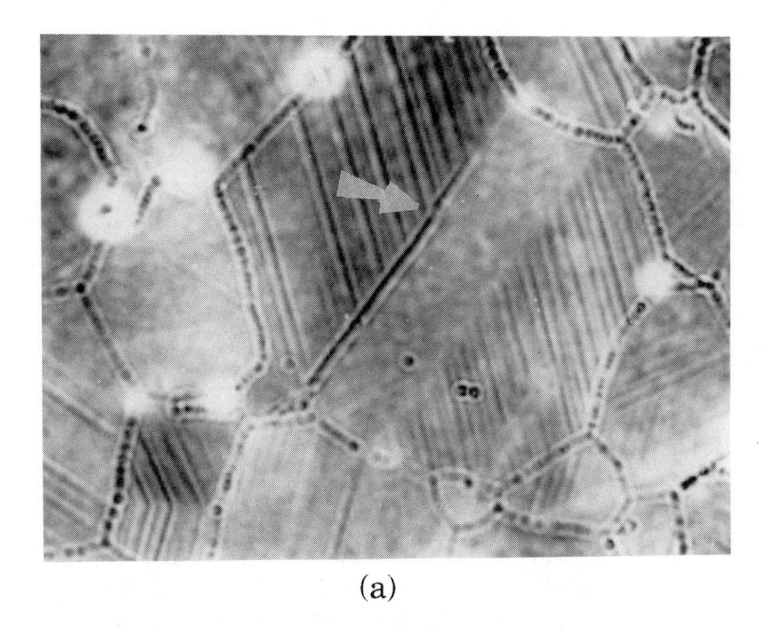

(a)

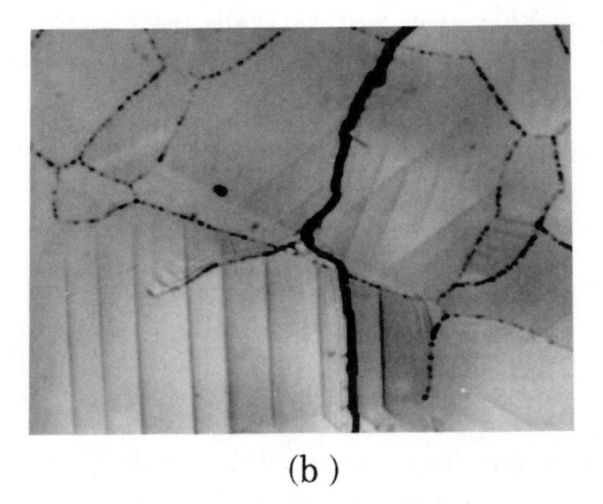

(b)

FIGURE 9.5. Acoustic micrographs of Waspaloy showing grain boundaries and persistent slip bands: (a) a fatigue crack within a single grain, following a twin boundary; and (b) a fatigue crack across two grains and including a grain boundary segment.

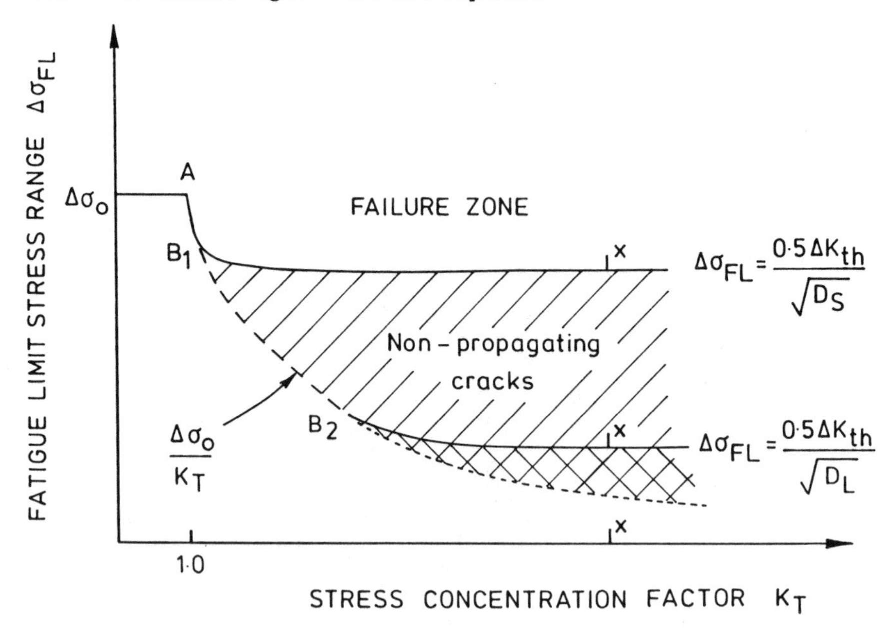

FIGURE 9.6. The effect of notch depth (D) shape (K_T) and threshold stress intensity range (ΔK_{th}) on fatigue limit conditions.

Possibly the first study to illustrate the retardation behavior of cracks at notches at values of ΔK less than ΔK_{th} (the threshold value of the LEFM stress intensity range) was that of Hammouda and Miller (1979). This followed earlier work of Smith and Miller (1977, 1978) who showed that if the LEFM ΔK_{th} value was eventually reached, the crack could not become non-propagating and fatigue failure must occur. Figure 9.6 collates this information. Of interest are the following features:

1. The boundary condition between failure and non-failure of sharp notches is not dependent on the value of the elastic stress concentration factor, K_T (Frost and Phillips, 1956).

2. The fatigue limit condition of notches with a stress concentration $K_T = X$ in Fig. 9.6 is dependent on the depth of the notch which will be different for large components (D_L) and small notches (D_S) that simulate the component in the laboratory.

3. In zone A–B_1 (small specimens) or zone A–B_2 (large specimens) the fatigue limit is dependent on material behavior, while at higher stress concentrations it is mainly dependent on geometrical conditions.

4. When designing large rotors, especially those subjected to high mean stresses which reduce the value of ΔK_{th}, the lowest value of the fa-

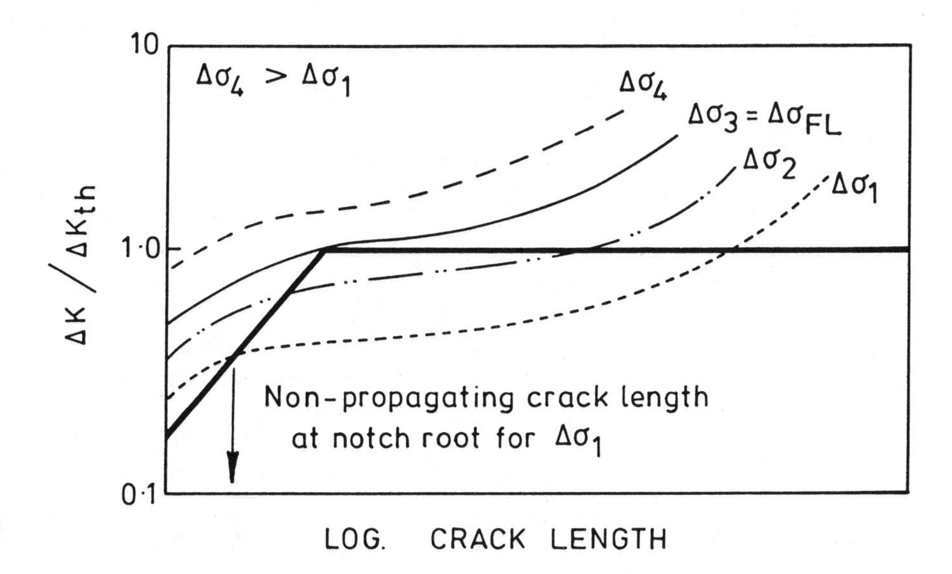

FIGURE 9.7. The variation of the crack tip stress intensity factor with crack length for various levels of cyclic stress applied to a notched specimen. The thick black line represents the fatigue limit condition of a given material.

tigue limit calculated by the two equations shown in Fig. 9.6 should be employed.

In order to understand the fatigue limit conditions of notches the difference between the response of a specific material and the general response of notches to cracks of different length is shown in Fig. 9.7. The fatigue limit condition ΔK_{FL} of the material, i.e. $da/dN = 0$, is shown by the thick black line in Fig. 9.7 in terms of a reformulation of the Kitagawa-Takahashi (1976) diagram. The other curves indicate the increase of applied values of the stress intensity factor range, ΔK_{app} as the crack length at the notch root increases, for various values of constant cyclic stress ranges. When $\Delta K_{app} > \Delta K_{FL}$ of the material, the crack will propagate and when ΔK_{app} attains the value of ΔK_{FL} of the material, the crack will stop. Initially cracks will propagate at notch roots but if the applied stress is too low, they will eventually be arrested. Note that the higher the applied stress, the longer will be the non-propagating crack. However, at $\Delta K = \Delta K_{th}$ the reverse applies, i.e. the length of the non-propagating crack decreases as the applied stress level increases.

Good correlation of experimental data with this theoretical appreciation of the behavior of cracks at notches for a variety of notch shapes has been provided by Yates and Brown (1987).

9.5 The Fatigue Limit of a Material

From the foregoing sections, we can now consider how the fatigue limit is influenced by both the material and defect size.

Since the fatigue failure process is entirely related to cracks, let us first examine Fig. 9.8 which shows how cracks grow throughout the lifetime. In Fig. 9.8(a), three different starter crack lengths are shown, namely a_o equals 10mm, 0.1mm and 0.001mm. The cyclic driving forces which can just propagate these cracks to failure are shown as ΔF_1, ΔF_3, and ΔF_5, requiring linear elastic fracture mechanics, elastic plastic fracture mechanics and microstructural fracture mechanics respectively to describe their growth which are related to the extent of plasticity at the crack tip and in the latter case the involvement of relevant microstructural features.

The fatigue limit for each case is equated to the cyclic driving forces ΔF_2, ΔF_4, and ΔF_6 which cannot quite cause the respective crack size to propagate.

Here we want to pursue the important effect of cyclic stress levels $\Delta\sigma_1$, $\Delta\sigma_2$, $\Delta\sigma_3$, $\Delta\sigma_4$, $\Delta\sigma_5$, introduced in Fig. 9.8(b), and their ability or inability to overcome various microstructural barriers such as twin boundaries, grain boundaries and strong second phase constituents as can be found e.g. in pearlite, these being represented by distance b_1, b_2, and b_3 respectively. As can be seen in Fig. 9.8(b), the stress level $\Delta\sigma_1$ cannot quite propagate the crack through the barrier at distance b_1 and hence represents a fatigue limit. A similar argument applies for $\Delta\sigma_2$ and $\Delta\sigma_3$. However, $\Delta\sigma_4$ can just overcome the strength of the strongest barrier in the microstructural system while $\Delta\sigma_5$ propagates the crack at a faster rate.

Of particular note in the response types presented in Fig. 9.8 are the following points:

1. In the LEFM regime, the crack growth rate is of the conventional form which is normally expressed in terms of the range of the elastic stress intensity factor.

2. In the Elastic-Plastic Fracture Mechanics (EPFM) region, crack growth invariably follows the type of equation first represented by Tomkins (1968).

3. In the microstructurally-influenced zone the crack grows immediately but at a decreasing rate as it approaches various microstructural barriers. Additionally, an exceedingly large proportion of the fatigue lifetime is spent overcoming the strength of the most effective microstructural barrier to crack propagation.

The fundamental $\underline{a}$ versus $\underline{N}$ curves of Fig. 9.8(b) can be translated into diagrams of the type shown in Fig. 9.9. In Fig. 9.9(a) the fatigue limit is shown as a function of crack size versus cyclic stress range and is given

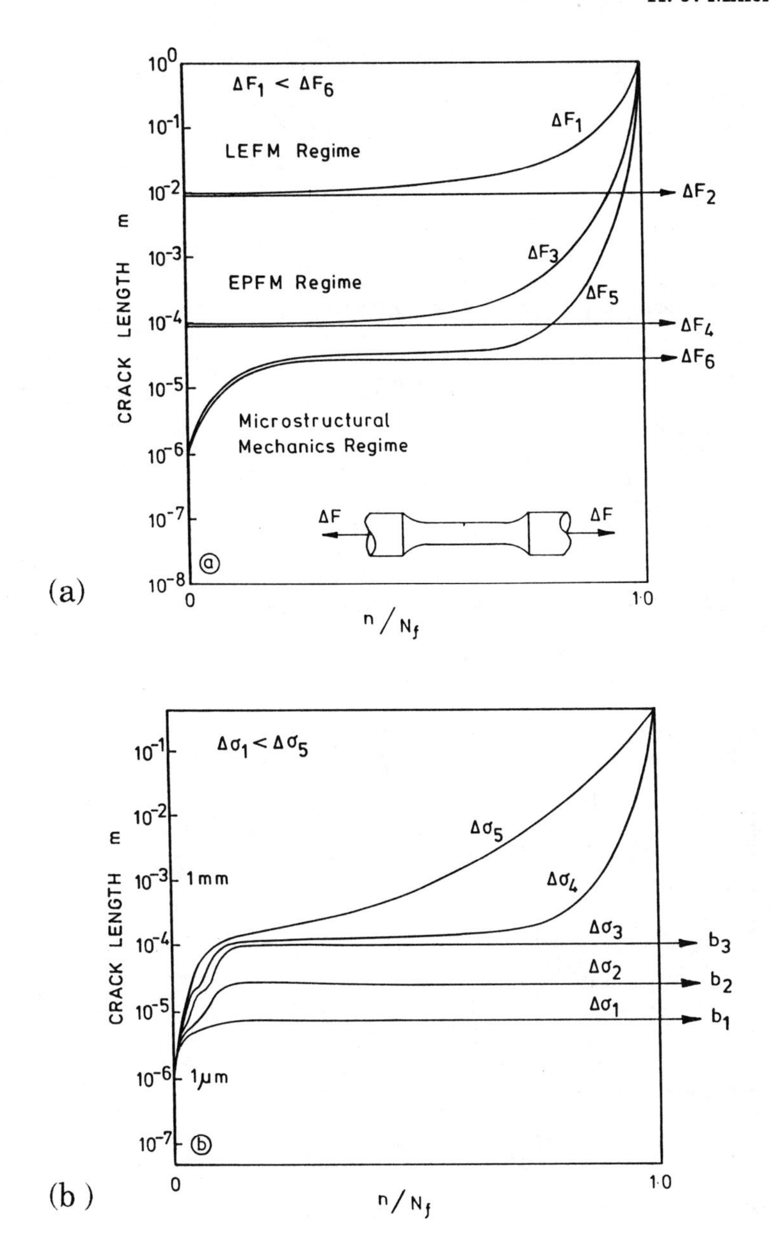

FIGURE 9.8. Crack growth behavior as a linear function of applied cycles normalized with respect to the number of cycles to failure: (a) Long, low-stress and short, high-stress cracks and their associated fracture mechanics regimes. (b) Short crack growth behavior as influenced by microstructural barriers of different strengths and spacing.

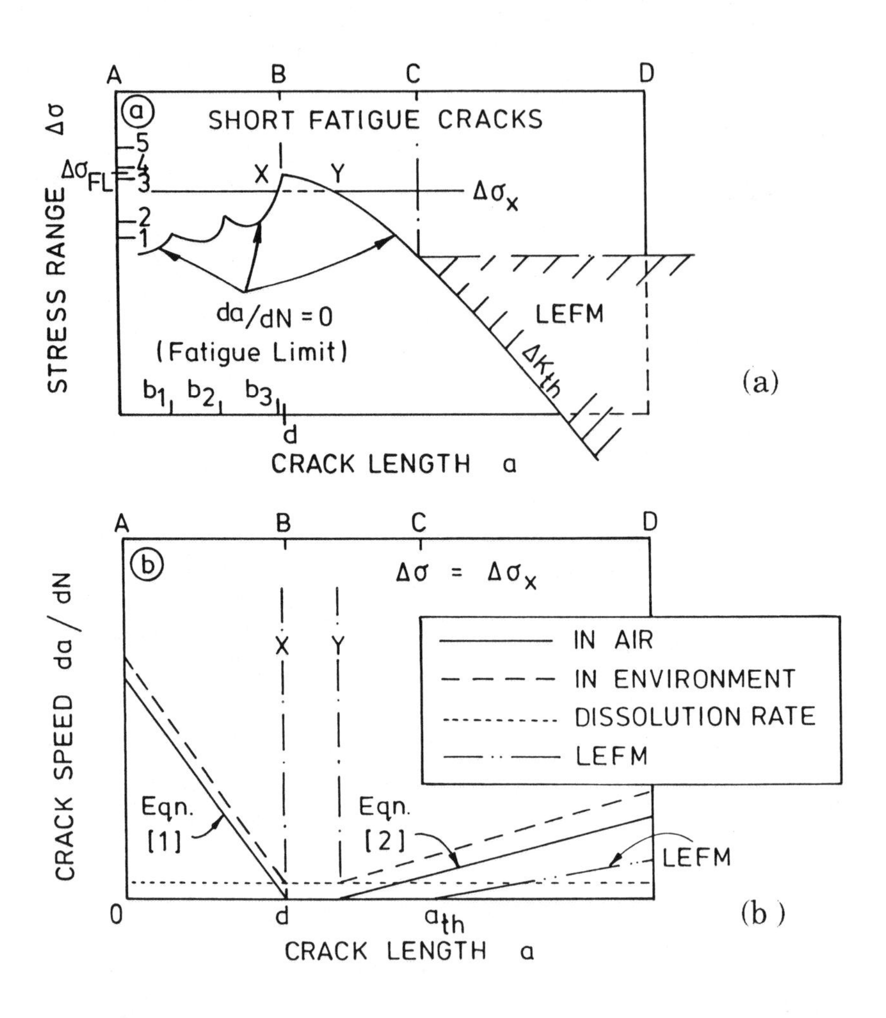

FIGURE 9.9. Short fatigue crack behavior: (a) the fatigue limit as a function of crack length and stress range, and (b) crack growth rates in different environments.

by the contour $da/dN = 0$, while in Fig. 9.9(b) the curves exhibiting the growth rate behavior of short cracks at a stress level $\Delta\sigma_x < \Delta\sigma_{FL}$ are presented. For a more detailed explanation of these curves see the studies of Miller (1985, 1991); Miller and de los Rios (1986,1991). Barriers b_1, b_2 and b_3 for cyclic stress ranges $\Delta\sigma_1$, $\Delta\sigma_2$ and $\Delta\sigma_3$ correspond to those of Fig. 9.8(b). The fatigue limit expressed by ΔK_{TH} has an upper bound limit of approximately 0.7 $\Delta\sigma_{FL}$ where $\Delta\sigma_{FL}$ is approximately equal to the cyclic yield stress range $\Delta\sigma_{cy}$. Above this level the LEFM assumption of small scale yielding no longer applies.

The equations of fatigue crack growth depicted in Fig. 9.9(b) are given for the cyclic stress range $\Delta\sigma_x$. The general form however may be expressed as

$$da/dN = A\Delta\gamma^{\alpha}\,(d - a) \tag{9.1}$$

for the microstructurally influenced Stage I crack growth phase. In equation 9.1, a is the crack length, A and α are material constants and d is the barrier distance associated with the fatigue limit. Clearly, one can substitute for the shear strain range $\Delta\gamma$ the stress range (tension or shear) using the appropriate constitutive relationships.

When the crack at $\Delta\sigma_x$ becomes non-propagating, i.e. $a = d$, if it becomes possible, in some way for the crack to extend by the distance X–Y shown in Fig. 9.9, then the crack will have escaped the environment of the barrier and would be able to propagate anew according to the following expression:

$$da/dN = B\Delta\gamma^{\beta}a - C \tag{9.2}$$

where B and β are material constants and C represents a threshold condition for the eventual production of a Stage II crack. Two models (Hobson, 1985; Navarro and de los Rios, 1987) are currently employed to characterize this regime in the work described here although alternatives are being examined, e.g. the work of Tanaka et al (1991). It will be noted that the LEFM curve provides a lower bound solution and Fig. 9.9(b) explains why any threshold condition can be found in the range $d \leq a \leq a_{th}$.

The model derived and developed by Brown and Hobson, (see Hobson, 1985), is based on experimental results and hence averages out experimental scatter due to microstructural variations, while that due to Navarro and de los Rios (1987) is theoretical and is based on the mobility of dislocations. The latter, although more complex, explains the reasons for the experimental scatter of short fatigue crack growth rates; as explained in Fig. 9.10. For example this Figure shows that the short fatigue crack can continue to slow down and re-accelerate as it crosses successive microstructural barriers following the penetration of the strongest barrier, but with a decreasing influence of microstructure as the crack grows longer. The interesting point of this model is that the experimental scatter of short crack growth data due to differences in grain size is accommodated between the upper and lower limit curves of the theory. The regime of decreasing microstructural

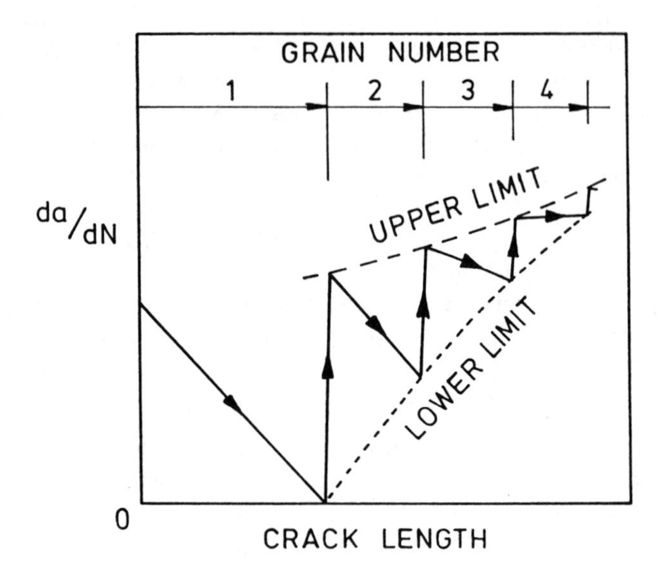

FIGURE 9.10. Schematic of the Navarro-de los Rios model, showing crack speed variations due to microstructural barriers to crack growth.

influence is the zone BC of Fig. 9.9(a) and is sometimes referred to as the physically small crack growth phase in which the crack, in axial tension loading changes to Stage II. Sun et al (1991) have successfully modelled experimental data due to Lukas et al (1989) to match the non-linear slope of this part of the fatigue limit characteristic.

9.6 Environmental Effects

It is well known that by introducing an aggressive environment, the fatigue limit in air can be eliminated, see Fig. 9.11. Since the fatigue limit is associated with short crack growth behavior, it is necessary to understand how microstructural barriers to crack growth are influenced by an aggressive environment. In this respect, both the Brown-Hobson (B–H) model and the Navarro-de los Rios (N–R) model have been used to quantify environmental effects on short crack growth behavior (see Akid, 1987; and Akid and Miller 1990a, b).

Figure 9.12 compares short crack growth data for two environments. Tests on shallow hour-glass shaped specimens were conducted under reversed torsion control at 4 Hz on a BS 4360 50B steel having a chemical composition (% wt) of 0.18 C, 0.11 Si, 0.83 Mn, 0.024 P, 0.036 S and Nb/V less than 0.020, remainder being ferrite. All fatigue tests were performed at

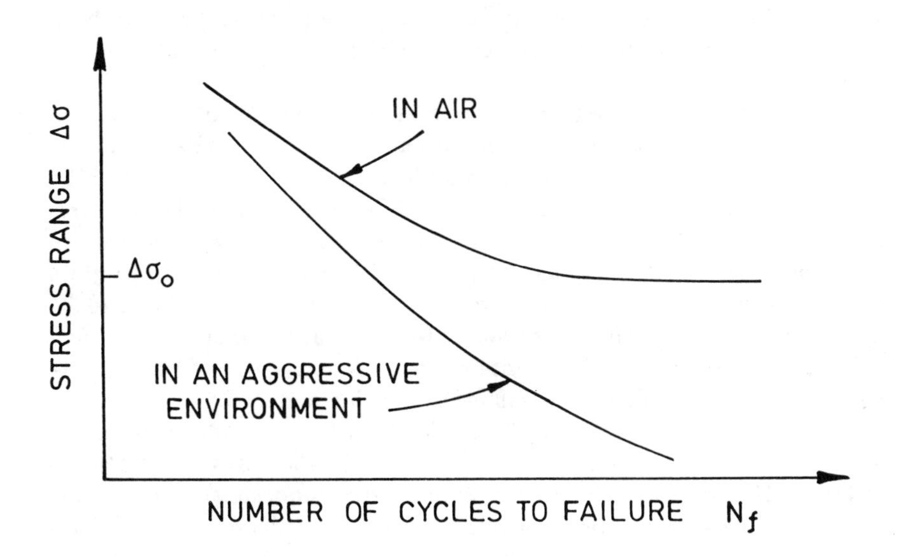

FIGURE 9.11. The effect of environment on fatigue endurance.

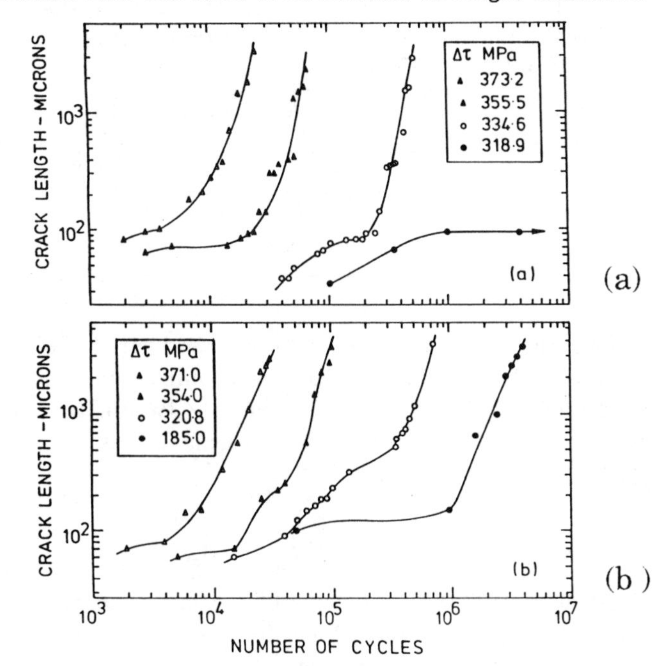

FIGURE 9.12. The effect of environment on short fatigue crack growth at various shear stress ranges (a) in laboratory air, and (b) in an 0.6M NaCl environment.

19–20°C with crack growth being monitored via a plastic-replication technique, using a microscope plus camera attachment and a pre-calibrated TV screen.

In Fig. 9.12(a) the arrest of the crack in air at $\Delta\tau$ (shear stress range) equal to 318.9 MPa is because the fatigue limit is 325 MPa. From Fig 9.12(b) the crack is seen to grow at a cyclic stress level as low as 185 MPa because of the influence of the 0.6M NaCl environment. In both figures the effect of the barrier is clearly shown with cracks slowing down and sometimes arresting at lengths in the range 70 to 110 microns. The grain size of the material was 15 microns in the transverse direction and 35 microns in the longitudinal direction. It follows that the short cracks could easily propagate through the ferrite grains only to be arrested at pearlite zones which had a volume fraction of 0.14 in this material.

Of interest is the fact that the data shown in Fig. 9.12 illustrates negligible environmental effects at the higher stress levels of 373/371 MPa and 355/354 MPa, probably due to corrosion deposits limiting the cyclic crack opening displacement which controls the rate of growth of cracks, or perhaps by not providing enough time for the corrosive environment to have its effect.

TABLE 9.1. Comparison of Calculated and Experimental Lifetimes in a 0.6M NaCl environment (after Akid and Miller 1990a)

Shear Stress	Actual Life	Calculated Life	
Range, $\Delta\tau$, MPa	Cycles	B–H	N–R
371.0	41,500	32,365	33,795
352.7	144,000	108,654	129,859
330.7	561,250	532,060	553,225
320.8	915,500	1,270,940	992,425

Full details of these and other tests are given elsewhere (Akid, 1987, Akid and Miller, 1990a,b) but Table 9.1 provides a comparison of actual fatigue lifetimes and predicted fatigue lifetimes using the two models, of B–H and N–R, previously described. The calculation assumed that the crack growth rates could be added together, i.e.

$$\left(\frac{da}{dN}\right)_{\text{environment}} = \left(\frac{da}{dN}\right)_{\text{air}} + \left(\frac{da}{dN}\right)_{\text{dissolution}} \tag{9.3}$$

as illustrated in Fig. 9.9(b). The value of the last term of Equation 9.3 was calculated to be of the order of 1×10^{-3} microns/cycle which agrees with the development rates of surface pits on unstressed material after immersion in 0.6M NaCl solution.

With reference to Fig. 9.9, it is clear that at the stress level below the fatigue limit e.g. $\Delta\sigma_x$, it is possible to extend the crack from X to Y (by the conjoint action of chemical and mechanical driving forces. Indeed, it is only necessary to apply the aggressive environment at the instant the crack has reached the barrier and to remove it after the barrier has been breached in order to eliminate the fatigue limit and obtain a "corrosion-fatigue" endurance (S–N) curve.

9.7 Concluding Remarks

The importance of cracks, whatever their size, has been shown to be of crucial importance in metal fatigue studies. This is particularly true at cyclic stress levels around and below the fatigue limit in unnotched specimens where cracks are of microstructural dimensions.

The crack size, its orientation with respect to the applied stress field, and the orientation of the surface with respect to the crack growth planes are factors that control fatigue lifetimes. Indeed, it is recognized that fatigue failure can result from tests conducted in such a manner that the value of the stress level does not change throughout the test, i.e. $\Delta\sigma = 0$. This can be achieved in experiments, where the three principal stresses (or strains) are maintained constant but rotated throughout a cycle; failure being dictated by the state of stress on the plane of the propagating crack.

Microstructural barriers to crack growth, in terms of their spacing and strength within the metal matrix, are crucial to the development of fatigue limits, which can be removed by a variety of techniques, e.g. introduction of an aggressive environment for a short time, or a few high-level stress cycles. Hence the fatigue limit is seen as being only a consequence of performing tests at a constant maximum stress range.

Effects such as surface finish and grain size on the smooth specimen fatigue limit can be recognized (Miller, 1985, 1991) as a consequence of Equation 9.1 in which the classical fatigue initiation period is transformed into a microstructural short fatigue crack propagation phase.

Finally with reference to Fig. 9.9, the zones (A–B) (B–C) and (C–D) require more individual attention in research. Specifically in A–B an increase in fatigue resistance can be achieved by improving the microstructural texture, e.g. by decreasing grain size and increasing the number and strength of barriers. In C–D an increase in fatigue resistance can be achieved at low mean stress levels by increasing the grain size which decreases the cyclic crack opening range. In the intermediate zone B–C, the effect of spacing and strength of various microstructural barriers on small fatigue crack propagation requires further investigation.

References

Akid, R. (1987). *The initiation and growth of short fatigue cracks in an aqueous saline environment.* PhD thesis, University of Sheffield. 28.

Akid, R. and Miller, K. J. (1990a). The initiation and growth of short fatigue cracks in an aqueous saline environment. In Scott, P. and Cottis, R. A., editors, *Environment Assisted Fatigue*, EGF publication 7. Mechanical Engineering Publications, London.

Akid, R. and Miller, K. J. (1990b). The effect of solution ph on the initiation and growth of short fatigue cracks. In *Fracture Behavior and Design of Materials and Structures*, pages 1753–1758. EMAS, London.

Allen, N. P. and Forrest, P. G. (1956). The influence of temperatures on the fatigue of metals. In *Fatigue of Metals*, pages 327–340, London. Institution of Mechanical Engineers, IMechE/ASME.

Brown, M. W. and Miller, K. J. (1973). A theory for fatigue failure under multiaxial stress-strain conditions. In *Proc Instn Mech Engrs*, pages 745–755.

Brown, M. W. and Miller, K. J., editors (1989). *Biaxial and Multiaxial Fatigue*, EGF publication 3. Mechanical Engineering Publications. .

de los Rios, E. R., Mohamed, H.J., and Miller, K. J. (1985). A micromechanics analysis for short fatigue crack growth. *Fatigue Fract Engng Mater Struct*, 8:49–63.

Forsyth, P. J. E. (1961). A two-stage process of fatigue crack growth. In *Symposium on "Crack Propagation"*, pages 76–94, Cranfield.

Frost, N. E. and Phillips, C. E. (1956). Studies in the formation and propagation of cracks in fatigue specimens. In *Metal Fatigue*, pages 520–526, London. Institution of Mechanical Engineers, IMechE/ASME.

Hammouda, M. M. and Miller, K. J. (1979). Elastic-plastic fracture mechanics analyses of notches. In *ASTM STP 668*, pages 703–719. American Society for Testing and Materials.

Heyes, P. J. (to be published, 1991). *Fatigue cracks at notch roots under high mean stress.* PhD thesis, University of Sheffield.

Hobson, P. D. (1985). *The growth of short fatigue cracks in a medium carbon steel.* PhD thesis, University of Sheffield.

Jacquet, P. A. (1956). Observations on the microstructure of a brass containing 67 percent of copper, subjected to alternating bending stresses. In *Fatigue of Metals*, pages 506–509, London. Institution of Mechanical Engineers, IMechE/ASME.

Kitagawa, H. and Takahashi, S. (1976). Applicability of fracture mechanics to very small cracks or the cracks in the early stage. In *Proc Second International Conference on Mechanical Behavior of Materials (ICM2)*, pages 627–631. American Society of Metals.

Kussmaul, K. F., McDiarmid, D. L., and Socie, D. F., editors (1991). *Fatigue under Biaxial and Multiaxial Loading*, ESIS publication 10. Mechanical Engineering Publications.

Lukas, P., Kunz, L., Weiss, B., and Stickler, R. (1989). Notch size effect in fatigue. *Fatigue Fract Engng Metals Struct*, 12:175–186.

McClintock, F. A. (1956). The growth of fatigue cracks under plastic torsion. In *Proceedings of the International Conference on Fatigue of Material*, pages 538–542, London. Institution of Mechanical Engineers, IMechE/ASME.

Miller, K. J. (1985). Initiation and growth rates of short fatigue cracks. In Bilby, B. A., Miller, K. J., and Willis, J. R., editors, *Fundamentals of Deformation and Fracture*. IUTAM Eshelby Memorial Symposium, Cambridge University Press.

Miller, K. J. (1991). Metal fatigue — past, current and future. *The 27th John Player Lecture, Proc Instn Mech Engs.*

Miller, K. J. and Brown, M. W., editors (1985). *Multiaxial Fatigue*, STP 853. American Society for Testing and Materials.

Miller, K. J. and de los Rios, E. R., editors (1986). *The Behaviour of Short Fatigue Cracks*. Mechanical Engineering Publications, London, egf publication 1 edition.

Miller, K. J. and de los Rios, E. R., editors (to be published, 1991). *Short Cracks*. Mechanical Engineering Publications, London, an ESIS (EGF) publication.

Navarro, A. and de los Rios, E. R. (1987). A model for short fatigue crack propagation with an interpretation of the short-long crack transition. *Fatigue Fract Engng Mater Struct*, 10:169–186.

Smith, R. A. and Miller, K. J. (1977). Fatigue cracks at notches. *Int J Mech Sci*, 19:11–22.

Smith, R. A. and Miller, K. J. (1978). Prediction of fatigue regimes in notched components. *Int J Mech Sci*, 20:201–206.

Sun, Z., de los Rios, E. R., and Miller, K. J. (1991). Modelling small fatigue cracks interacting with grain boundaries. *Fatigue Fract Engng Mater Struct*, 14:277–291.

Tanaka, K., Kinefuchi, M., and Yokomaku, T. (to be published, 1991). Modelling of statistical characteristics of propagation of small fatigue cracks. In *Short Cracks*, ESIS (EGF) publication 12. Mechanical Engineering Publications, London.

Tomkins, B. (1968). Fatigue crack propagation — an analysis. *Phil Mag*, 18:1041–1066.

Yates, J. R. and Brown, M. W. (1987). Prediction of the length of non-propagating fatigue cracks. *Fatigue Fract Engng Mater Struct*, 10:187–201.

Zhang, W. (1991). *Short fatigue crack behavior under different loading systems.* PhD thesis, University of Sheffield.

10

Reflections on Contributions to Deformation and Fracture

F. A. McClintock

ABSTRACT These reflections are based on some of my publications over the years, mostly in the fields of deformation and fracture. My apologies to those co-authors whose work I have omitted. To them and to others whose work *is* mentioned, I am deeply indebted for inspiration, discussion, dedication, and often tedious work. In the end, I hope these associates have shared the thrill of new-found insight that may be of service to engineers. Needless to say, I am also deeply indebted to the many colleagues in the field, whose findings have often been greater, and with whom discussion have been exciting and instructive. Finally, the essential support of this work by many agencies over the years is very much appreciated. The topics are in order of their first major publication.

10.1 Fatigue in Mode I

Thesis work at CalTech on the fatigue of single crystals grew out of a common, but in part misguided, attempt to understand fatigue (of gas turbines) from tests on "simple" microstructures. It showed that fatigue crack growth, as well as initiation, is often associated with primarily single slip (McClintock 1953). Further, a statistical analysis verifies that macroscopically normal cracks in polycrystals (Stage II) can consist microscopically of shear cracks (Stage I); there need be no essentially different mechanism (McClintock 1952). Paul Paris' recognition that macroscopic fatigue crack growth rates depend on the range of stress intensity factor led to a theoretical model (McClintock 1963) that, while predicting a threshold before it was reported experimentally, also depended on damage ahead of the crack tip, which microscopic studies fail to support. Further, the observed crack growth rates are far *lower* than predicted by the crack tip displacement that sometimes governs in the fully plastic case (McClintock 1969). In fact Barsom and then Markus Speidel showed that the growth rate in different materials correlates with the modulus of elasticity E but is relatively independent of yield strength. This led to an unpublished correlation of crack growth by normalizing with the atomic or molecular diameter b to give $da/bdN = [\Delta K/E\sqrt{Cb}]^m$, where the empirical constant C is *of the order* of 2 for a wide range of metals and polymers. The physical basis of such a relation, including the effect of microstructure and the rough crystalline facets of the growing crack, is an important unsolved problem.

10.2 Uncertainties and Choosing Working Stresses

When statistical data are insufficient, as is usually the case, how does the engineer use all the knowledge and lore about fracture that has been and is still being gathered? Perhaps by expressing a judgement about the data in the form of uncertainty intervals with specified odds (Kline and McClintock 1953) as applied to the design of meter-sized magnets for fusion reactors. (McClintock, Feng, and Vieira 1991)

10.3 Statistics of Fatigue and Fracture

Noting variation in position of fracture along the classical hourglass fatigue bending specimens gave rise to the correlation between that and the size effect (McClintock 1955 and 1955a), made far simpler and more elegant by Wallodi Weibull's unfortunately unacknowledged suggestion that the extreme-value distributions should be tried. A more general, three-dimensional stress field has been treated (McClintock 1971).[1] As Gumbel pointed out, these distributions are ones with constant form as the sample size changes, but are not necessarily approached in the limit. An extreme-value distribution of crack lengths does not lead to an extreme-value distribution of strengths (McClintock 1974).

Two-dimensional boundary element modeling of polycrystals with random grain boundary strengths gives insight into the effect of variability on macroscopic strength (McClintock 1976 and 1979), and toughness (McClintock and Bassani 1981). The extension to three-dimensional grains and to trans-granular cleavage, with skewed intersections, is a challenge, but is becoming feasible. Again, macrocrack roughness is a key problem.

10.4 Studies in Mode III: Fatigue and Fracture

The insight into stress and strain fields around cracks came to me suddenly on noting my accidental omission of the fillets in the re-entrant corners of an I-beam I had assigned in a plasticity class. I thought I had given the students an impossible problem, but it *was* possible to find the stress

[1] In McClintock 1971, read $1 - \Phi$ for Φ after Eq. 50 and delete "no" before Eq. 54).

and strain fields, as it was for the corresponding problem with the flanks folded flat on themselves to form a crack. (Meanwhile Max Williams was seeing similar possibilities in his studies of external corners in plane strain elasticity). Then there was the excitement of cutting a groove in a rod and twisting it back and forth to find that indeed the cracks generated surfaces parallel to the axis, not helices, so the analysis had relevance (McClintock 1956). A quantitative theory for fracture depended on critical strain over a finite distance. Only years later when H. Neuber was visiting, did I suddenly realize that the idea had come to me from his work, and was not my own creation.

The membrane-sand-hill analogies for torsion suggested that the elastic-plastic solution would be possible, but Jan Hult and I found it elusive until in a crucial five-minute discussion after a seminar, William Prager suggested we try a hodograph transformation. Jan Hult still had to prompt me on some of the complex variable analysis before I could present the results at the International Congress of Applied Mechanics in Brussels in 1957 (Hult and McClintock 1957), where George Irwin was presenting his strain energy release rate analysis and Max Williams his stress singularity, both for plane elasticity. Later James Rice pointed out, as I should have seen, that some of our numerical integrals could be regarded as elliptic integrals. It turned out, however, that we needed such accuracy in a small region of the tables that numerical integration was still about as good.

This work led to a number of analytical and experimental studies (see McClintock 1971)[2] including by Ithan Kayan on the size effect in longitudinally grooved bars, computation by Joseph Walsh on crack transients in such bars and on the strain distribution and plastic zones in circumferentially grooved bars (including studies of what later became known as the Dugdale-Barenblatt model of collapsing the plastic zone to a plane), and by Alexander Mackenzie on strain measurements in rounded roots. Ani Chitaley (1970) analyzed dynamically running Mode III cracks.

This work found application to cracks in turbo-generator shafts subject to electrical faults. Ritchie, Nayeb-Hashemi, and Tschegg (1983) considered plastic zones, micromechanisms of growth that here *do* produce damage ahead of the crack, and effects of friction and loading sequences. The final report by McClintock, Nayeb-Hashemi, Ritchie, Wu, and Wood (1984) included results from a computer program for damage accumulation within the plastic zones caused by any arbitrary torque history, as well as a preliminary risk analysis of a fault clearance, related to Section 10.2 above.

---

[2]Here and elsewhere "see McClintock 1971" refers to summaries and references in that review article, to save space in spite of inconvenience to the reader.

10.5 Atomic and Dislocation Studies

The possibility that couple stresses affect dislocation energies, suggested by some earlier work with sophomores in a mechanics class, was ruled out (McClintock 1960), but that does not mean that it would not affect the core transitions with dilatation found by William O'Day's meticulous bubble raft experiments (McClintock and O'Day, 1965). The same reservation holds for the general exponential atomic force-separation law that is finding application to cleavage. O'Day's work showed that dislocations and grain boundaries both reduced the dilatational strength by about a factor of two, for a reasonable range of interatomic hardnesses (bubble sizes).

The idea of using dislocation arrays or dipoles with the boundary element method to represent plasticity seemed like a good suggestion to make to Jehuda Tirosh and James Joyce. Although it seems to work well in some cases (Tirosh and McClintock 1978), with higher dislocation densities instabilities set in. Thinking of dislocation structures in electron micrographs shows that this is a natural occurrence. One case in which this can be followed is a hexagonal mono-crystal of aluminum with a $\{111\}$ plane as a cross-section. X-ray observations confirm that initial loops from the faces react to form the more stable dislocation network (McClintock and Prinz 1983). This case provides one more example of how difficult it will be to achieve the dream of deriving flow and evolutionary equations from actual dislocation motions and reactions.

10.6 Elastic-Plastic Fracture in Plane Stress

The first paper (McClintock 1958) should have been titled "Ductile Fracture Instability in Plane Stress" rather than "in Shear", because the experiments were for the more practical case of plane stress. They included simulated corrosion by cutting the crack as well as stable growth on loading. Even though the underlying equations are different, the extending of the plastic zone more ahead of the crack than to either side in both cases leads to remarkably similar behavior. The similar final behavior on different loading paths shows that the R-curve is a special case. The analogy led to several papers on the significance of the analysis for testing, on root radius, and on loading rate, as well as a call from the ASTM to George Irwin and me to reach a common ground, which was a pleasure (see McClintock 1971).

10.7 Rigid-Plastic Fracture Mechanics

For structures involved in earthquakes and accidents, this is a seriously neglected field. For early work on unequal or anti-symmetric grooves and the effects of strain-hardening, (see McClintock 1971). Perhaps most important was the recognition arising from a study of crack growth beyond the initial stage of Rice and Johnson (both papers stimulated by lunch together) that hole growth only on the plane ahead of the crack gave an unrealistically large crack opening angle, and that zig-zagging was required, which helps to explain the crack roughness as well (McClintock 1969). The slip line analysis of Carson showing the microscopically unstable, but macroscopically stable nature of the plane strain solutions was shown in McClintock and Bassani (1981). James Joyce found that sliding off, rather than blunting, occurred at crack tips of sufficiently ductile material (see McClintock 1971). With typically dense hole nuclei, he turned to the HRR singularity to correlate initial crack growth (Joyce and McClintock 1976).

In Mode II of a single groove in a plate, Margherita Clerico (McClintock and Clerico 1980) found the slip line field from a modification of the Green and Hundy field. Very likely that field, wrapped somewhere around the arc of James Rice's upper bound solution, will give the field for a grooved plate under combined bending and torsion. For practical tests at high triaxiality with double grooves, Sarah Wineman's (McClintock and Wineman 1987) wedge tests, symmetrical and asymmetrical, give insight.

10.8 Elastic Fracture Mechanics

Work on drilling of oil wells led to findings with Joseph Walsh (McClintock and Walsh 1962) that friction on Griffith cracks predicted the fracture data on rocks under pressure. The possibly high coefficients of friction suggest that roughness plays a role, as studied by others.

In rail webs, McClintock and Wineman (1990) showed that the U-shaped residual stresses from roller-straightening railroad rails could drive cracks spontaneously, especially with high-strength steels. How to reduce the U-shaped stresses by inexpensive roller-straightening is still a puzzle.

10.9 Hole Growth Analysis

Since the work of Constance Elam, vitalized by the much later electron microscopy, it was known that even with structurally brittle aluminum alloys, the micro-mechanism was the growth and coalescence of holes. The trick in getting an analysis to provide some insight lay in considering generalized plane strain, so the triaxial tension only had to bias the hole growth. The resulting paper (McClintock 1968) was almost rejected (I thought complimented) as a mixture of analysis, experiment, and conjecture. Even early on, Harland Alpaugh's experiments showed the rather limited hole growth before localization or fine cracking set in between the holes (McClintock 1968a). In spite of Alan Needleman, Viggo Tvergaard, and others' good work on localization in sheets from fine holes, the irregular fine cracks that are often found remain to be explained. Saul Kaplan and Charles Berg extended the model to hole growth in shear bands, using Berg's thorough solution and experimental verification of plane hole growth in a viscous medium (see McClintock 1971).

In view of these and more precise studies of various aspects of hole growth, it is surprising to see damage characterized, without apology, by using only one or two parameters. A description of even the holes as regularly-spaced ellipsoids requires a triad for their orientation, a triad for their principal axes, and another pair of triads for the lattice representing their spacing. Remember Al Backofen's wolf's ear fractures on tension after monotonic torsion, but not after reversed torsion!

For insight into spall fracture, dynamics and strain rate effects were included (McClintock 1973).

10.10 Surface Roughening and Coatings

Under stress reversal Charles Berg (1966) showed that the mechanics solutions to be surprisingly kinematically reversible, although cyclic experiments show surface roughening and increasing acuity of initially round holes (McClintock 1963). The cause appears to be still unexplained.

John Holmes (1990) recognized the importance of surface coatings and changed a research program in that direction, away from creep. His experiments showed that in nickel aluminide coatings on nickel-based superalloys the coating failure was typically, but not always, due to surface roughening through the depth of the coating. Later Esteban Busso (1988) studied the stress history in harder nickel-cobalt-chromium-aluminum-ytrium coatings and found that the critical event was the formation of Kirkendall voids by diffusion across the interface.

10.11 Mixed-Mode and Asymmetric Configurations

The first mixed-mode study used the boundary element method for a horizontal crack in a railroad rail, finding the history of closure, friction, and plasticity on a few planes radiating from the crack tip (McClintock 1977). Later George Kardomateas simulated a part-through stress corrosion crack in the weld along the base plate of oil storage tanks, forcing non-hardening plasticity into a single 45° shear band from the crack tip. The band is subject to tensile loading but shear deformation. The crack growth ductility was reduced below that for a symmetrical singly grooved specimen by a factor of about 2, in a variety of alloys (Kardomateas and McClintock 1987, McClintock 1990).

10.12 Creep

John Bassani showed the progressive development of elastic, plastic, primary creep, steady state creep, and steady crack growth fields around suddenly loaded cracks (McClintock and Bassani 1981), and compared them with computation results (Bassani and McClintock 1981). The domains of validity change not only with time but also with stress level and hence life, so that laboratory tests may not be representative of service. While some are skeptical about the Hui-Riedel fields for steady state growth, the form of their singularities is evocative of the constant crack growth rate at varying stress intensity factor that environmental effects do cause. These results must be reconciled with micro-mechanical modelling and with the failure of the singularity concept in some computational and experimental cases discussed by others.

10.13 Closure

The chase goes on. There seem to exist upper and lower bound theorems for power-law creep, and hence for strain-hardening plasticity. It appears possible to derive Andrade's one-third power transient creep from a simple set of flow and evolutionary equations. The physicists can use the Grüneisen constants to make a simple prediction of the temperature dependence of the bulk modulus. How well does it work? For shear? The distortion of the yield locus of a square beam under an arbitrary history of two-plane bending gives interesting insights into a generalized Bauschinger effect. True stress-strain curves from video data. Fold-bend tests of plates with high ductility. The necking of a plate subject to bending superimposed on a constant tension has applications to the panels of a ship.

"The search for truth is one way hard and another easy, for no one can grasp it fully nor miss it wholly. Each adds to our store of knowledge, and from it arises a certain grandeur."

....Aristotle (ca 335 B.C.)

It is great to be in the game, to have a little more time to learn what others are doing, and to have the prospect of leaving good problems to good people.

REFERENCES

Aristotle (ca 335 B.C.) Freely quoted through several sources from *Metaphysics, Book II, Chapter 1*, 993a, 1.30–40. 40.

Bassani, J. and McClintock, F. A. (1981). Creep relaxation of stress around a crack tip, Int. J. Solids Struct., 17:479–492. 39.

Berg, C.A. and McClintock, F. A. (1966). Zeit. Angew. Math. Phys., 17:453–456. 33.

Busso, E. P. and McClintock, F. A. (1988). Stress-strain histories in coatings on single crystal specimens of a turbine blade alloy, Int. J. Solids Struct., 24:1113-1130. 35.

Chitaley, A. D. and McClintock, F. A. (1970). Elastic-plastic mechanics of steady crack growth under anti-plane shear, J. Mech. Phys. Solids, 19:147–163. 16.

Holmes, J. W. and McClintock, F. A. (1990). The chemical and mechanical processes of thermal fatigue degradation of an aluminide coating, Met. Trans. A., 21A:1209-1222. 34.

Huff, H. W., Joyce, J. A., and McClintock, F. A. (1969). Fully plastic crack growth under monotonic and repeated bending. In *Fracture*, Proc. 2nd Int. Conf. on Fracture, Chapman and Hall, London, 83–94. 4.

Hult, J. A. H. and McClintock, F. A. (1957). Elastic-plastic stress and strain distributions around sharp notches under repeated shear. *9th Int. Cong. Appl. Mech.*, 8:51–58. 15.

Joyce, J. A. and McClintock, F. A. (1976). Predicting ductile fracture initiation in large parts. In Miyamoto, H., Kunio, T., Okamura, , Weiss, V., Williams, M. and Liu, S. H., editors, *Strength and structure of solid materials*. Noordhoff, Leyden. 157–181. 25.

Kardomateas, G. A. and McClintock, F. A. (1987). Tests and interpretation of mixed mode I and II fully plastic fracture from simulated weld defects, Int. J. Fract. 35:103–124. 37.

Kline, S. J. and McClintock, F. A. (1953). Describing uncertainties in experiments, Mech. Eng., 75:3–8. 5.

McClintock, F. A., Feng, J. and Vieira, R. (1991). Using statistical single-sample and uncertainty analyses in design, applied to a Tokamak central solenoid. In *Proceedings of the 14th IEEE Symposium on Fusion Engineering*, San Diego. 6.

McClintock, F. A. (1990). Reduced crack growth ductility due to asymmetric configurations, Int. J. Fract. 42:357–370. 38.

McClintock, F. A. and Wineman, S. J. (1990). Residual stress near a rail end, Theo. and Appl. Fract. Mech., 13:29–37. 22.

McClintock, F. A. Nayeb-Hashemi, H. Ritchie, R. O., Wu, E. and Wood, W. T. (1984). Assessing expired fatigue life in large turbine shafts. Oak Ridge Nat. Lab. Rep. ORNL/Sub/83-9062/1, Nat. Tech. Ing. Serv. U.S. Dept. of Comm., Springfield, VA. 18.

McClintock, F. A. and Prinz, F. (1983). A model for the evolution of a twist dislocation network, Acta Met. 31:827–32. 22.

McClintock, F. A. and Bassani, J. (1981). Three-Dimensional Constitutive Relationships and Ductile Fracture. In Nemat-Nasser, S. editor, North Holland, Amsterdam, 123–145. 13.

McClintock, F. A. and Clerico, M. (1980). The transverse shearing of single-grooved specimens, J. Mech. Phys. Solids, 28:1–16. 26

McClintock, F. A. and Zaverl, F., Jr. (1979). An analysis of the mechanics and statistics of brittle crack initiation, Int. J. Fract., 15:107–118. 12.

McClintock, F. A. (1977). Plastic flow around a crack under friction and combined stress. In Taplin, D.M.R., editor, *Fract. 77*, University of Waterloo Press, Waterloo. 49–64. 36.

McClintock, F. A. and Mayson, H. J. (1986). Stress effects on brittle crack statistics. In Cowin, S.C. and Carrol, M.M., editors, *Effects of voids on Materials Deformation*, AMD-16, ASME, New York 31–45. 11.

McClintock, F. A. (1974). *Statistics of brittle fracture.* In Bradt, R.D., Hasselman, D.D.H. and Lange, F.F., editors, *Fract. Mech. of Ceramics VI*, Plenum Press, New York. 93–114. 10.

McClintock, F. A. (1973). Models of spall fracture by hole growth. In Rohde, R.W., Butcher, B.M., Holland, J.R., Karnes, C.H., editors, *Metallurgical Effects at High Strain Rates*, Plenum Press., New York. 415–427. 32.

McClintock, F. A., (1971). Plasticity Aspects of Fracture, Academic Press In Liebowitz, H., editor, *Fracture* 3:47–225. 9.

McClintock, F. A. (1968). Crack growth in fully plastic grooved tensile specimens. In Argon, A.S., editor, *Physics of strength and plasticity*, MIT Press, Cambridge, MA, 307–326. 24.

McClintock, F. A. (1968). A Criterion for ductile fracture by the growth of holes, J. Appl. Mech., 35:363–371. 30.

McClintock, F. A. (1968a). On the mechanics of fracture from inclusions. In *Ductility*, ASM, 255–277. 31.

McClintock, F. A. and Wineman, S. (1987). A wedge test for quantifying fully plastic fracture, Int. J. Fract., 33:285–295. 27.

McClintock, F. A. and O'Day, W. R., Jr. (1965). Biaxial tension, distributed dislocation cores and fracture in bubble rafts. In *Proc. 1st Intl. Conf. on Fract. 3*, A-5, Jap. Soc. Strength, Fract. Matrls., Sendai, Japan.75–981. 20.

McClintock, F. A. (1963). On the Plasticity of the Growth of Fatigue Cracks In Drucker, D.C. and Gilman, J.J., editors, *Fracture of Solids. Met. Soc. AIME Conf. Ser.* 20: Interscience, New York 65–102. 3.

McClintock, F. A. and Walsh, J. B. (1962). Friction on Griffith cracks in rocks under pressure. In *Proc. 4th U.S. Natl. Cong. Appl. Mech.*, ASME 2: 1015–1021. 28.

McClintock, F. A. (1960). Contribution of interface couples to the energy of a dislocation. Letter to Ed., Acta Met., 8: 127. 19.

McClintock, F. A. (1958). Ductile fracture instability in shear, J. Appl. Mech., 25: 581–588. 23.

McClintock, F. A. (1956). The growth of fatigue cracks under plastic torsion. In *Proc. Int. Conf. Fatigue Metals*, Inst. Mech. Eng., London. 538–542. 14.

McClintock, F.A. (1955). The statistical theory of size and shape effects in fatigue, J. Appl. Mech., 22:421–426. 7.

McClintock, F.A. (1955a). A criterion for minimum scatter in fatigue testing, J. Appl. Mech., 22:427–431. 8.

McClintock, F.A. (1953). Fatigue tests of single crystals of ingot iron. In *Proc. 1st Nat. Cong. Appl. Mech.*, ASME, N.Y., 653–659. 1.

McClintock, F. A. (1952). On the direction of fatigue cracks in polycrystalline ingot iron, J. Appl. Mech., 19:54–56. 2.

Nayeb-Hashemi, H., McClintock, F. A. and Ritchie, R. O. (1983). Influence of overload and block loading sequences on mode III fatigue crack propagation in A469 rotor steel, Eng. Fract. Mech., 18:763-783. 17.

Tirosh, J. and McClintock, F. A. (1978). A numerical approach for solving 2D in anti-plane elasto-plastic problems by assembling dislocations. In Armen, H. and Jones, R.F., editors, *Applications of numerical methods to forming processes*, AMD-28, ASME, New York, 155–161. 21

Index